Advances in Behavioral Based Safety

N. A. Siddiqui · Faisal Khan · S. M. Tauseef ·
Waddah S. Ghanem · Vikram Garaniya
Editors

Advances in Behavioral Based Safety

Proceedings of HSFEA 2020

Springer

Editors
N. A. Siddiqui
Health Safety, Environment and Civil Engineering
University of Petroleum and Energy Studies
Dehradun, Uttarakhand, India

S. M. Tauseef
University of Petroleum and Energy Studies
Dehradun, Uttarakhand, India

Vikram Garaniya
Australian Maritime College
Newnham, TAS, Australia

Faisal Khan
MKO Process Safety Centre
College Station, TX, USA

Waddah S. Ghanem
Emirates National Oil Company Limited
Dubai, United Arab Emirates

ISBN 978-981-16-8269-8 ISBN 978-981-16-8270-4 (eBook)
https://doi.org/10.1007/978-981-16-8270-4

This Springer imprint is published by the registered company Springer Nature Singapore Pte Ltd.
The registered company address is: 152 Beach Road, #21-01/04 Gateway East, Singapore 189721, Singapore

Contents

About the Editors

Dr. N. A. Siddiqui is currently associated with University of Petroleum & Energy Studies, Dehradun, as Professor & Program Director - Health, Safety & Environment and civil Engineering. He pursued his postgraduate degree in Environmental Science and Doctorate in Environmental Biology where he focused on environmental impact assessment. Dr. Siddiqui specializes in the area of safety management, occupational health & safety, environmental pollution, and environmental monitoring and control techniques. He has worked closely with major research centres both nationally and internationally. He has more than 170 research papers to his credit and has participated in several national and international conferences. He has authored 14 books. Dr. Siddiqui has guided more than 200 M.Tech thesis and 21 Ph.D. thesis.

Dr. Faisal Khan is Professor, Director, MKO Process Safety Centre, College Station, Texas, United States . He is Founder of Centre for Risk Integrity and Safety and Engineering (C-RISE). His areas of research interest include offshore safety and risk engineering, inherent safety, risk management, and risk-based integrity assessment and management. He is actively involved with multinational oil and gas industries on the issue of safety and asset integrity. He continues to serve subject matter expert to several international organizations. He is the recipient of President Outstanding Research Award of 2012–13, CSChE National Award on Process Safety Management of 2014, President Outstanding Research Supervision Award of 2013–14, and recently Society of Petroleum Engineer award for his contribution in Health, Safety and Risk Engineering. He has authored over 400 research articles in peer-reviewed journals and conferences on safety, risk, and reliability engineering.

Dr. S. M. Tauseef is Professor and Associate Dean (R&D) at the University of Petroleum and Energy Studies, India. He holds an M.S. in Chemical Engineering (Process Design) from Sharif University of Technology, Iran, and a Ph.D. in Environmental Engineering from Pondicherry University. Dr. Tauseef's research interests include process safety, especially forecasting and consequence assessment of accidents in chemical process industries. He has developed methodologies for application of computational fluid dynamics (CFD) to dense gas dispersion, vapor cloud explosions, and single and multiple pool fires. His other interest is in solid waste management, especially the design of continuously operable vermireactor systems for assorted wastes, and high-rate anaerobic digesters. He has published more than 90 journal articles, co-authored/edited eleven books, five patents, and presented papers at 30 conferences.

Dr. Waddah S. Ghanem is currently appointed to as Senior Director for Logistics and Marine Assurance for the ENOC Group in Dubai, UAE. He is Hon. Chairman for the Energy Institute—Middle East and is also Chairman of the Platform for Connected Leadership, Dubai, UAE. Dr. Waddah holds several academic and vocational qualifications including two PG Diplomas (British Safety Council), an M.Sc. in Environmental Sciences (UAE University), an Executive MBA, and a Doctorate from the University of Bradford in Governance and Leadership. He has received several awards, most notably, the Rashid Prize for Scholarship Excellence in 2007 (Dubai), and the IEMA Sustainability Leader of the Year (UK) in 2019. He is Visiting Professor at the University of Technology Petronas, Malaysia and the University of Petroleum and Energy Studies, India. He is Fellow of several institutions such as the Energy Institute (EI), Institution of Chemical Engineers (IChemE), Institute of Environmental and Management Assessment (IEMA), and the GCC Board of Directors Institute of which he is also appointed Mentor for directors.

Dr. Vikram Garaniya is currently serving as Acting Director, National Centre for Maritime Engineering and Hydrodynamics (NCMEH) and Associate Head of Research at the Australian Maritime College (AMC)—a specialised institute of the University of Tasmania (UTAS). He is also the Research Leader for the Environment, Energy and Safety Engineering (EES) Research Group—a key focus of this group is coastal environmental research, including air and water pollution. Over the last eight years, in collaboration with industry and other institutions, he has secured many competitive research grants. He has led many externally funded projects, including consultancies with the Australian Maritime Safety Authority (AMSA) and multi-institutional, multi-country projects funded by the International Association of Maritime Universities (IAMU). He has published papers in journals of national and international repute.

Risk Assessment of a NHT Heat Exchanger Using Bow Tie Analysis: An Intuitionistic Fuzzy Approach

Arjun P. Kumar and V. R. Renjith

1 Introduction

Risk assessment plays a vital role as far as an organisation aims to maintain the risk associated with its operation on a tolerable level. Many forms of risk assessment are there in practice today, of which fault tree analysis (FTA) is deductive technique including logic gates where all the potential causes to a top event (TE) are identified. If the probability of occurrence of these causes are known, obtaining the probability of the undesired event is attainable using Boolean logic. Event tree analysis (ETA) is an inductive technique where consequences arising out of/in the course of the undesired event can be identified. Once if the probability of occurrence of pivot events (PE) or failure probability of safety barriers is known, consequence probability can also be identified.

Heat exchangers are very important part as far as the refinery operations are concerned. It helps in decreasing the expenditure for heating purposes by utilising the heat given out by the product of different operation. Heat exchanger is sometimes given as a single one or as a series of the same. Although it requires less maintenance and attention, there is possibility that unexpected causes can create unwanted and undesired events in the operations. There are so many examples that can be pointed out relating to the catastrophic failure of heat exchanger and corresponding accidents. Some of those accidents that occurred due to the same are:

- The Esso Longford gas explosion was a catastrophic and major industrial accident that occurred at the Esso natural gas plant in Longford, Victoria, Australia, on

A. P. Kumar (✉)
Division of Safety and Fire Engineering, School of Engineering, CUSAT, 682022 Kochi, Kerala, India
e-mail: arjunpk@cusat.ac.in

V. R. Renjith
Faculty of Division of Safety and Fire Engineering, School of Engineering, CUSAT, 682022 Kochi, Kerala, India

N. A. Siddiqui et al. (eds.), *Advances in Behavioral Based Safety*,
https://doi.org/10.1007/978-981-16-8270-4_1

25 September 1998, killing two workers and injuring eight. Victoria State's gas supply severely affected for two weeks.

- The Tesoro Anacortes refinery disaster was an industrial accident at the Tesoro Anacortes refinery in Anacortes, Washington, on 2 April 2010. Seven workers sustained fatal burns in an explosion and fire that followed when a heat exchanger violently ruptured after a maintenance restart.
- The Williams Olefins Plant blast happened on 13 June 2013 at a petrochemical plant situated in Geismar, an unincorporated and to a great extent modern zone 20 miles southeast of Baton Rouge, Louisiana. Two labourers were killed, and 114 were harmed. A reserve heat exchanger had loaded up with hydrocarbon and was disengaged from its pressure relief; not long after the heat exchanger was heated up with boiling water, the hydrocarbon blazed to fume, cracked the warmth exchanger, and detonated.

Risk assessment of the vital parts of refinery operations can contribute to decrease in these undesirable events. The main problem in carrying out risk assessment is the unavailability of failure data. The proposed methodology makes use of expert opinions in linguistic terms for finding out failure probabilities and can be efficiently applied to the above-mentioned problem. Expert opinions collected are converted to crisp possibility and probability values using mathematical methods.

The methodology is applied in BT analysis where fault tree analysis and event tree analysis are carried out to identify the basic events and consequences related to the release prevention barrier failure. FTA method is a deductive technique where Boolean logic is used to illustrate the logical relations between undesired top event and its causes propagating through intermediate events. ETA method allows to obtain the occurrence probability of consequence by considering the failure probabilities of safety barriers or occurrence probability of pivot events related to initiating event. The basic event probabilities and pivot event probabilities are identified to obtain top event and consequence probabilities.

A naphtha hydrotreating unit is a part of refinery operations where the naphtha cut is introduced into the reactor after heating to remove the sulphur and nitrogen compounds. The processes associated with the same are always carried out at temperatures ranging between 205 and 260 °C. This heating is achieved by different means like preheating, reheater rotation of heated output feed through heat exchangers to transfer residual heat. Figure 1 shows a typical layout of NHT. The heat exchangers shown can either be a single one or be a series based on the heating needs. Since isolation of different components in the input feed is taking place at different temperature, different sections of equipment are present which in turn consist of heat exchangers operating at different temperatures which result in a complex system.

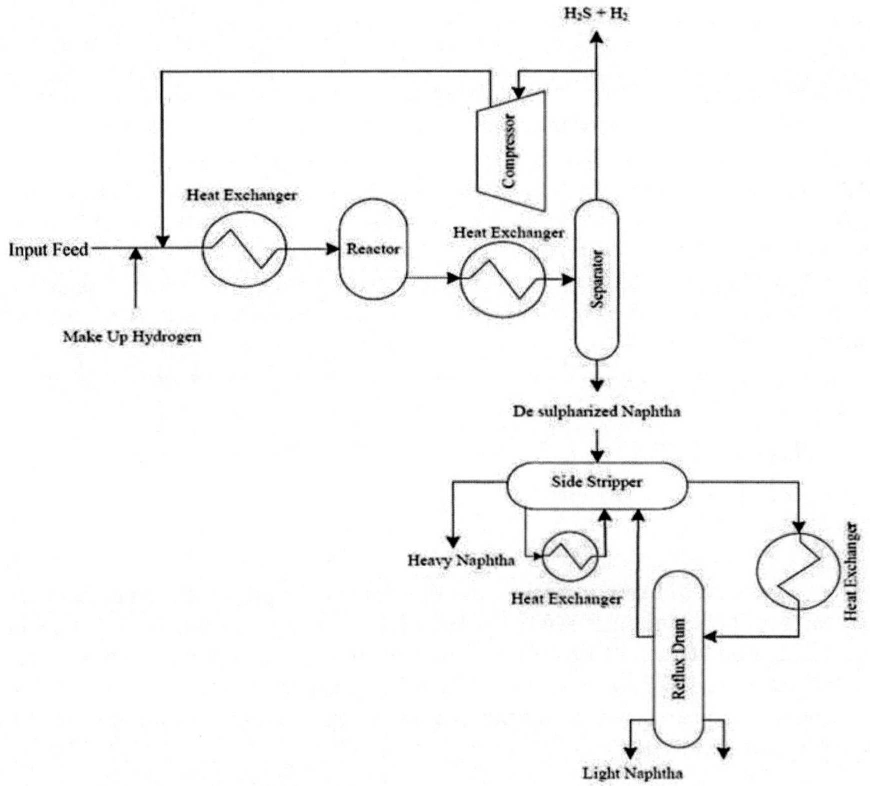

Fig. 1 Layout of a naphtha hydrotreater with heat exchangers

2 Materials and Methods

2.1 *Basics of FTA*

Conventional FTA

FTA makes use of logic gates to identify the possible paths and propagation of causes of a specific event to it undesired happening. Identification of the TE is the primary part in FTA, and afterwards, the path is drawn until the BEs are obtained. The AND and OR gates are the logic gates used in the analysis where AND gate represents the need of both the input events for happening of the output event where OR gate represents the need of only one of the events. The Boolean operations can be applied to estimate the probability of the undesired TE. Here, the probability values of the BE are either obtained from previous literatures.

FFTA

Zadeh [8] introduced the concept of fuzzy set theory and always been a prominent tool in risk assessment where sufficient data were not available. Fuzzy sets are defined on a universal set (X) characterised by a membership degree denoted by $\mu(x)$ in the interval [0, 1]. Here, $\mu(x)$ provides a measure of the degree of similarity of an element in X to the fuzzy subset. Fuzzy sets are defined for specific linguistic variables. The linguistic variables can be represented by different types of fuzzy numbers such as triangular, trapezoidal, or Gaussian shape membership function. In this paper, we make use of nonlinear triangular intutionistic fuzzy numbers to represent our linguistuc variables.

2.2 *Basics of ETA*

Conventional ETA

ETA is a forward-thinking logical method to identify the probabilities of outcomes by considering the failure of safety barriers (SB) or the pivot events (PE) that could occur as a result of the initiating event. The analysis progresses forwards considering the failure as well as success of the SB and PE considered. ETA is used as a tool to identify outcomes before occurring and setting up extra measures to prevent the possibility of occurrence.

FETA

FETA makes use of fuzzy set theory where proper data are not available in the case of failure probabilities. Expert elicitation with fuzzification, and further defuzzification enables to find the needed failure probabilities of SB's and occurrence probability of PEs.

2.3 *Proposed Model*

Concepts

Intuitionistic Fuzzy Method

Intuitionistic fuzzy sets were introduced by Atanassov [2] and are represented as follows:

Let A be an intuitionistic fuzzy set (IFS) in the universal set X defined as,

$$A = \{\langle x, \mu_A(x), \upsilon_A(x)\rangle | x \in X\}$$

Here, $\mu_A : X \to [0, 1]$ and $\upsilon_A : X \to [0, 1]$ are the membership and non-membership functions of an element $x \in A \subset X$ and for every $x \in X\ 0 \leq \mu_A(x) + \upsilon_A(x) \leq 1$

The method here involves the utilisation of fuzzification methods to convert linguistic terms to nonlinear triangular form of the fuzzy set, aggregation method to aggregate different fuzzy numbers, and final defuzzification methods to obtain the crisp possibility and probability scores.

Application of Methodology

The basic steps of the methodology applied by Kumar and Kaushik [4] can be summarised as follows:

- *Step 1:* FT and ET formulation
- *Step 2:* linguistic expert data collection
- *Step 3:* intuitionistic fuzzy failure probability (IFFP) evaluation
- *Step 4:* aggregation of opinions
- *Step 5:* defuzzification of IFFP to possibility and probability values
- *Step 6:* top event probability estimation.
- *Step 7:* pivot event probability estimation
- *Step 8:* consequence probability estimation

Step 1: FT and ET formulation

The FT was adopted from [1], and based on the same, the ET was constructed considering the PEs that could occur due to the TE considered. Fourteen possible consequences are identified with the help of ET (Fig. 2) (Tables 1, 2).

Step 2: Linguistic Expert Data collection

Qualitative data sets including seven linguistic terms, viz. VL, L, RL, M, RH, H, VH were introduced to the experts from different reas who have knowledge in the respective field, and opinions were collected. Both opinions about failure probability of BE and occurrence probability of PE are collected. Even though they may give dissimilar judgements, steps are taken in the following part to ensure weightage to experts and give importance to the more valuable opinion. Table 3 shows the details of the experts selected, and Table 4 is the opinions given by the experts w.r.t the BE and PE considered.

Step 3: Intuitionistic Fuzzy Failure Probability (IFFP) evaluation

The experts' opinions are converted to IFFP values with the help of previous literature. Huang et al. [3] and Liu et al. [5] developed the inductive approach, and it is used to represent the non-membership and membership values related to the failure probability of basic events. Initially, the extreme right and left sides' membership

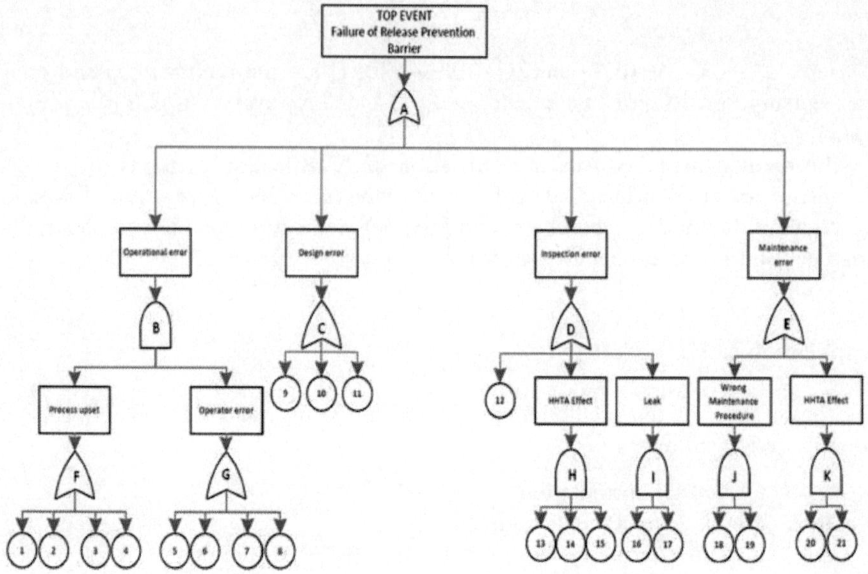

Fig. 2 Fault tree related to the top event

Table 1 Basic events adopted from [1]

Basic event	Basic event description
1	High temperature hydrogen attack (HTHA)
2	Difficulty with valve operation during start up
3	No report on leaks from heat exchanger during start up
4	Hydrogen induced cold cracking
5	Inexperience
6	No permit on job carrying
7	Failure of external supervision
8	Incorrect procedure
9	Poor construction material for NHT heat exchanger
10	High mechanical stress
11	Insufficient instrumentation to measure process conditions
12	Long delay in inspection schedule
13	Inadequate methods for detecting HTHA

(continued)

Table 1 (continued)

Basic event	Basic event description
14	Inadequate training of the inspectors to detect HTHA easily
15	Failure of HTHA inspection of heat exchanger
16	Failure of detection of leaks from heat exchanger flanges
17	Failure of minor release detection
18	Wrong maintenance procedure
19	Delay maintenance operations
20	HTHA degradation monitoring performed but failed to detect
21	HTHA degradation monitoring specified but not performed

Table 2 Identified consequences

Consequence	Consequence description	Outcome
1	Safe detection and rectification	Safe
2	Explosion and fire hazards causing minimal damage and casualties	Minimal damage and casualties
3	Explosion and fire hazards causing minimal damage with possible casualties	Minimal damage and possible casualties
4	Explosion and fire spread causing increased damages and possible casualties	Increased damage and possible casualties
5	Explosion and fire spread causing increased damages and casualties	Increased damage and casualties
6	Explosion and fire hazards causing moderate damages and possible casualties	Moderate damage and possible casualties
7	Explosion and fire hazards causing moderate damages and casualties	Moderate damage and casualties
8	VCE and fire hazards causing minimal damage and casualties	Minimal damage and casualties
9	VCE and fire hazards causing minimal damage with possible casualties	Minimal damage and possible casualties
10	VCE with fire spread causing increased damages and possible casualties	Increased damage and possible casualties
11	VCE and fire spread causing increased damages and casualties	Increased damage and casualties

(continued)

Table 2 (continued)

Consequence	Consequence description	Outcome
12	VCE and fire hazards causing moderate damages and possible casualties	Moderate damage and possible casualties
13	VCE and fire hazards causing moderate damages and casualties	Moderate damage and casualties
14	Formation of hazardous mixture with air	Hazardous atmosphere

Table 3 Details of selected experts

Expert	Professional position	Job experience (years)	Education
E1	Professor	15	Ph.D
E2	Manager	10	M.Tech
E3	Engineer	8	B.Tech
E4	Engineer	4	M.Tech
E5	operator	6	ITI

and non-membership functions of the TIFN are identified, and the two are selected to represent "very high" and "very low". Later to define the five other linguistic variables in between, the area in between the two IFN selected first is divided. Table 5 shows the IFFP values corresponding to the linguistic variables. These are the values which will be given to expert opinions and afterwards aggregated w.r.t the different opinions given related to BE and PE.

Step 4: Aggregation of opinions

In this step, the opinions given by the experts are then aggregated to obtain the aggregated IFFP. This happen in 7 steps and for understanding the calculations for a single BE is also given. BE 1 is selected, and the steps are applied to find out the aggregated IFFP of the event.

(a) *Calculation of Expectancy Evaluation*
The expectancy evaluation, $EE(A_i)$, for ith Expert (E_i) opinion's triangular IF number represented as $p_{ij} = \left(a_i, b_i, c_i; a_i', b_i, c_i'\right)$ can be found out using the formula,

$$EE(A_i) = \frac{(a_i + a_i{}') + 4b_i + (c_i + c_i')}{8}$$

Note that, TIFN numbers are taken as separate in the calculation and tabulation for the ease of calculation and should be considered as in the form represented above.

Table 6 shows the expectancy evaluations calculations corresponding to each of the experts utilising the formula.

Table 4 Opinions of selected experts

Opinions					
BE	*Expert 1*	*Expert 2*	*Expert 3*	*Expert 4*	*Expert 5*
1	RL	RL	L	L	RL
2	L	VL	L	L	VL
3	RL	VL	RL	L	L
4	VL	L	L	L	VL
5	VL	L	VL	L	VL
6	L	VL	VL	L	L
7	L	M	RL	M	M
8	L	VL	L	VL	L
9	VL	VL	L	L	L
10	L	VL	VL	L	L
11	VL	VL	VL	VL	L
12	VL	VL	VL	VL	L
13	M	RH	L	RL	L
14	RL	L	RL	L	VL
15	RL	L	M	L	L
16	VL	L	L	L	VL
17	VL	VL	VL	L	VL
18	VL	VL	VL	L	RL
19	L	VL	L	VL	L
20	RL	M	M	RL	L
21	L	RL	RL	RL	VL
PE	*Expert 1*	*Expert 2*	*Expert 3*	*Expert 4*	*Expert 5*
1	H	RH	RH	M	M
2	VH	VH	VH	H	H
3	H	VH	H	H	VH
4	M	RL	H	H	M
5	VH	H	VH	H	VH
6	M	M	RL	RL	RL

(b) *Calculation of Similarity degree and matrix*
The similarity between the opinions of the experts A_i and A_j selected is found out using the equation given, and the matrix is then formulated.

$$S(A_i, A_j) = \begin{cases} EE(A_i)/EE(A_j);\ for EE(A_i) \le EE(A_j) \\ EE(A_j)/EE(A_i);\ for EE(A_i) \ge EE(A_j) \end{cases}$$

Table 5 IFFP values of opinions

Opinion	IFFP values of opinions (general)						Probability
	a	b	c	a′	b′	c′	
VL	0	0.04	0.08	0	0.04	0.08	2.31E−07
L	0.07	0.13	0.19	0.06	0.13	0.2	4.61E−05
RL	0.17	0.27	0.37	0.15	0.27	0.39	6.23E−04
M	0.35	0.5	0.65	0.32	0.5	0.68	5.00E−03
RH	0.63	0.73	0.83	0.61	0.73	0.85	2.23E−02
H	0.81	0.87	0.93	0.79	0.87	0.95	6.01E−02
VH	0.92	0.96	1	0.92	0.96	1	1.59E−01

Table 6 Expectancy evaluations of BE 1

Basic Event 1 expectancy evaluation $EE(A_i)$					
Expert	1	2	3	4	5
Opinions	RL	RL	L	L	RL
Expert 1 (A_1) p_{ij}					
a	b	c	a′	b′	c′
0.17	0.27	0.37	0.15	0.27	0.39
$EE(A_1) = 0.27$					
Expert 2 (A_2) p_{ij}					
a	b	c	a′	b′	c′
0.17	0.27	0.37	0.15	0.27	0.39
$EE(A_2) = 0.27$					
Expert 3 (A_3) p_{ij}					
a	b	c	a′	b′	c′
0.07	0.13	0.19	0.06	0.13	0.2
$EE(A_3) = 0.13$					
Expert 4 (A_4) p_{ij}					
a	b	c	a′	b′	c′
0.07	0.13	0.19	0.06	0.13	0.2
$EE(A_4) = 0.13$					
Expert 5 (A_5) p_{ij}					
a	b	c	a′	b′	c′
0.17	0.27	0.37	0.15	0.27	0.39
$EE(A_5) = 0.27$					

For n experts, the similarity matrix can be formed as follows. Note that, for $i = j$, $S(A_i, A_j) = 1$

$$\begin{bmatrix} 1 & S_{(A_1,A_2)} & S_{(A_1,A_3)} & . & S_{(A_1,A_m)} \\ S_{(A_2,A_1)} & 1 & . & . & S_{(A_2,A_m)} \\ . & . & 1 & . & . \\ . & . & . & 1 & . \\ S_{(A_m,A_1)} & S_{(A_m,A_2)} & . & . & 1 \end{bmatrix}$$

Figure 3 shown below is the similarity degree matrix corresponding to BE 1 in the FTA. This is a 5 * 5 matrix since the number of experts selected is 5, and for all similar opinions, the term corresponding in the matrix will be equal to 1 (Fig. 4).

(c) *Calculation of Average Agreement Degree*
For each of the Experts E_i (for i = 1,2…, m) selected, the average agreement degree $AAD(E_i)$ can be found out using,

$$AAD(E_i) = \frac{1}{m-1}\sum_{j=1}^{m} S(A_i, A_j);\ i = (1, 2, \ldots, m),\ j \neq i$$

Table 7 shows the $AAD(E_i)$ w.r.t basic event selected, and $\sum AAD(E_i)$ is the sum of all degrees.

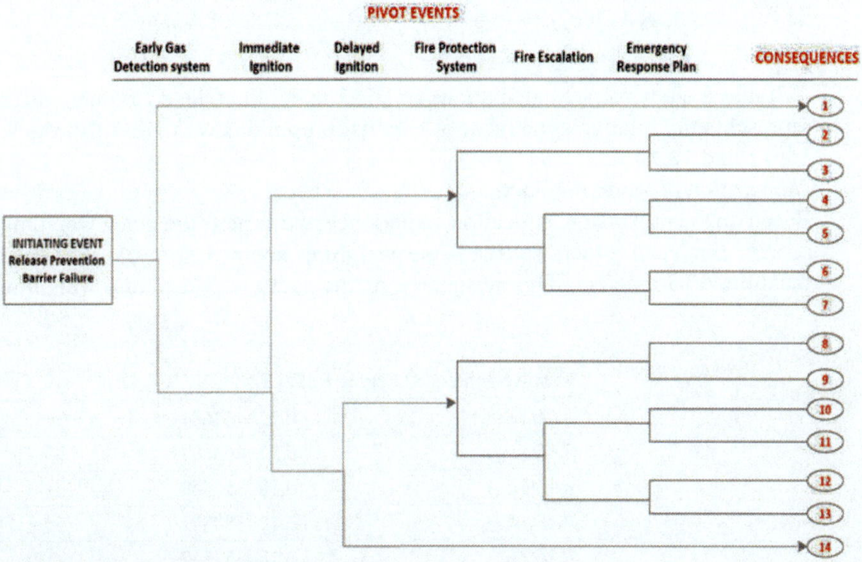

Fig. 3 Event tree related to the initiating event

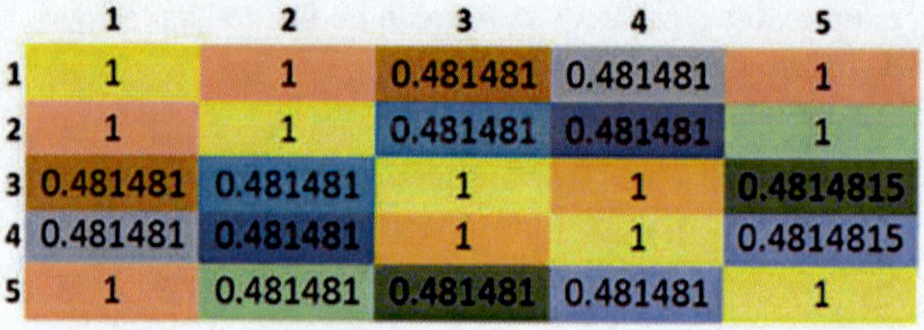

	1	2	3	4	5
1	1	1	0.481481	0.481481	1
2	1	1	0.481481	0.481481	1
3	0.481481	0.481481	1	1	0.4814815
4	0.481481	0.481481	1	1	0.4814815
5	1	0.481481	0.481481	0.481481	1

Fig. 4 Similarity degree matrix for BE 1

Table 7 Average agreement degrees of BE 1

Average agreement degree $AAD(E_i)$ for BE1	
$AAD(E_1)$	0.740740741
$AAD(E_2)$	0.740740741
$AAD(E_3)$	0.611111111
$AAD(E_4)$	0.611111111
$AAD(E_5)$	0.611111111
$\Sigma\, AAD(E_i)$	3.314814815

(d) *Calculation of Relative Agreement Degree*
For each of the Experts E_i (for $i = 1, 2, \ldots, m$) selected, the relative agreement degree $RAD(E_i)$ can be found out using,

$$RAD(E_i) = \frac{AAD(E_i)}{\sum_{i=1}^{m} AAD(E_i)} i = (1, 2, \ldots, m)$$

Table 8 shows RAD calculations of BE 1 using the above formula which represents the relative agreement of experts in opinions w.r.t BE 1 (Tables 9, 10, 11 and 12).

(e) *Calculation of Weighing Factor*
Based on the profession, education, experience, the experts are given weighing scores. For each expert selected, the weighing score differs based on the mentioned parameters. This weightage of the score w.r.t the total weighing

Table 8 Relative agreement degrees of BE 1

Relative agreement degree $RAD\ (E_i)$ for BE 1	
$RAD\ (E_1)$	0.223463687
$RAD\ (E_2)$	0.223463687
$RAD\ (E_3)$	0.184357542
$RAD\ (E_4)$	0.184357542
$RAD\ (E_5)$	0.184357542

Table 9 Weighing factor criteria and scores for experts

Classification	Score
1. Professional position	
Professor, GM/DGM, Chief Engineer, Director	5
Assistant Professor, Manager, Factory Inspector	4
Engineer, Supervisors	3
Foreman, Technician, Graduate apprentice	2
Operator	1
2. Job experience (years)	
>=20	5
15 to 19	4
10 to 14	3
5 to 9	2
<5	1
3. Education	
Ph.D	6
M.Tech	5
MSc or B.Tech	4
Diploma or B.Sc.	3
ITI	2
Secondary school	1

Table 10 Weighing factors of selected experts

Expert	Position score	Experience score	Education score	Weighing score (WS)	Weighing factor (WF)
E1	5	4	6	15	0.3
E2	4	3	5	12	0.24
E3	3	2	4	9	0.18
E4	3	1	5	9	0.18
E5	1	2	2	5	0.1
	Total weighing score			50	

Table 11 Aggregated weights for BE 1

Aggregated weights (w) for BE_1	
w_1	0.261731844
w_2	0.231731844
w_3	0.182178771
w_4	0.182178771
w_5	0.142178771

Table 12 Aggregated IFFP for BE 1

Aggregated IFFP of BE 1 P_j ($\Sigma w_i * p_{ij}$)						
$w_i * p_{ij}$	a	b	c	a′	b′	c′
E1	0.044494	0.070668	0.096841	0.0392598	0.070668	0.102075
E2	0.039394	0.062568	0.085741	0.0347598	0.062568	0.090375
E3	0.012753	0.023683	0.034614	0.0109307	0.023683	0.036436
E4	0.012753	0.023683	0.034614	0.0109307	0.023683	0.036436
E5	0.02417	0.038388	0.052606	0.0213268	0.038388	0.05545
$\Sigma w_i * p_{ij}$	0.133564	0.21899	0.304416	0.1172078	0.21899	0.320772

score gives the weighing factor of the expert and is the indication of the importance of opinion given by the expert.

The weighing factor of the *i*th expert is as follows,

$$WF(E_i) = \frac{WS(E_i)}{\sum_{i=1}^{m} WS(E_i)}; i = (1, 2, \ldots, m)$$

(f) *Calculation of Aggregated Weights*
Aggregated weight w_i is the aggregation of the relative agreement degree $RAD(E_i)$ and weighing factor $WF(E_i)$ with and importance factor β where ($0 \leq \beta \leq 1$). β shows the relative importance given to RAD and WF of experts.

Here, equal weightage is given to both RAD and WF so that β is assigned a value of 0.5, and the aggregated weight $w_i$is obtained as follows,

$$w_i = \beta.RAD(E_i) + (1 - \beta).WF(E_i); i = 1, 2, \ldots, m)$$

(g) *Calculation of Aggregated IFFP*
The calculation of aggregated IFFP is done by combining the opinions of the selected experts and can be done by using the equation given below,

$$P_j = \sum_{i=1}^{m} w_i \otimes p_{ij}; j = 1, 2, \ldots, n$$

Applying the method to the opinions given for other basic events, we get the IFFP for each as given in Table 13.

Step 5: Defuzzification of IFFP to possibility and probability values

The aggregated IFFP obtained thereby is converted to crisp possibility score and then to probability values by using centroid method of defuzzification proposed by Vargheese [7] and logarithmic function proposed by Onisawa and Nishiwaki [6], respectively. Therefore, for an IFFP represented as $(a_i, b_i, c_i; a_i', b_i, c_i')$, the possibility score and probability are,

Table 13 Aggregated IFFP of basic event

Basic event	Intuitionistic fuzzy failure probability					
	a	b	c	a′	b′	c′
1	0.133564	0.21899	0.304416	0.117208	0.21899	0.320772
2	0.048065	0.101798	0.155531	0.041199	0.101798	0.162397
3	0.104432	0.179549	0.254665	0.091493	0.179549	0.267604
4	0.043174	0.095509	0.147845	0.037006	0.095509	0.154012
5	0.026213	0.073703	0.121192	0.022468	0.073703	0.124937
6	0.04443	0.097125	0.149819	0.038083	0.097125	0.156166
7	0.259442	0.381772	0.504101	0.235388	0.381772	0.528156
8	0.043787	0.096297	0.148808	0.037532	0.096297	0.155063
9	0.039587	0.090897	0.142208	0.033932	0.090897	0.147863
10	0.04443	0.097125	0.149819	0.038083	0.097125	0.156166
11	0.007942	0.050211	0.09248	0.006807	0.050211	0.093614
12	0.037983	0.09037	0.142757	0.033132	0.09037	0.147607
13	0.270858	0.369019	0.46718	0.251939	0.369019	0.486099
14	0.108077	0.184149	0.260221	0.094585	0.184149	0.273713
15	0.136525	0.21981	0.303094	0.121082	0.21981	0.318537
16	0.043174	0.095509	0.147845	0.037006	0.095509	0.154012
17	0.009279	0.05193	0.094581	0.007953	0.05193	0.095906
18	0.033783	0.08497	0.136157	0.029532	0.08497	0.140407
19	0.043787	0.096297	0.148808	0.037532	0.096297	0.155063
20	0.228582	0.343114	0.457646	0.20596	0.343114	0.480268
21	0.126585	0.210189	0.293794	0.11129	0.210189	0.309088

$$S = \frac{1}{3}\left[\frac{(c' - a')(b - 2c' - 2a') + (c - a)(a + b + c) + 3(c'^2 - a'^2)}{(c' - a' + c - a}\right]$$

$$P = \begin{cases} \dfrac{1}{10^{\left(\left[\frac{1-S}{S}\right]^{\frac{1}{3}} x2.301\right)}}; S \neq 0 \\ . \\ 0; S = 0 \end{cases}$$

$S =$ crisp probability score
$P =$ probability
Table 14 shows the defuzzified values of possibility scores and probability values of the 21 BE.

Table 14 Possibility and probability of basic events

Basic event	Intuitionistic fuzzy failure probability						Possibility(S)	Failure probability of basic events
	a	b	c	a′	b′	c′		
1	0.133564	0.21899	0.304416	0.117208	0.21899	0.320772	0.218989944	0.000305101
2	0.048065	0.101798	0.155531	0.041199	0.101798	0.162397	0.101797902	1.75891E−05
3	0.104432	0.179549	0.254665	0.091493	0.179549	0.267604	0.179548503	0.000151939
4	0.043174	0.095509	0.147845	0.037006	0.095509	0.154012	0.095509317	1.35453E−05
5	0.026213	0.073703	0.121192	0.022468	0.073703	0.124937	0.073702632	4.4687E−06
6	0.04443	0.097125	0.149819	0.038083	0.097125	0.156166	0.097124845	1.4515E−05
7	0.259442	0.381772	0.504101	0.235388	0.381772	0.528156	0.381771689	0.001985725
8	0.043787	0.096297	0.148808	0.037532	0.096297	0.155063	0.096297368	1.40125E−05
9	0.039587	0.090897	0.142208	0.033932	0.090897	0.147863	0.090897368	1.10258E−05
10	0.04443	0.097125	0.149819	0.038083	0.097125	0.156166	0.097124845	1.4515E−05
11	0.007942	0.050211	0.09248	0.006807	0.050211	0.093614	0.05021066	7.39707E−07
12	0.037983	0.09037	0.142757	0.033132	0.09037	0.147607	0.09036997	1.07603E−05
13	0.270858	0.369019	0.46718	0.251939	0.369019	0.486099	0.369018953	0.001772071
14	0.108077	0.184149	0.260221	0.094585	0.184149	0.273713	0.184149176	0.000166274
15	0.136525	0.21981	0.303094	0.121082	0.21981	0.318537	0.219809784	0.000309064
16	0.043174	0.095509	0.147845	0.037006	0.095509	0.154012	0.095509317	1.35453E−05
17	0.009279	0.05193	0.094581	0.007953	0.05193	0.095906	0.051929787	8.73243E−07
18	0.033783	0.08497	0.136157	0.029532	0.08497	0.140407	0.08496997	8.29313E−06
19	0.043787	0.096297	0.148808	0.037532	0.096297	0.155063	0.096297368	1.40125E−05
20	0.228582	0.343114	0.457646	0.20596	0.343114	0.480268	0.343113982	0.001389456
21	0.126585	0.210189	0.293794	0.11129	0.210189	0.309088	0.210189384	0.000264674

Step 6: Top event probability Estimation.

(a) *Cut Set Analysis*
Cut set analysis was done on FT to find out the most significant basic events or their combinations which when occurred will propagate and reach the TE. The probabilities of the cut sets were also calculated and shown in Table 15.

(b) *Calculation of Top Event Probability*
Making use of the simple Boolean algebra, the top event probability can be calculated. The basic operations on gates used here are as follows,

For an OR gate with two events as input, the probability of the output event P_{a+b} is,

$$P_{a+b} = P(a) + P(b) - P(a) * P(b)$$

Table 15 Cut sets of fault tree

Serial No	Cut set	Probability
1	BE 10	1.45E−05
2	BE 9	1.10E−05
3	BE 12	1.08E−05
4	BE 11	7.40E−07
5	BE 1, BE 7	6.06E−07
6	BE 20, BE 21	3.68E−07
7	BE 3, BE 7	3.02E−07
8	BE 2, BE 7	3.49E−08
9	BE 4, BE 7	2.69E−08
10	BE 1, BE 6	4.43E−09
11	BE 1, BE 8	4.28E−09
12	BE 3, BE 6	2.21E−09
13	BE 3, BE 8	2.13E−09
14	BE 1, BE 5	1.36E−09
15	BE 3, BE 5	6.79E−10
16	BE 2, BE 6	2.55E−10
17	BE 2, BE 8	2.46E−10
18	BE 4, BE 6	1.97E−10
19	BE 4, BE 8	1.90E−10
20	BE 18, BE 19	1.16E−10
21	BE 13, BE 14, BE 15	9.11E−11
22	BE 2, BE 5	7.86E−11
23	BE 4, BE 5	6.05E−11
24	BE 16, BE 17	1.18E−11

If the events are considered independent and $P(a) \cdot P(b)$ is very small, then the above equation can be approximated as

$$P_{a+b} = P(a) + P(b)$$

Therefore, for an n input gate, the equation becomes $P_{a+b+\cdots+n} = P(a) + P(b) + \cdots + P(n)$

For an AND gate with two independent events as input, the probability of the output event $P_{a.b}$ is,

$$P_{a*b} = P(a) * P(b)$$

Therefore, for an n input gate, the equation becomes $P_{a*b\ldots*n} = P(a) * P(b) * \cdots * P(n)$

Applying the above algebra on the FT using the obtained probability value BE, the probability of the top event P_T which is the *failure of release prevention barrier is estimated to be* ***2.65E−3***.

Step 7: Pivot Event Probability Estimation

The same method of aggregation is applied here for the opinions by experts, and final IFFP values are obtained. Afterwards, defuzzification procedures are done in order to get the possibility and probability values of PE.

Tables 16 and 17 show aggregated IFFP of PEs and list of PEs with estimated probability.

Step 8: Consequence Probability Estimation

Let

P_T = probability of occurrence of the TE.

P_1 = probability of occurrence/failure of PE 1.

Then, the success and failure probability of the first branch will be calculated as follows,

$$P(failure) = P_T * P_1$$

$$P(success) = P_T * (1 - P_1)$$

Similarly, successive failure and success probabilities of the upcoming branches are calculated until the ET reaches the undesired outcome/consequence. Table 18 shows the probability values of consequences obtained.

Table 16 Aggregated IFFP of pivot events

Pivot events	Intuitionistic fuzzy probability						Possibility score(S)	Crisp probability of pivot events
	a	b	c	a′	b′	c′		
1	0.5780511	0.6854301	0.7928091	0.5546432	0.6854301	0.816217	0.68543013	1.67928E−02
2	0.8827631	0.9295334	0.9763038	0.8759928	0.9295334	0.9830741	0.929533443	1.06208E−01
3	0.8502058	0.9028957	0.9555855	0.837516	0.9028957	0.9682754	0.902895674	8.04913E−02
4	0.4943334	0.5997112	0.705089	0.4701459	0.5997112	0.7292765	0.599711174	9.75180E−03
5	0.8752278	0.9233682	0.9715086	0.8670874	0.9233682	0.979649	0.923368212	9.91566E−02
6	0.2511691	0.373716	0.496263	0.2266597	0.373716	0.5207724	0.373716022	1.84873E−03

Table 17 List of pivot events with and probability

Pivot event	Pivot event description	Probability
1	Probability of failure of early gas detection system	1.67928E−02
2	Probability for immediate ignition	1.06208E−01
3	Probability for delayed ignition	8.04913E−02
4	Probability of failure of fire protection system	9.75180E−03
5	Probability of fire escalation	9.91566E−02
6	Probability of failure of emergency response	1.84873E−03

Table 18 Pivot events with estimated probability

Consequence	Consequence description	Outcome	Probability
1	Safe detection and rectification	Safe	2.60550E−02
2	Explosion and fire hazards causing minimal damage and casualties	Minimal damage and casualties	3.93140E−04
3	Explosion and fire hazards causing minimal damage with possible casualties	Minimal damage and possible casualties	7.28155E−07
4	Explosion and fire spread causing increased damages and possible casualties	Increased damage and possible casualties	3.48769E−06
5	Explosion and fire spread causing increased damages and casualties	Increased damage and casualties	6.45973E−09
6	Explosion and fire hazards causing moderate damages and possible casualties	Moderate damage and possible casualties	3.83892E−07
7	Explosion and fire hazards causing moderate damages and casualties	Moderate damage and casualties	7.17075E−09
8	VCE and fire hazards causing minimal damage and casualties	Minimal damage and casualties	4.29559E−05
9	VCE and fire hazards causing minimal damage with possible casualties	Minimal damage and possible casualties	7.05813E−10
10	VCE with fire spread causing increased damages and possible casualties	Increased damage and possible casualties	3.81077E−07
11	VCE and fire spread causing increased damages and casualties	Increased damage and casualties	7.05813E−10

(continued)

Table 18 (continued)

Consequence	Consequence description	Outcome	Probability
12	VCE and fire hazards causing moderate damages and possible casualties	Moderate damage and possible casualties	4.19455E−08
13	VCE and fire hazards causing moderate damages and casualties	Moderate damage and casualties	7.76894E−11
14	Formation of hazardous mixture with air	Hazardous atmosphere	3.80431E−06

3 Results and Discussions

Using the IFFP method, the failure probabilities of BEs, PEs, and consequences were identified. Tables 19, 20, and 21 show, respectively, the probability values of BEs,

Table 19 Basic events with probability in descending order

Basic event	Basic event description	Probability
7	Failure of external supervision	0.001985725
13	Inadequate methods for detecting HTHA	0.001772071
20	HTHA degradation monitoring performed but failed to detect	0.001389456
15	Failure of HTHA inspection of heat exchanger	0.000309064
1	High-temperature hydrogen attack (HTHA)	0.000305101
21	HTHA degradation monitoring specified but not performed	0.000264674
14	Inadequate training of the inspectors to detect HTHA easily	0.000166274
3	No report on leaks from heat exchanger during start up	0.000151939
2	Difficulty with valve operation during start up	1.75891E−05
6	No permit on job carrying	1.4515E−05
10	High mechanical stress	1.4515E−05
8	Incorrect procedure	1.40125E−05
19	Delay maintenance operations	1.40125E−05
4	Hydrogen-induced cold cracking	1.35453E−05
16	Failure of detection of leaks from heat exchanger flanges	1.35453E−05
9	Poor construction material for NHT heat exchanger	1.10258E−05
12	Long delay in inspection schedule	1.07603E−05
18	Wrong maintenance procedure	8.29313E−06
5	Inexperience	4.4687E−06
17	Failure of minor release detection	8.73243E−07
11	Insufficient instrumentation to measure process conditions	7.39707E−07

Table 20 Pivot events with probability in descending order

Pivot event	Pivot event description	Probability
2	Probability for immediate ignition	0.106208
5	Probability of fire escalation	0.099157
3	Probability for delayed ignition	0.080491
1	Probability of failure of early gas detection system	0.016793
4	Probability of failure of fire protection system	0.009752
6	Probability of failure of emergency response	0.001849

Table 21 Consequence with probability in descending order

Consequence	Consequence description	Outcome	Probability
1	Safe detection and rectification	Safe	0.02605499
2	Explosion and fire hazards causing minimal damage and casualties	Minimal damage and casualties	0.00039314
8	VCE and fire hazards causing minimal damage and casualties	Minimal damage and casualties	4.29559E−05
14	Formation of hazardous mixture with air	Hazardous atmosphere	3.80431E−06
4	Explosion and fire spread causing increased damages and possible casualties	Increased damage and possible casualties	3.48769E−06
3	Explosion and fire hazards causing minimal damage with possible casualties	Minimal damage and possible casualties	7.28155E−07
6	Explosion and fire hazards causing moderate damages and possible casualties	Moderate damage and possible casualties	3.83892E−07
10	VCE with fire spread causing increased damages and possible casualties	Increased damage and possible casualties	3.81077E−07
12	VCE and fire hazards causing moderate damages and possible casualties	Moderate damage and possible casualties	4.19455E−08
7	Explosion and fire hazards causing moderate damages and casualties	Moderate damage and casualties	7.17075E−09
5	Explosion and fire spread causing increased damages and casualties	Increased damage and casualties	6.45973E−09
9	VCE and fire hazards causing minimal damage with possible casualties	Minimal damage and possible casualties	7.05813E−10

(continued)

Table 21 (continued)

Consequence	Consequence description	Outcome	Probability
11	VCE and fire spread causing increased damages and casualties	Increased damage and casualties	7.05813E−10
13	VCE and fire hazards causing moderate damages and casualties	Moderate damage and casualties	7.76894E−11

PEs, and consequences in the descending order. As compared to other method, this method allows a clear ranking since all the probability values obtained will be have more significant figures.

4 Conclusions

This study shows that FTA and ETA along with the use of IF method yield a model which is simple as well as reliable for assessing the risk associated with events with unknown probability and uncertainties. Even though the risk associated with an instrument/process is less, it is always good to have the understanding about the same. For such a purpose, this model can be effectively and effortlessly applied. The method successfully identifies the probability associated with each of BE, PE, and consequences and reduces vagueness in the obtained values which helps in effective differentiation and ranking. However, the application and feasibility across different processes and equipment shall have to be found out by applying the method. Reducing the gap in the proposed approach by sensitivity analysis for importance factor, different importance analysis and its comparison, etc. is the modifications that have to be done in the approach.

References

1. Adedigba, S. A., Khan, F., & Yang, M. (2016). Dynamic safety analysis of process systems using nonlinear and non-sequential accident model. *Chemical Engineering Research and Design, 111*, 169–183. https://doi.org/10.1016/j.cherd.2016.04.013
2. Atanassov, K. T. (1986). Intuitionistic fuzzy sets. *Fuzzy Sets and Systems, 20*(1), 87–96. https://doi.org/10.1016/S0165-0114(86)80034-3
3. Huang, D., Chen, T., & Wang, M. J. J. (2001). A fuzzy set approach for event tree analysis. *Fuzzy Sets and Systems, 118*(1), 153–165. https://doi.org/10.1016/S0165-0114(98)00288-7
4. Kumar, M., & Kaushik, M. (2020). System failure probability evaluation using fault tree analysis and expert opinions in intuitionistic fuzzy environment. *Journal of Loss Prevention in the Process Industries, 67*, 104236.https://doi.org/10.1016/j.jlp.2020.104236
5. Liu, J., et al. (2008). Self-tuning of fuzzy belief rule bases for engineering system safety analysis. *Annals of Operations Research, 163*(1), 143–168. https://doi.org/10.1007/s10479-008-0327-0

6. Onisawa, T., & Nishiwaki, Y. (1988). Fuzzy human reliability analysis on the Chernobyl accident. *Fuzzy Sets and Systems, 28*(2), 115–127. https://doi.org/10.1016/0165-0114(88)90194-7
7. Varghese, A. (2012). Centroid of an intuitionistic fuzzy number. *Notes on Instuitionistic Fuzzy Sets, 18*(1), 19–24.
8. Zadeh, L. A. (1965). Fuzzy sets. *Information and Control*. https://doi.org/10.1016/S0019-9958(65)90241-X

Mathematical Modelling Approach to Estimate COVID-19 Susceptibility and Rate of Transmission

Ankit Dasgotra, Vishal Kumar Singh, S. M. Tauseef, Ravi Kumar Patel, Sunil Kumar Tiwari, and Bikarama Prasad Yadav

1 Introduction

SARS-CoV was first detected in humans in 2002 and then the strains of the coronavirus family emerged at various times such as H1N1 followed by H5N1 and H5N7 during 2009. After a few years, MERS-CoV emerged in the Middle East in 2015 which was again a respiratory syndrome, whereas now, from the same family of a novel coronavirus, COVID-19 has emerged as the most contagious of all [1, 2]. COVID-19 has been stated pandemic, and the virus which causes it has been found to have profound variations in its arrangement and epidemiology. They have zoonotically enclosed RNA respiratory viruses that barely spread in their original form among humans, but they replicate to permit effective human transmission. Similar properties were shown by SARS in 2002–2003 and H1N1 influenza in 2009 [3, 4].

Common and recognized routes of transmission were identified as droplet transmission and airborne transmission. In droplet transmission, the droplet diameter is greater than 5 μm and travelling distance is less than 1 m. Droplets act as a viable virus carrier, and on encountering facial parts such as nose, lips, or eyes, the virus enters the upper respiratory tract. Accordingly, in airborne transmission, the droplets with diameter ≤ 5 μm can travel a distance more than 1 m and enter through the nasal passage of individuals [4, 5]. The two different modes (i) absolute contact transmission (contaminated surface not involved) and (ii) indirect contact transmission (contaminated surface involved) in the transmission of pandemic potential viruses have been provocative [5–7]. Nevertheless, a range of studies and models has proposed that

A. Dasgotra (✉) · R. K. Patel · S. K. Tiwari
School of Engineering, University of Petroleum & Energy Studies, 248007 Bidholi, Dehradun, India

A. Dasgotra · V. K. Singh · S. M. Tauseef · B. P. Yadav
Department of Health, Environment and Civil Engineering, University of Petroleum and Energy Studies, Safety, Dehradun 248007, India

N. A. Siddiqui et al. (eds.), *Advances in Behavioral Based Safety*,
https://doi.org/10.1007/978-981-16-8270-4_2

indirect contact transmission would be the primary mode of transmission for certain respiratory viruses, in certain situations [8].

Furthermore, some reports show that coronavirus like SARS, MERS, and influenza (H1N1) are capable of survival on dry surfaces for long duration that allows further spread [9–11]. SARS-CoV and influenza virus can also live for prolonged periods in the environmental reservoirs such as soil, food, and sewage [12–17]. The significance of transmission of such viruses can be seen in the study of influenza and human coronavirus experiments, which were conducted on dry surfaces, field investigations, and surface sampling for viruses.

Several studies, involving viruses from CoV family on dry surfaces, have highlighted the role of infected surfaces in the spread of viruses, and consideration of the role of dry surfaces in the spread of the virus may further improve findings in this area [1]. Comparing transmission paths, droplets and airborne routes are a significant source of direct contact, whereas transmission source of indirect contact is still unknown. Influenza virus, SARS-CoV, and possibly MERS-CoV when released into the atmosphere in amounts well above the contagious dosage, they could live on surfaces for long periods, and several authorities or studies have reported contamination of hospital surfaces [18]. The infected surface may further contaminate hands or instruments, thereby causing inoculation by interaction with facial parts such as eyes, nose, and mouth. Pandemic potential viruses like influenza, MERS-CoV, and SARS-CoV could live on dry surfaces for prolonged periods, causing contamination in open locations which requires improved disinfection ensuring prevention and successful control of infection [1]. Research is done to validate empirical mathematical model, various models have been developed to estimate the rate of virus suppression, etc., and majorly, it was implemented on coronavirus family viruses like Middle East respiratory syndrome (MERS) and severe acute respiratory syndrome (SARS) [1, 18].

Mathematical modelling can be used as an appropriate tool in these types of virus susceptibility and rate of transmission effecting, susceptible-infected-recovered (S-I-R) is a basic infection dynamic approach for mathematical modelling, and it has been implemented in SARS, MERS, Hong Kong flu, etc. Advance approach such as susceptible-infected-recovered-susceptible (S-I-R-S), susceptible-exposed-infected-recovered (S-E-I-R), and bats-hosts-reservoir-people (B-H-R-P) transmission network model can be implemented in COVID-19 computational research. Resulting data can be used to validate the concepts of social distancing, quarantine, lockdown, etc. with a realistic scenario.

The outbreak caused by COVID-19 has challenged the researchers and health officials due to the exponential increase in the number of infected cases along with significant escalation in the number of deaths. Detail analysis is still undergoing, but the scientific data are yet to be revealed, whereas the public is interested in the duration of the outbreak and the expected number of infected cases to be reached at its highest level. Since the outbreak, various mathematical models have been proposed by various researchers to predict the number using calculation [19]. In this paper, we have reviewed models that can be used to predict the spread of this novel coronavirus (COVID-19) and the effect of mitigate measures to contain its spread.

2 COVID-19

The novel coronavirus (COVD-19) has affected approximately 213 countries including small territories also. Given the magnitude and the extent of its spread, the WHO has declared it a pandemic, and globally, it has been treated as a notified disaster. Apart from human suffering, it is also causing major economic disruptions. As per WHO, coronavirus (COVID-19) belongs to the coronavirus family such as the Middle East respiratory syndrome coronavirus (MERS-CoV) and severe acute respiratory syndrome coronavirus (SARS-CoV) which causes illness ranging from the common cold to respiratory infections [19, 20]. It has been predicted that the emergence of this virus has been from bats which were primarily from the Wuhan market, China [19, 20]. Symptoms of COVID-19 show from common cold further leading to respiratory disease, which is being spread after people, encounter the symptomatic person [18–20]. Figure 1 represents some common transmission modes of COVID-19.

As on 10 September 2020, 28,238,792 cases of novel coronavirus have been detected globally resulting in 911,500 deaths, and 20,238,900 have recovered. The active number of cases remains 7,027,675, and 60,717 patients are reported to be in critical condition. United States of America, India, Brazil, Russia, and Peru are five major impacted countries combined having approximately 61% of global coronavirus cases [20]. Table 1 represents the top 10 countries with their COVID-19 data.

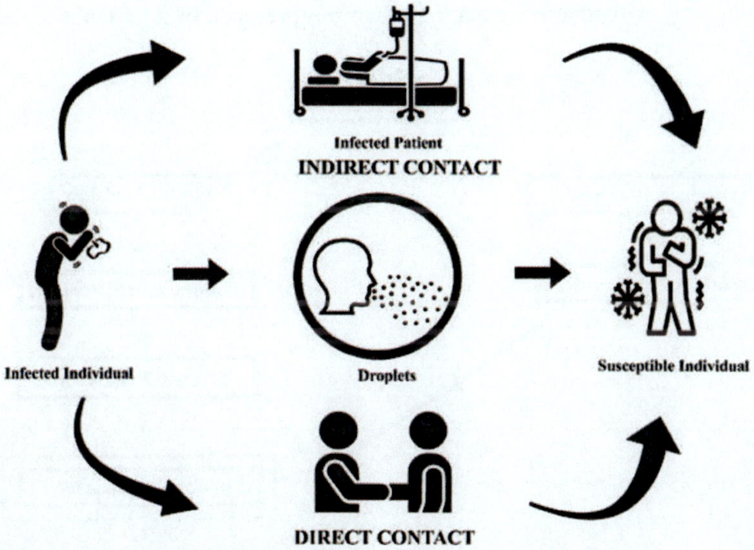

Fig. 1 Common transmission modes of COVID-19 [1, 18–20]

Table 1 Top 10 countries COVID-19 data till 27th November 2020 [20]

S. No	Country	Total cases	Total deaths
1	United States of America	13,615,593	272,284
2	India	9,406,995	136,878
3	Brazil	6,290,272	172,637
4	Russia	2,269,316	39,527
5	France	2,208,699	52,127
6	Spain	1,646,192	44,668
7	UK	1,605,172	58,030
8	Italy	1,564,532	54,363
9	Argentina	1,413,375	38,322
10	Colombia	1,299,613	36,401

3 Mathematical Modelling Methodology

Modelling is considered as an important tool to study epidemics worldwide, and it is useful for studying of infection dynamics approach in world emerging problems like COVID-19 nowadays. In Fig. 2, we can signify the route map of the mathematical approach, and it is implementing connectivity with the real scenario.

Mathematical modelling can be utilized to study the contribution of the diverse elements to the empirical observations. This relation can be stated with ease not containing any mathematical equation when the presence of a variable represents

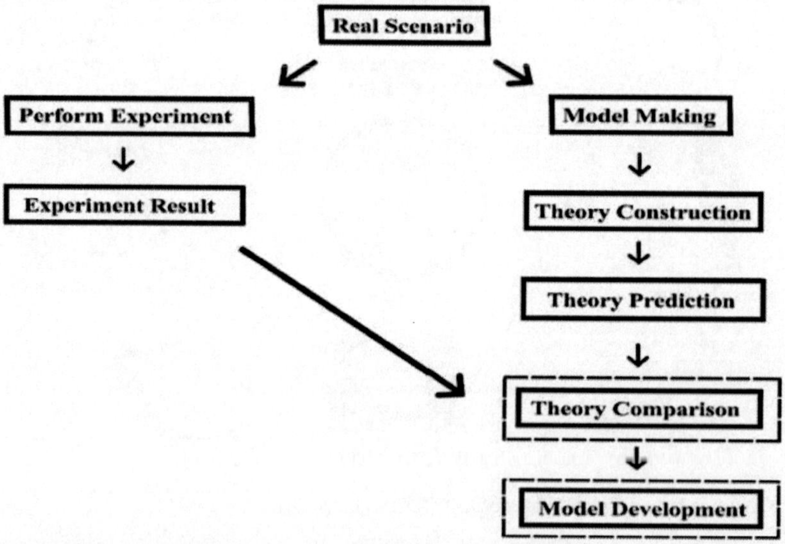

Fig. 2 Approach and connectivity of mathematical modelling [21]

linearity to various variables. In these types of modelling, variables raising the rate of infection are not adding up; therefore, these models play an important role in the identification of disease/virus [21]. The following are some major utilized infectious dynamics and advanced models which can be implemented in predicting future situations.

3.1 *Susceptible-Infected-Recovered (S-I-R) Modelling*

S-I-R model proposed by William Ogilvy Kermack can be used to detect the spread of disease and is based on a simplified approach that supposes that spread of the virus is mainly due to person-to-person transmission [21, 22]. In S-I-R type of modelling, very few infected individuals were considered that can pass the virus/disease to many people. Three groups are mainly categorized here passing through stages of check-ups, etc. Firstly, they are susceptible, then they become infectious, and finally, they recover [23]. This type of methodology is applied in a situation that occurs for a short period, and it takes the form of a disease that infects or kills a major part of the society before it vanishes [24].

$$\mathrm{S} \rightarrow \mathrm{I} \rightarrow \mathrm{R}$$

Modelling starts with the identification of independent and dependent variables. t = independent variable, i.e., period in days.

As each variable is a function of time, S-I-R variables are stated as:

$$\begin{aligned} \mathrm{S} &= \mathrm{S(t)}, \\ \mathrm{I} &= \mathrm{I\,(t)}, \\ \mathrm{R} &= \mathrm{R\,(t)} \end{aligned}$$

If P is the total population in a specific area, the fraction of the total population concerning all three variables can be represented as

$$\begin{aligned} \mathrm{s(t)} &= \mathrm{S(t)/P}, \\ \mathrm{i(d)} &= \mathrm{I(t)/P}, \\ \mathrm{r(t)} &= \mathrm{R(t)/P} \end{aligned}$$

Using fractions makes interpretation easy as each of the variables susceptible, infected, and recovered is reported as a fraction of the total population. The following are some assumptions to be considered while defining this differential equation model approach.

I. At each time, $\mathbf{s(t) + i(t) + r(t) = 1}$ and $\mathbf{S + I + R = P}$

II. Here, new born and immigrants are ignored; therefore, no addition to the susceptible category.
III. A person leaves the susceptible group by entering the infected category.
IV. If every infected person contacts a specified range of people (i.e., b) per day that is a sufficient number for virus supersession.
V. On average, an infected person infects several new people every day is represented by bs(t).
VI. Fixed fraction (i.e., k) number of infected persons recovers on a specific given day ("infected" is "infectious" that can transmit the virus to a susceptible individual. A "recovered" individual can still feel miserable and might even die later from virus. With these assumptions, the authors arrived at the following differential equations.

(a) The susceptible equation:

$$\frac{ds}{dt} = -bs(t)I(t)$$

This equation leads to the following differential equation for s(t).

$$\frac{ds}{dt} = -bs(t)i(t)$$

(b) The recovered equation:

$$\frac{dr}{dt} = ki(t)$$

(c) The infected equation:

$$\frac{ds}{dt} + \frac{di}{dt} + \frac{dr}{dt} = U$$

Therefore,

$$\frac{di}{dt} = bs(t)i(t) - ki(t)$$

Figure 3 represents the trajectory of the epidemic in the I-S graph, and it is clear that threshold effect can be observed from the existent curve.

$$R_0 = \frac{Pb}{K} > 1$$

The equation states that the number of infected will increase until the susceptible is reduced to k/b and will decrease thereafter.

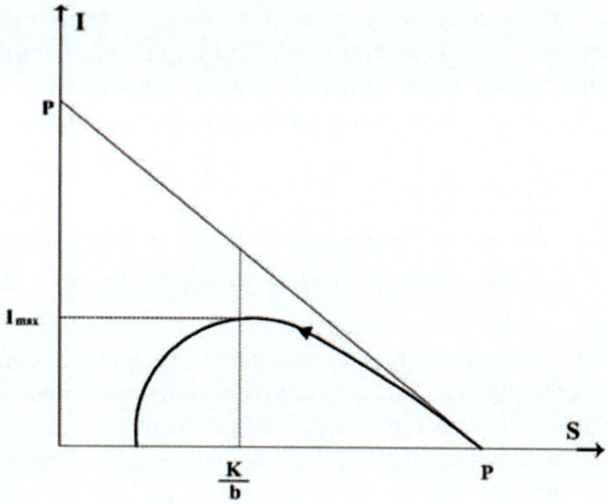

Fig. 3 Graphical representation of epidemic [21–24]

Here, R_0 = basic reproductive ratio represents the threshold for an epidemic to occur. It also represents the average value of susceptible contaminated by an infected person. Now, dividing the infected equation by the susceptible equation,

$$\frac{di}{ds} = -\frac{bs(t)i(t) - ki(t)}{bs(t)i(t)}$$

Therefore,

$$\frac{di}{ds} = -\left(1 + \frac{ki(t)}{bs(t)i(t)}\right) = \frac{k}{bs(t)} - 1$$

Integrating the above equation,

$$i = \frac{k}{b}\log s(t) - s(t) + c$$

Here, the approximation calculation for the value of constant, i.e., c as

$$c \approx P - \frac{k}{b}\log P$$

From this integrated equation, the maximum number of the infections can be computed as

$$I_{max} = P\left(1 - \frac{1 + \log R_0}{R_0}\right)$$

The above equation states that infection (I) must disappear somewhere near the positive value of susceptible (s). This implies that the epidemic disappears. Therefore, the epidemic terminates before the condition where all susceptible moves to the infected category and few infected recovers completely.

3.2 Susceptible-Infected-Recovered (S-I-R) Modelling Considering Demographic Effects

Basic susceptible-infected-recovered (S-I-R) modelling is elaborated in this section with the addition of demography, i.e., new-born babies, immigrants, and deaths taking place are considered here [23, 25]. This model closely represents the real scenario. The model is represented with the help of a block diagram in Fig. 4. Here, births (or immigrants) addition is represented at the rate "m" and deaths/emigrants at the rate "d".

(d) The susceptible equation with demography:

$$\frac{ds}{dt} = mP - bs(t)i(t) - ds(t)$$

(e) The recovered equation with demography:

$$\frac{dr}{dt} = ki(t) - dr(t)$$

(f) The infected equation with demography:

$$\frac{di}{dt} = bs(t)i(t) - ki(t) - di(t)$$

Total population, P = S + I + R (and dP/dt = 0).

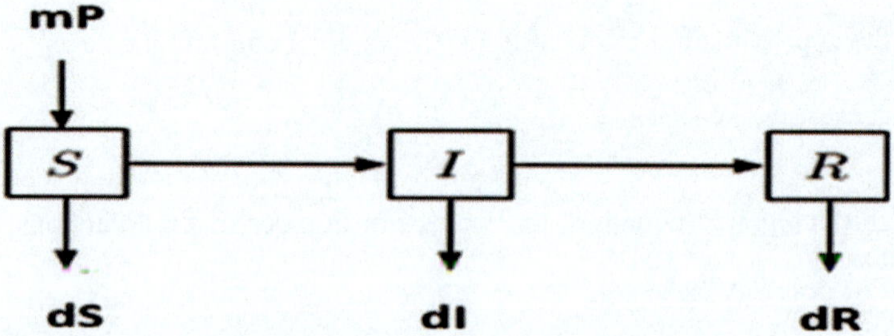

Fig. 4 Representation of S-I-R modelling with demography [23, 25]

Here, an assumption on birth rate and death rate is the same, i.e., m = d. There exist two equilibrium stages, i.e., no disease (represented by superscript *) and endemic (represented by superscript **). This endemic scenario exists with the assumption that the basic reproductive ratio remains positive (i.e., $R_0 > 1$).

$$S^* = P,\ I^* = 0,\ R^* = 0$$

And,

$$S^{**} = \frac{k+d}{b}$$

Dividing both sides by P,

$$\frac{S^{**}}{P} = \frac{k+d}{Pb} = \frac{1}{R_0},$$

$$I^{**} = \frac{d(P - S^{**})}{bS^{**}} = \frac{d(R_0 - 1)}{b}$$

Endemic equilibrium exists only if $S^{**} < P$ and $I^{**} > 0$ which means $R_0 > 1$. Here, $R_0 = \frac{Pb}{k+d}$

3.3 *Susceptible-Exposed-Infected-Recovered (S-E-I-R) Modelling*

Susceptible-exposed-infected-recovered (S-E-I-R) modelling approach is a modified S-I-R modelling approach that includes the category of people who are exposed to the virus but cannot transfer it to possible contacts. This type of occurrence delays and minimizes the chances of virus transmission via contact, etc. [21, 24]. Mainly, when a virus carrier is infectious without symptoms, there is a possibility that he/she cannot transmit it to another person. Here, term h is introduced that specifies the latent period in an exposed individual. The latent period specifies a period of virus/disease, where a person is exposed to infection, but he/she is not yet infectious [26]. In simple words, the rate at which an exposed person becomes infective, i.e., period of transmission from E to I.

As each variable is a function of time, exposed variable E is stated as E = E (t).

P is the total population in a specific area, and the fraction of the total population concerning exposed term can be represented as e(t) = E(t)/P (Fig. 5).

Total population, P = S + E + I + R.

And, basic reproductive ratio (R_0) can be calculated here assuming $m = d$ and represented as

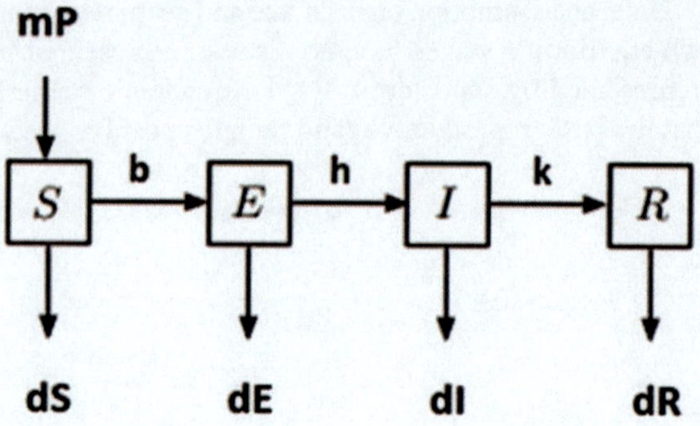

Fig. 5 Compartmental diagram representing of S-E-I-R modelling [21, 24, 26]

$$R_0 = \frac{Pb}{k+d}\frac{h}{h+d}$$

(a) The susceptible equation in S-E-I-R model:

$$\frac{ds}{dt} = m(P - s(t)) - b\frac{s(t)i(t)}{P} - ds(t)$$

(b) The exposed equation in S-E-I-R model:

$$\frac{de}{dt} = \frac{bs(t)i(t)}{P} - (m+h).e(t)$$

(c) The recovered equation in S-E-I-R model:

$$\frac{dr}{dt} = ki(t) - dr(t) + ms(t)$$

(d) The infected equation in S-E-I-R model:

$$\frac{di}{dt} = he(t) - (m+k).i(t)$$

3.4 *Susceptible-Infected-Recovered-Susceptible (S-I-R-S) Modelling*

S-I-R-S modelling is mainly considered where immunity exists for a short period and the recovered person re-enters the susceptible categorized group. Therefore, it is known as an epidemic model having the existence of temporary immunity [26, 27]. A term z is considered here to specify the immunity losing rate. Mathematical modelling of S-I-R-S is represented as follows (Fig. 6).

Three main differential equations represented as

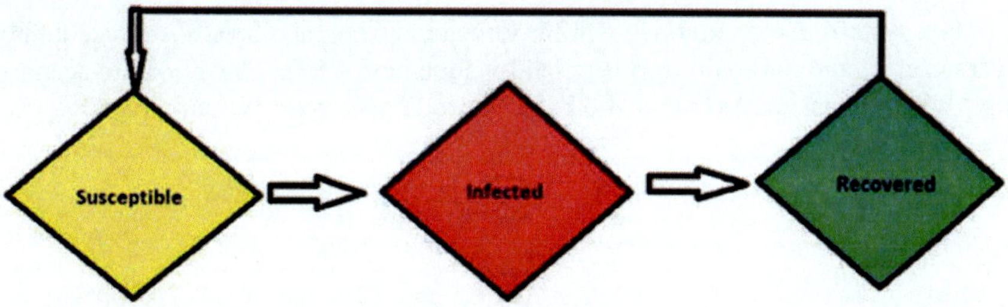

Fig. 6 S-I-R-S compartmental representation (it supposes source is saved from re-entering of infection temporally after recovery) [26, 27]

$$\frac{ds}{dt} = -bs(t)i(t) + zr(t),$$

$$\frac{di}{dt} = bs(t)i(t) - ki(t),$$

$$and \frac{dr}{dt} = ki(t) - zr(t)$$

Total population, P = S + I + R and dP/dt = 0.

In this system, there exist two equilibrium stages, i.e., no disease (represented by superscript *) and endemic (represented by superscript **). This endemic scenario exists with the assumption that the basic reproductive ratio remains positive (i.e., Pb/k > 1).

$$S^* = P, \ I^* = 0, \ R^* = 0$$

And,

$$S^{**} = \frac{k}{b}, I^{**} = \frac{P - \left(\frac{k}{b}\right)}{1 + \left(\frac{k}{z}\right)}, R^{**} = \frac{P - \left(\frac{k}{b}\right)}{1 + \left(\frac{z}{k}\right)} \tag{1}$$

Three main differential equations represented as

$$\frac{ds}{dt} = -bs(t)i(t) + zr(t),$$

$$\frac{di}{dt} = bs(t)i(t) - ki(t),$$

$$and \frac{dr}{dt} = ki(t) - zr(t)$$

Total population, P = S + I + R and dP/dt = 0.

In this system, there exist two equilibrium stages, i.e., no disease (represented by superscript *) and endemic (represented by superscript **). This endemic scenario exists with the assumption that the basic reproductive ratio remains positive (i.e., Pb/k > 1).

$$\mathrm{S}^{*} = \mathrm{P},\ \mathrm{I}^{*} = 0,\ \mathrm{R}^{*} = 0$$

And,

$$S^{**} = \frac{k}{b},\ I^{**} = \frac{P - \left(\frac{k}{b}\right)}{1 + \left(\frac{k}{z}\right)},\ R^{**} = \frac{P - \left(\frac{k}{b}\right)}{1 + \left(\frac{z}{k}\right)} \tag{2}$$

3.5 *Bats-Hosts-Reservoir-People Transmission Network Model*

B-H-R-P transmission network model is a newly developed mathematical model that was proposed by bioRxiv on 19 January 2020 [25]. It is specified on COVID-19 occurrence, developed using some known sources/parameters of transmission. Here, it is stated that the virus transverses between the main sources, i.e., bats, and then it enters into unknown hosts, i.e., wild animals, fishes. The hosts were sent to market in the form of seafood, and other sources were stated as the reservoir of the novel coronavirus. People get exposure through buying these products in the market, and thus, virus is further transmitted. Calculation of basic reproduction number (R0) is mainly from the reservoir-people model for accessibility of COVID-19 transmission [25].

B-H-R-P transmission network model is developed by networking connectivity between four sources of transmission, i.e., bats, hosts, reservoir, and people. S-E-I-R modelling is approached in bats and hosts. Infected bats are transmitting the virus hosts susceptible group, and further infected hosts transfer it to reservoirs (i.e., seafood). Reservoirs are categorized in a single group, i.e., do not follow any modelling approach because they are haunted form of hosts and remain infected. Further, people caught this virus via reservoirs and people which are categorized into five groups, i.e., susceptible, exposed, symptomatic infected people, asymptomatic infected people, and removed. Therefore, S-E-I-R model is applied in people transmission part with the addition of an extra group as asymptomatic infected people.

4 Validation of Mathematical Modelling

Day to day data are required to validate any type of infectious mathematical modelling for certain virus/disease. S-E-I-R is having exposed parameter; that is why, it requires data of people who are exposed to the virus but not infected. S-I-R-S requires people data who are recovered and again suspected. So, a simple approach to validate it is via S-I-R modelling requiring only three parameters, i.e., number of suspected individuals from the total population, infectious, and recovered (including deaths also).

Suppose, we have an area having a total population of 5000. Out of the total population, 2500 are suspected individuals, 2000 are confirmed with infection, and 500 are recovered (i.e., recovered from infection or dead).

Here, First condition is implemented, i.e., $P = S + I + R$.

Suppose, average period of infectiousness at thirteen days, i.e., $k = 1/13 = 0.077$.

And, if each infected would make possibility infecting contact every four days, then $b = 1/4 = 0.25$. Then, $R0 = 0.25/0.077 = 3.25$.

$R0 > 1$, therefore, it is epidemic-type situation.

Suppose, we want to report the situation for the next five days implementing source data. Initially, $S(0) = 2500$

$I(0) = 2000$

$R(0) = 500$

Calculating the rate of susceptibility,

$$dS/dt = -bs(t)I(t) = -0.25 \times (2500/5000) \times 2000 = -250$$

Similarly, $dI/dt = bs(t)I(t) - kI(t) = 250 - (0.077 \times 2000) = 96$ and $dR/dt = kI(t) = 154$

Therefore, for $t = 1$ (first-day data) (Fig. 7):

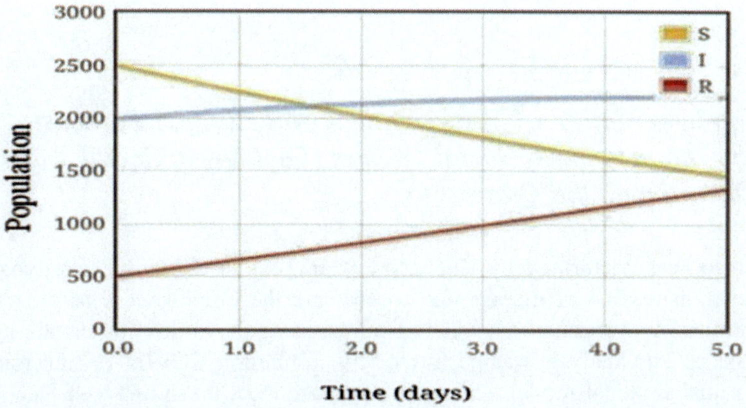

Fig. 7 Graphical representation of the example

$$S(1) = 2500 - 250 = 2250$$
$$I(1) = 2000 + 96 = 2096$$
$$R(1) = 500 + 154 = 654$$

For second day,

$$S(2) = 2250 - (0.25 \times 0.45 \times 2096) = 2250 - 235.8 = 2014.2$$
$$I(2) = 2096 + (235.8 - 161.392) = 2170.408$$
$$R(2) = 654 + 161.392 = 815.392$$

For third day,

$$S(3) = 2014.2 - (0.25 \times 0.4028 \times 2170.408) = 1795.64$$
$$I(3) = 2170.408 + (218.56 - 167.1214) = 2221.85$$
$$R(3) = 815.392 + 167.1214 = 982.5134$$

For fourth day,

$$S(4) = 1795.64 - (0.25 \times 0.3591 \times 2221.85) = 1596.1835$$
$$I(4) = 2221.85 + (199.4665 - 171.0824) = 2250.23$$
$$R(4) = 982.5134 + 171.0824 = 1153.6$$

For fifth day,

$$S(5) = 1596.1835 - (0.25 \times 0.3192 \times 2250.23) = 1416.6$$
$$I(5) = 2250.23 + (179.57 - 173.27) = 2256.53$$
$$R(5) = 1153.6 + 173.27 = 1326.87$$

4.1 *Implementation of Mathematical Modelling to Forecast the Importance of Social Factors (Lockdown, Social Distancing, and Quarantine)*

An analysis was performed by Kucharski et al. [28] to determine the dynamics of initial transmission of disease and to evaluate the effectiveness of prevention and control in new areas which was critical to assess the ability for sustaining the transmission. The analysis was performed by combining COVID-19 transmission with the mathematical model where four datasets were taken between December 2019 and February 2020 within the outside and inside Wuhan. If more cases were introduced from outside Wuhan, then the observed estimation will be observed to assess the potential for sustained human-to-human transmission. It was observed that one week before the travel restriction, the median daily reproduction number (Rt) from 2.35 (1.15–4.77) started to decline after one week of restriction on 23 Jan 2020 to 1.05 (0.41–2.39). Based on the observation for Rt, the transmission potential was calculated in various locations in Wuhan for SARS-like variation. Understanding the model if at least four independent cases are introduced in an area than there are 50% of symptoms of this disease will establish within that population. The results predict that the transmission of COVID-19 would decline in later weeks in January 2020 after the introduction of control measures. With the invasion of more numbers in China from international borders with identical symptoms of disease before the implementation of protective measures, initially, the transmission chain might not establish but eventually, it will outbreak [28].

A data-driven susceptible-exposed-infectious-quarantine-recovered (SEIQR) model was developed by Yuzhen Zhang et al. which can simulate the epidemic with interventions of social distancing and epicentre lockdown. To estimate the model parameters, the officially reported data were combined with the population migration data by estimating the daily infected ration and daily susceptible population size. On 1st January 2020, the estimated latent infected individual initially was 380.1 (379.8–381.0), and after inclusion of thirty days of social distancing, the number of infected people was reduced from 2.2 to 1.58, whereas in other provinces, it reduced from 2.56 to 1.65, and with the implementation of social distancing, it could reduce the impact of the pandemic in China with an estimated reduction of 98.9% of infected people and 99.3% of many deaths by 23rd February 2020. However, with the lockdown of point source or epicentre, it would partially reduce the effect and would lead to the improvement in the scenario. To make it more effective and minimize the impact, social distancing should be implemented stepwise where the epicentre should be targeted first followed by province and later the whole nation to minimize the economic loss [29].

Benjamin Ivorra and Angel M defined BE-CoDids (between countries disease spread) epidemiological model to study the transmission of COVID-19 between humans and within countries. It is a combination of the individual-based model (where countries are considered as an individual) which is simulated between different countries with compartment model (a system of ordinary differential equations), simulating the rate of spread of disease within countries. When the simulation completes, BE-CoDids produces an output of outbreak characteristics (the risk of disease introduces or spreads per country, the magnitude of epicentre, etc.) of COVID-19 of referring countries [30, 31] (Fig. 8).

5 Conclusion

The transmission of COVID-19 is modelled using different types of mathematical models. The S-I-R modelling is a basic approach that does not include births, migrants, deaths emigrants, and it can be used for a small community where variation

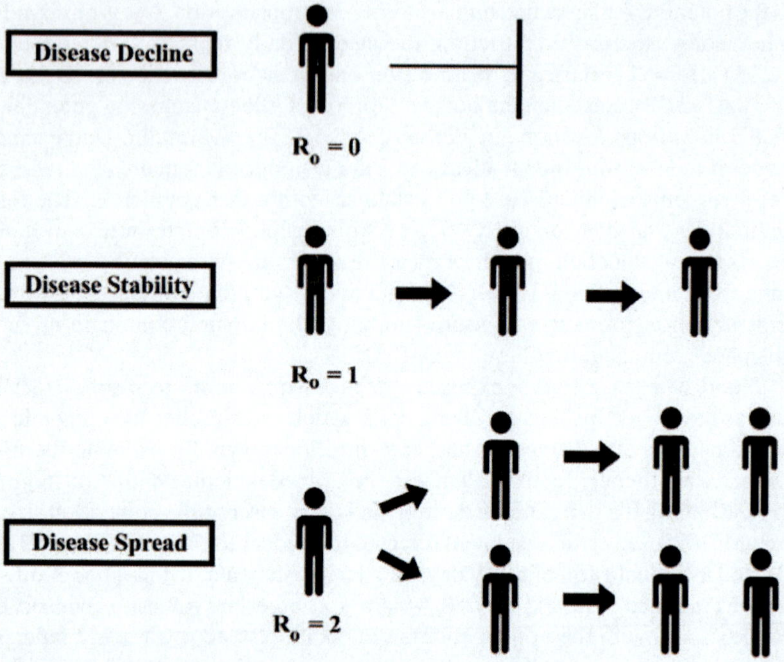

Fig. 8 Role of basic reproductive ratio in virus transmission [30–33]

is negligible, and S-I-R modelling can be updated concerning demography consideration, and it validates a realistic situation scenario in major areas. While S-E-I-R model explains epidemic situations, it can be implemented well in presently COVID-19 emerging situations. It includes an extra term, i.e., exposed that is infected with the virus but cannot transfer further to possible contacts. The S-I-R-S modelling is approached where recovered people are having temporary immunity, and again, it enters in the susceptible category. Immunity and medication factors play an important factor in S-E-I-R and S-I-R-S modelling. All mentioned mathematical modelling approaches can be arranged in a network model, and predictive results can be calculated for COVID-19 situation. B-H-R-P transmission network model is well-implemented and well-defined model using initial reported conditions of COVID-19. Using and updating these models, we can calculate and predict many parameters required as precautions that can be used for the welfare of society. The requirement of emergencies situations like social distancing, quarantine, and lockdown can be well determined using calculations from this mathematical approach. Further, it can also be used in computational tools for simulating epidemic or pandemic conditions.

Acknowledgements We would like to acknowledge University of Petroleum and Energy Studies for supporting in providing resources for this study.

References

1. Otter, J. A., Donskey, C., Yezli, S., Douthwaite, S., Goldenberg, S. D., & Weber, D. J. (2016). Transmission of SARS and MERS coronaviruses and influenza virus in healthcare settings: The possible role of dry surface contamination. *Journal of Hospital Infection, 92*(3), 235–250. https://doi.org/10.1016/j.jhin.2015.08.027
2. de Groot, R. J., et al. (2013). Middle East Respiratory Syndrome Coronavirus (MERS-CoV): Announcement of the Coronavirus Study Group. *Journal of Virology, 87*(14), 7790–7792. https://doi.org/10.1128/jvi.01244-13
3. Stalin Raj, V., et al. (2013). Dipeptidyl peptidase 4 is a functional receptor for the emerging human coronavirus-EMC. nature.com. https://doi.org/10.1038/nature12005.
4. Fineberg, H. V. (2014). Pandemic Preparedness and Response—Lessons from the H1N1 Influenza of 2009. *New England Journal of Medicine, 370*(14), 1335–1342. https://doi.org/10.1056/nejmra1208802
5. Annex, C. (2009). *Natural ventilation for infection control in health-care settings: Respiratory droplets*.
6. Boone, S. A., & Gerba, C. P. (2007). Significance of fomites in the spread of respiratory and enteric viral disease. *Applied and Environmental Microbiology, 73*(6), 1687–1696. https://doi.org/10.1128/AEM.02051-06
7. Bridges, C. B., Kuehnert, M. J., & Hall, C. B. (2003). Transmission of influenza: Implications for control in health care settings. *Clinical Infectious Diseases, 37*(8), 1094–1101. https://doi.org/10.1086/378292
8. Brankston, G., Gitterman, L., Hirji, Z., Lemieux, C., & Gardam, M. (2007). Transmission of influenza A in human beings. *Lancet Infectious Diseases, 7*(4), 257–265. https://doi.org/10.1016/S1473-3099(07)70029-4
9. Kramer, A., Schwebke, I., & Kampf, G. (2006). How long do nosocomial pathogens persist on inanimate surfaces? A systematic review. *BMC Infectious Diseases, 6*. https://doi.org/10.1186/1471-2334-6-130.
10. Van Doremalen, N., et al. (2014). Host species restriction of Middle East respiratory syndrome coronavirus through its receptor, dipeptidyl peptidase 4. *American Society of Microbiology*. https://doi.org/10.1128/JVI.00676-14
11. Chan, K. H., Peiris, J. S. M., Lam, S. Y., Poon, L. L. M., Yuen, K. Y., & Seto, W. H. (2011). The effects of temperature and relative humidity on the viability of the SARS coronavirus. *Advances in Virology, 2011*. https://doi.org/10.1155/2011/734690.
12. Coulliette, A. D., Perry, K. A., Edwards, J. R., & Noble-Wang, J. A. (2013). Persistence of the 2009 pandemic influenza a (H1N1) virus on N95 respirators. *Applied and Environment Microbiology, 79*(7), 2148–2155. https://doi.org/10.1128/AEM.03850-12
13. Casanova, L., Rutala, W. A., Weber, D. J., & Sobsey, M. D. (2009). Survival of surrogate coronaviruses in water. *Water Research, 43*(7), 1893–1898. https://doi.org/10.1016/j.watres.2009.02.002
14. Mullis, L., Saif, L. J., Zhang, Y., Zhang, X., & Azevedo, M. S. P. (2012). Stability of bovine coronavirus on lettuce surfaces under household refrigeration conditions. *Food Microbiology, 30*(1), 180–186. https://doi.org/10.1016/j.fm.2011.12.009
15. Yépiz-Gómez, M. S., Gerba, C. P., & Bright, K. R. (2013). Survival of respiratory viruses on fresh produce. *Food and Environmental Virology, 5*(3), 150–156. https://doi.org/10.1007/s12560-013-9114-4
16. Guan, W., et al. (2020). Clinical characteristics of coronavirus disease 2019 in China. *New England Journal of Medicine*, NEJMoa2002032. https://doi.org/10.1056/NEJMoa2002032.
17. Shigematsu, S., et al. (2014). Influenza A virus survival in water is influenced by the origin species of the host cell. *Influenza and Other Respiratory Viruses, 8*(1), 123–130. https://doi.org/10.1111/irv.12179
18. Choi, B. C. K., & Pak, A. W. P. (2003). A simple approximate mathematical model to predict the number of severe acute respiratory syndrome cases and deaths. *Journal of Epidemiology and Community Health, 57*(10), 831–835. https://doi.org/10.1136/jech.57.10.831

19. Nesteruk, I. (2020). Statistics-based predictions of coronavirus epidemic spreading in Mainland China. *Innovative Biosystems and Bioengineering, 4*(1), 13–18. https://doi.org/10.20535/ibb.2020.4.1.195074
20. World Health Organisation. (2020). Coronavirus (COVID-19) events as they happen. Retrieved 07 Feb, 2021. https://www.who.int/emergencies/diseases/novel-coronavirus-2019/events-as-they-happen.
21. Jit, M., & Brisson, M. (2011). Modelling the epidemiology of infectious diseases for decision analysis: A primer. *PharmacoEconomics, 29*(5), 371–386. https://doi.org/10.2165/11539960-000000000-00000
22. Smith, D., Moore, L. (2016). The-SIR-model-for-spread-of-disease-the-differential-equation-model. *Journal of Online Mathematics its Applications.* Accessed Feb 05, 2021, from http://www.maa.org/press/periodicals/loci/joma/.
23. Bjørnstad, O. N., Finkenstädt, B. F., & Grenfell, B. T. (2002). Dynamics of measles epidemics: Estimating scaling of transmission rates using a Time series SIR model. *Ecological Monographs, 72*(2), 169–184. https://doi.org/10.1890/0012-9615(2002)072[0169:DOMEES]2.0.CO;2
24. Murphy, T. F., Brauer, A. L., Grant, B. J. B., & Sethi, S. (2005). Moraxella catarrhalis in chronic obstructive pulmonary disease: Burden of disease and immune response. *American Journal of Respiratory and Critical Care Medicine, 172*(2), 195–199. https://doi.org/10.1164/rccm.200412-1747OC
25. Chen, T. M., Rui, J., Wang, Q. P., Zhao, Z. Y., Cui, J. A., & Yin, L. (2020). A mathematical model for simulating the phase-based transmissibility of a novel coronavirus. *Infectious Diseases of Poverty, 9*(1). https://doi.org/10.1186/s40249-020-00640-3.
26. Elie, R., Hubert, E., & Turinici, G. (2020). Contact rate epidemic control of COVID-19: An equilibrium view. *Mathematical Modelling of Natural Phenomena, 15*, 35. https://doi.org/10.1051/mmnp/2020022
27. Ellner, S. P., & Guckenheimer, J. (2011). *Dynamic models in biology, 9781400840.*
28. Kucharski, A. J., et al. (2020). Early dynamics of transmission and control of COVID-19: A mathematical modelling study. *The Lancet Infectious Diseases, 3099*(20), 1–7. https://doi.org/10.1016/S1473-3099(20)30144-4
29. Zhang, Y., Jiang, B., Yuan, J., & Tao, Y. (2020). The impact of social distancing and epicenter lockdown on the COVID-19 epidemic in mainland China: A data-driven SEIQR model study. *medRxiv*. https://doi.org/10.1101/2020.03.04.20031187.
30. Ivorra, B., & Ramos, Á. (2020). Application of the Be-CoDiS mathematical model to forecast the international spread of the 2019–20 Wuhan coronavirus outbreak. *researchgate.net*. https://doi.org/10.13140/RG.2.2.31460.94081.
31. Chaves, L. F., Hurtado, L. A., Rojas, M. R., Friberg, M. D., Rodríguez, R. M., & Avila-Aguero, M. L. (2020). COVID-19 basic reproduction number and assessment of initial suppression policies in Costa Rica. *Mathematical Modelling of Natural Phenomena, 15*. https://doi.org/10.1051/mmnp/2020019.
32. Ciarochi, J. (2020). How COVID-19 and other infectious diseases spread: Mathematical modelling. *Medium*, 1–16. Retrieved Feb 07, 2021, from https://triplebyte.com/blog/modeling-infectious-diseases.
33. Tengaa, P. E., Mwalili, S., & Orwa, G. O. (2020). Deterministic modeling for HIV and AIDS epidemics with viral load detectability. *The Journal of Mathematics and Computer Science, 10*(4), 728–757. https://doi.org/10.28919/jmcs/4474

Date Palm Waste and Attempts to Use it as an Energy Source: State-of-the-Art

Tabassum-Abbasi, M. Nayeem-Shah, S. A. Abbasi, and Tanseem Abbasi

1 Introduction

1.1 The Date Palm

Date palm *(Phoenix dactylifera)* is, arguably, the most ancient of the world's cultivated trees [1], with a history going back 10,000 years. Conquerors introduced date palm to the conquered lands—for example Alexander took it to Western India (now in Pakistan) and Moors to Spain. In turn, traders and explorers spread it to other lands, including Mexico and on to Northern America. At present, date palm is cultivated across a large belt encompassing numerous countries of the Old World, stretching ±8,000 km from east to west and ±2000 km from north to south [1]. The arid and semi-arid regions of the Americas are also significant repositories of the date palm, with over 600,000 trees. All-in-all, about 3% of the world's cultivable surface is believed to be occupied by the date palm [2].

The genus *Phoenix* has 12 different species which are all native to tropical/subtropical Africa/Asia. Of these, six species provide edible fruits [3], but it is the fruit of *P.dactylifera* (date) that has won over the test-buds of nearly everyone across the world. This has led to the planting of over 105 million date trees [4, 5], and the production of marketable dates is now reaching 8 million metric tonnes (mmt) per year. The global demand for dates is increasing at a fast rate: it jumped from 2.8 mmt in 1985 to 5.4 mmt in 2001 and 7.6 mmt in 2010 [6]. Hence, the extent of date palm cultivation is expected to increase in coming years. Ambitious plans to do so are on

Tabassum-Abbasi (✉)
School of Engineering, University of Petroleum and Energy Studies, Bidholi, Dehradun 248007, India

M. Nayeem-Shah · S. A. Abbasi · T. Abbasi
Centre for Pollution Control & Environmental Engineering, Pondicherry University, Chinakalapet, Puducherry 605014, India
e-mail: abbasi.cpee@gmail.com

N. A. Siddiqui et al. (eds.), *Advances in Behavioral Based Safety*,
https://doi.org/10.1007/978-981-16-8270-4_3

the envil in many countries. Even in some regions of India which do not have arid climate, date palm cultivation is being done.

Given the necessity that it must survive even in harsh arid environments, date palm tree has a very rugged body, so rugged that it can re-grow after being damaged by fire. It is this ability which, perhaps, has prompted the taxonomical naming of date palm after the legendry bird Phoenix which is supposed to have lived to be 500 years old and had risen with renewed vigour after having been burnt to ashes [7, 8]. It is also possible that the generic name is derived from the region—Phoenicia—where date palm was initially common during the Biblical times. The name of the species *dactylifera* has its origins either in *dachel*, a Hebrew world, or in the Greek word *daktulos*—both seek to represent date's finger-like shape. Date features prominently in the scriptures of all the three Abrahmic religions—Judaism, Christianity, and Islam—especially in the *Quran*. There is even belief, hotly contested, that it was a date palm and not an apple tree which was the "tree of knowledge" in the Garden of Eden described in the Testaments and in the *Quran* [9, 10].

1.2 The Date Palm Waste

In pre-modern times, especially before large reservoirs of oil were discovered in West Asia and Africa, the number of date palms standing in the world was much lesser than what it is at present. And, whatever residue that used to be generated in the form of dead leaves or dead trees was utilized in a myriad of ways. For example the leaves were used in making mats, screens, baskets, brooms, hats, fans bags, etc. The trunk, with its great tensile strength, was used to make poles, beams, rafters, lintels, girders, jetties, and light foot bridges. Crates and furniture were also shaped out of date palm residue [11]. Indeed, the palm was as useful in its life as it was after death.

Then, the economies of several of the date-producing regions experienced a sharp upswing. Better alternatives to the products based on date palm tree residues became available and affordable. These happenings led to lesser and lesser use of date palm residues in the making of traditional, hand-made, products. Alongside this, a sharp increase in the cultivation of date palm has occurred. The cumulative impact is the generation of increasingly large quantities of date palm waste. By now, it is one of the biggest streams of lignocellulosic solid wastes in the world, exceeding 12 mmt per year [12]. It consists of the following sub-streams (Fig. 1):

(i) Leaves or fronds (including the midrib called rachi and attached leaflets):

An adult date palm has 100–125 green leaves approximately. It grows 10–26 new leaves annually. The leaves have a life of 3–7 years and grow to an average of 4 m, depending on the variety of the tree, its age, and the environmental conditions. A frond midrib (rachi) has a maximum width of 0.5 m, while the rest of the leaf is 0.25 m and narrows towards the top. The cross-section of a frond midrib or petiole is relatively triangular and has two lateral angles and one dorsal. It does not have spines

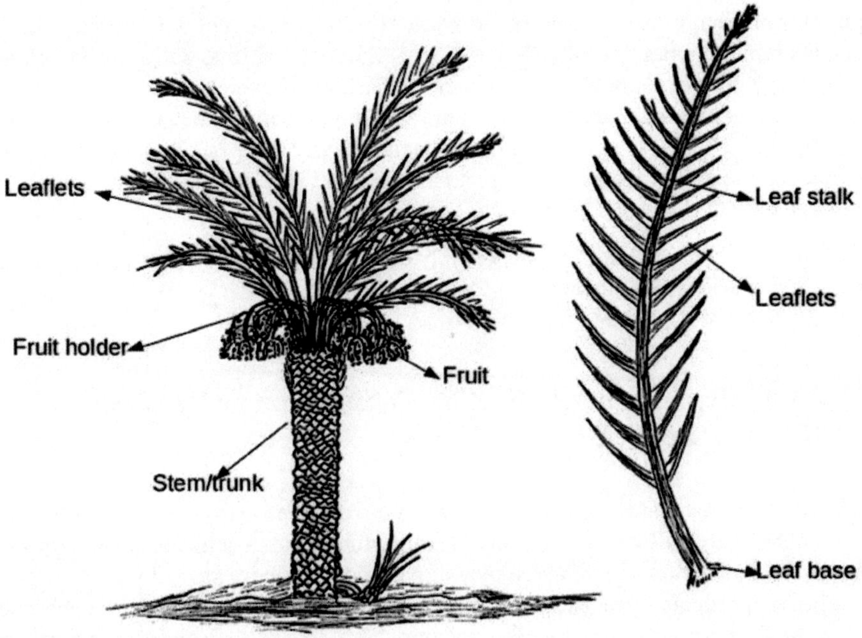

Fig. 1 Typical date palm and its components

for a short distance but is full of spines on both sides thereafter. After the leaves die, they do not fall and have to be plucked off. Each palm is estimated to generate about 1.4 kg of waste leaves per Kg of dates [13]. By another estimate, 20 kg of waste leaves is generated annually per date palm [14].

(ii) *Trunk:* the date palm trunk, also called stem or stipe, is vertical, cylindrical, and columnar. It is of nearly uniform girth (1–1.1 m circumference) all the way up and can rise to 20 m or higher. It is composed of tough, fibrous vascular bundles held together in a matrix of cellular tissue. Some date palms have branches, but it does not adversely affect the tree's productivity. Every year, a sizeable number of palm trees complete their life and are removed, contributing to the waste.

(iii) *Date seeds (also called stones, pits, or kernels):* date seeds are oblong, ventrally grooved, with a small embryo and a hard endosperm. Depending on the variety of the date, its seed can have weight ranging from 0.5 to 4 g, length 12 to 36 mm, and breadth 6 to 13 mm. About 10–15% of the weight of a date is taken up by its seed; the fleshy pericarp is taking up the rest [14, 15].

(iv) *Date flesh:* substantial losses of date fruits occur in the course of picking, storage, and transportation (Al-Jasser, 2010). At the stage of packaging, quality control leaves out dates that have been infected, damaged, malformed, or affected in some way which may compromise their marketability. The flesh of these dates constitutes the fourth major stream of date-based waste. Fruits of different date varieties have weights ranging from 2 to 60 g, length 8 to 110 mm, width 8 to 32 mm, and colour yellow to black. In their dry form, which is the form in which dates can be easily preserved and marketed, date flesh contains about 80% sugar, 2.5% vitamin C, 2.1% protein, and 1% fat. It also contains appreciable quantities of other vitamins and minerals and is regarded as a "complete food" [16, 17]. Currently, a small fraction of waste date flash is used as animal feed [16].

(v) *Date fruit holder:* it is a proportionately thinner waste steam than the other three but is, nevertheless, significant.

2 Production of Energy/Fuel by Combustion or Pyrolysis of Date Palm Waste

The first systematic study on the combustion of date seeds (DS) and palm stalks is due to Al-Omari [18] who used a small furnace fitted with a conical fuel bed for the purpose. Combustion was initiated by using liquid petroleum gas (LPG).

The hot products released by the combustion of LPG were allowed to penetrate the conical fuel bed from below for 2–3 min. This caused the solid fuel present in the bed to undergo gasification and pyrolysis. The resulting combustible gases were mixed with air and ignited using an external ignition source. The effect of varied fuel feed conditions and various airflow rates was then studied. The energy generated was then assessed in terms of the heat transferred to the cooling water (flowing in a water jacket around the furnace) as a function of time.

In a follow-up study, Al-Omari [19] compared combustion and heat transfer occurring in furnaces fuelled with DS and coal. It was seen that much more volatile material was released from DS than from coal during the pyrolysis process. The material subsequently burned under diffusion or partially premixed flame conditions.

In more recent years, there has been a spurt in the attempts to generate fuels from date palm waste [5, 20–22]. These attempts reflect increasing global concerns to find carbon–neutral fuels. The ever-increasing quantities of date palm waste (DPW) have also prompted more intense efforts to make them utilizable. Sait et al. [20] have argued that the bulk density of the DS (560 kg/m^3), as found by them, being much higher than the bulk density of leaf and leaf stem (298 and 216 kg/m^3, respectively) indicates that DS can be processed without densification. This has the potential of reducing pre-processing cost of date seeds in comparison with the other two waste streams.

Sait et al. [20] also found that "DS showed the highest calorific value (CV) of the three DPW streams studied by them. It was close to wood, straw pellets, and pine biomass. Leaves also had good CV, but the leaf stem's CV suffered due to its high moisture content. DS and leaf also had high volatile matter and low ash components compared to the leaf stem". This made them attractive for combustion process [23]. Sait et al. [20] also felt that "DS appeared suitable for bio-oil making

Table 1 Characteristic of date palm waste which lead to different streams of waste

Constituent, %	Leaves	Rachi	Trunk	Seeds	Flesh	Reference
Moisture	7.1	12.1	10	6.4	16.9	[28, 29]
Ash	15.2	6	4.2	1.2	1.7	[28, 29]
Fat	1.68	2.41	0.09	2	1.35	Juhami et al. (2013), Abed et al. (2012), [30]
Carbohydrates	76.9	84.5	79.5	83.1	52.6–88.6	Juhami et al. (2013), [16, 31]
Holo-cellulose	59.5	74.8	58.7	–	–	[32–34]
C	40.8	39.8	42.76	45.8–51.2	–	[28, 29]
N	0.63	0.19	0.21	0.73–0.81	–	[28, 29]
O	35.2	43	45.5	40.9–46.6	–	[28, 29]
H	6	5.7	5.8	6.3–6.4	–	[28, 29]

from pyrolysis, but no attempt to actually make bio-oil was made by them. Thermogravimetric analysis (TGA) indicated that the two substrates can be completely pyrolysed at 400 °C. Unburned char accounted for about 20 wt.% for seed and leaf and 28 wt.% for leaf stem. Kinetic studies showed activation energy in the range 9.7–43.6 kJ/mole in nitrogen atmosphere and 9.04–30.95 kJ/mole in air. The highest value was for the seed and lowest for the leaf stem". Another TGA-based study [21] has revealed that the "heating values of DS and PS (22.27 MJ/Kg and 19.66 MJ/Kg) were found to be higher than the energy content of other materials: rapeseed straw—17.64 MJ/Kg [24] pine chips—18.98 MJ/Kg [25], rice husk—14.42– 18.31 MJ/Kg [26], and wood—19.10 MJ/Kg [27]. The moisture related to the rough weight was found to be much lower—about 8% for DS and 5.8% for PS, when compared to dried wood (20%) and fresh wood (45%) (Table 1)".

Scanning electron micrographs (SEM) of raw DS and PS revealed considerable amounts of carbon and oxygen on the external surfaces. Other elements were found in trace amounts. They were magnesium, cobalt, iron, aluminium, silicon, chlorine, phosphorus, and X-ray diffraction patterns showed that in both substrates, the major part of the matter is amorphous. There are a few diffraction peaks which indicate that a small amount of crystalline matter lies interspersed.

All-in-all, the properties of DS and PS were found to be similar; the former possessing higher energy density". Rigorous TGA and differential thermal analysis of DS produced results similar to the ones reported by Sait et al. [20]. Additional insights were "(a) the most distinguished peak in fast pyrolysis stage is related to the complete pyrolysis of hemicellulose, which took place until 350 °C, while the pyrolysis of cellulose occurred from 250 to 500 °C. The pattern was similar to the one found earlier for rice husk by Manasary & Ghaly [35] and Varhegyi et al. [36]; (b) slow degradation of biomass continued until 900 °C, at which point complete devolatilization occurred. Evidently, and expectantly, lignin degraded more slowly than cellulose".

According to Abed et al. [21], "scanning electron microscopy (SEM) of the char particles revealed that pyrolysis temperature and heating rate influenced the size and shape of the particles due to a general increase in size and proportion of voids and a decrease in cell-wall thickness. It seemed favourable to pyrolyse DS at 500 °C, at which temperature the most highly porous and spherical particles were observed. At low pyrolysis temperature and low heating rate, no porous structure was seen and the particles were thick-walled and covered by tar agglomerates. Mass loss rates increased with increasing particle sizes during the stages of hemicellulose decomposition, as well as of lignin decomposition, but no effect over the total duration of the decomposition process was observed".

The weight loss in the first stage was less than 5% and was similar when pyrolysis was done in inert or air atmosphere. However, in the second stage of thermal degradation, there was much larger weight loss (>85%) in the air atmosphere as compared to the weight loss in the inert atmosphere (≈65%), possibly due to char oxidation. The authors also conducted kinetic study which may be helpful in designing DS pyrolysis systems.

In their combustibility tests, [5] found that fibrous stems of the waste (DPW)—date palm leaves (DPL), rachis (DPR), and trunk (DPT)—ignited much more quickly than DS at the temperatures studied by then (500–800 °C). Higher the temperature, greater was the combustibility of all the streams.

2.1 *Generation of Gaseous and Particulate Pollutants*

As noted by El May et al. [5], all the streams of DPW studied by them were "emitted carbon oxides, volatile organic compounds, tars, and particles when combusted at 500 °C. High molar proportions of carbon monoxide were emitted due to the very short residence time of the gases in the hot zone of the furnace. Samples with low bulk density were affected lesser by increase in temperature, and their total carbon emission factors all improved, (reduced), with the rise in temperature. But for DS, due to its higher bulk density, the carbon balance sheet did not change significantly when the temperature was increased from 500 to 800 °C. The emission of ultrafine particulate matter (PM), with the nanoparticle PM 0.1 being the dominant form, occurred with all forms of DPW. At 500 °C, particles were mainly produced during the char combustion and tar devolatilization phases, this got intensified at 800 °C, leading to higher amount of particles".

The emission factors were very close to the values reported earlier for other energetic crops and conventional biomass, thereby indicating that DPW has similar potential for use as direct fuel. El May et al. [5] also found that "at 800 °C, aerosols from tars were dependent on the initial amount of elemental chlorine in the DPW, and the number size distribution of the tar fraction was centred at a diameter of 74 nm. The total number of size fractions for PM10, PM1, and PM0.1 was also dependent on the content of chlorine in the residue. With regards to aerosol emissions, optimization of the combustion of date palm by-products appeared likely to be most efficient at high temperature and low air flow rate".

2.2 *Generation of Bio-Oil*

Joardder et al. [22] have generated bio-oil, besides activated carbon, by the pyrolysis of DS using an externally heated (400–600 °C) fixed bed reactor with nitrogen as the carrier gas. According to them, "a maximum liquid yield of 50 wt. % and char of 30 wt.% were obtained at a reactor bed temperature of 500 °C and a running time of 120 min. The oil was found to possess favourable flash point, density, and viscosity. Its calorific value (28.636 MJ/kg) was significantly higher than other biomass derived oils, as well as solid DS".

3 Generation of Bioethanol

The first, and so far the only, study towards the generation of bioethanol from DPW has come from Al-Zuhair et al., [37]. Using date palm funds or rachi (DPR) as raw material and the enzymes laccase, xylanase, and cellulose in a sequence of pretreatment and hydrolysis, they were able to obtain a 45.6% yield of reducing sugars. Hydrolysis alone with cellulose—without the sequential pretreatment with laccase and xylanase—provides a yield of only 5.6%. The study of Al-Zuhair et al., [37] provides proof-of-concept, with only an indication that further work may lead to potential application.

4 Production of Biohydrogen and Liquid Biofuels

Due to its high sugar content and its richness in terms of nutrients, date flesh has good potential to be fermented into liquid fuels or hydrogen [38–40].

The option of hydrogen production has been explored for the first time by Abd-Alla et al. [41]. According to their report, they have "carried out three-stage fermentation, using three different bacteria, for the purpose. A facultative anaerobe, *Escherichia coli* EGY, was used in the first stage to consume O_2 and maintain strict anaerobic conditions for a second-stage dark fermentative H_2 production by the strictly anaerobic Clostridium acetobutylicum ATCC 824. Subsequently, a third-stage photofermentation using Rhodobacter capsulatus DSM 1710 was conducted for the H_2 production. Overall, 7.8 mol of H_2 per mol of sucrose was obtained when 5 g/L of sucrose was supplemented as rotten date fruits". According to Abd-Alla et al. [41], "these findings indicate that this route may enable economically viable H_2 production as there is no need in it for addition of a reducing agent or flushing with argon, both of which are expensive".

Fermentation of date flesh-water slurry to obtain ethanol and vinegar in batch and continuous membrane reactors [42] and ethanol [43] has been attempted. According to Gupta & Kushwaha [44], "fermentation at a fixed temperature using a mutant strain *Saccharomyces cerevisiae* ATCC 36,858 and a commercial *S. cerevisiae* (STAR brand) activated at different periods has demonstrated that *S. cerevisiae* ATCC 36,858 could selectively convert glucose to ethanol and biomass with minimal fructose conversion". In a follow-up report, Gaily et al. [45] note that "high fructose yield above 91% of the original fructose was obtained with ATCC 36,858 in addition to the ethanol yield which was found to be 63% of the theoretical. Yields of the order of 136 g/L could be obtained under optimum conditions which consisted of an incubation period of 72 h, inoculum content of 4% (w/v), sugars concentration of 180 g/L, and ammonium phosphate concentration of 1 g/L [46]".

Spoiled date fruits have also been evaluated as a potential substrate for fermentation into acetone, butanol, and ethanol (ABE) by Abd-Alla & El-Enany [47]. They report that "mixed cultures of C. acetobutylicum ATCC 824 and Bacillus subtilis DSM 4451 and of C. cetobutylicum ATCC 824 and B. subtilis DSM 4451 were observed to consume sugar, amino acid, and NPK, as well as any available oxygen during growth. It reduced the need to develop anaerobic conditions to commence fermentation. The study further indicated that the aerobic B. subtilis DSM 5541 can be used to install and maintain anaerobic conditions for strictly anaerobic C. acetobutylicum ATCC 824, and the supplementation of yeast extract or ammonium nitrate to spoiled date fruits homogenate markedly increased the total ABE production. The results suggest that rotten date palm fruits homogenate can be efficiently used as substrate without the need for any pretreatment or hydrolysis and without requirement to add any costly reducing agent to the medium or flushing with N_2 to ensure anaerobic condition for commercial ABE production by mixed cultures". Use of date fruit to produce biobutanol by *C. acetobutylicum* NCIMB 13,357 has also been reported [48].

The success of these feasibility studies, notwithstanding, no attempt to take the technology to larger scale—even at pilot plant level—has been reported so far.

5 Generation of Biogas

Considering that nearly all potential avenues have been explored towards gainful utilization of date Palm waste [37, 49–57], it is surprising that there is no pre-existing report on the efforts to anaerobically digest any of the DPW streams. These authors have attempted to bridge this knowledge gap [12]. The resulting process involves soaking of DPW—in 0.75–1.25% aqueous NaOH solutions for a few days, followed by anaerobic digestion in semi-batch reactors. The process is able to yield 3.2–5.9 L of flammable biogas per Kg of waste per day. In terms of reactor space, the gas production is in the range 0.64–1.2 m^3 per m^3 reactor volume per day, indicating satisfactory space efficiency of the process. A significant aspect of this process is that the spent waste, which retains all nutrients (except parts of carbon, hydrogen, oxygen, besides traces of nitrogen which are released as biogas) that showed good potential for serving as feedstock for the production of compost/vermicompost.

6 Summary and Conclusion

The paper brings out that date palm (*Phoenix dactylifera*) is one of the most cultivated of the horticultural species in the world, with over 105 million trees standing across the arid and semi-arid regions of the world. These lead to the production of about 12 million tonnes of waste biomass per year. In absence of economically viable ways available for its utilization at present, most of this waste causes insanitation.

The paper reviews the present state-of-the-art of date palm waste (DPW) utilization, as a possible source of energy. It is brought out that the most of the work in this direction is of recent origin and indicates that DPW has the potential of being a viable feedstock for the generation of solid, liquid, as well as gaseous fuels.

Acknowledgements SAA thanks Indian National Science Academy [INSA], New Delhi, and PP thanks Council for Scientific and Industrial Research [CSIR], New Delhi, for support.

References

1. Zaid, A., & de Wet, P. F. (2002). Date palm cultivation. Abdelouahhab Zaid (ed.) Chapter II, FAO Plant production and protection.
2. Dowson, V. H. W. (1982). Date production and protection with special reference to North Africa and the Near East. *FAO Technical Bull No, 35*, 294.
3. Chevalier, A. (1952). Recherches sur les Phoenix africains. *International Journal of Applied Botany and Tropical Agriculture*, 355–356.
4. Agoudjil, B., Benchabaneb, A., Boudennec, A., Ibos, L., & Fois, M. (2011). Renewable Materials to Reduce Building Heat Loss: Characterization of Date Palm Wood. *Energy and Buildings, 43*, 491–497.
5. El May, Y., Dorge, S., Jeguirim, M., Trouvé, G., & Said, R. (2012). Measurement of gaseous and particulate pollutants during combustion of date palm wastes for energy recovery. *Aerosol and Air Quality Research, 12*, 814–825.
6. FAO STAT (2012). Food and Agriculture Organization of the United Nations. Retrieved Dec 2012, from http://faostat.fao.org/site/339/default.aspx.
7. Pliny, C. (1489). The Elder. Trans. Historia Naturale, Book XIII, Cap. III, 3 columns on the palmae. Translated into Italian by Christofore Landioro Fiorentino and published by Bartolamio de Zani de Portesio.
8. Van Zyl, H. J. (1983). Date cultivation in South Africa. Fruit and Fruit Technology Research Institute, Department of Agriculture, Stellenbosch, RSA. *Information Bulletin, 504*, 26.
9. Soussa, A. (1969). Irrigation and civilisation in Oued Rafi din. First Part. Adib ed. Baghdad. Iraq.

10. Zaid, A., & Arias-Jiménez, E. J. (2002). Date palm cultivation. Food and Agricultural Organization of the United Nations, Rome. Retrieved April 2013, from http://www.fao.org/docrep/006/Y4360E/Y4360E00.HTM.
11. Barreveld, W. H. (1993). Date palm products. Food and Agriculture Organization of the United Nations Rome, Fao Agricultural Services Bulletin No. 101. Retrieved April, 2013, from http://www.fao.org/docrep/t0681E/t0681e00.htm.
12. Al-Juhaimi, F.Y., Hamand, S.H., Al-Ahaideb, I.S., Al-Otaibi, M.M., Nayeem-Shah, M., Abbasi, T., & Abbasi, S.A. (2013). An assessment of the feasibility of extracting clean energy as biogas from date palm waste, *under review*.
13. Bchini, H., Hsayoui, S., & Aloui, S. (2002). Gestion de la Matière Organique et Compostage des Palmes Sèches Dans le Milieu Oasien. *Annales de l'INRAT, 75*, 299–312.
14. Ismail, Z. Z. (2012). Kinetic study for phosphate removal from water by recycled date-palm wastes as agricultural by-products. *International Journal of Environmental Studies, 69*, 135–149.
15. Briones, R., Serrano, L., Younes, R.B., Mondragon, I., & Labidi, J. (2011). Polyol production by chemical modification of date seeds.*Industrial Crops and Products, 34*, 1035–1040.
16. Besbes, S., Blecker, C., Deroanne, C., Drira, N.-E., & Attia, H. (2004). Date seeds: Chemical composition and characteristic profiles of the lipid fraction. *Food chemistry, 84*, 577–584.
17. Al Farsi, M. A., & Lee, C. Y. (2008). Nutritional and functional properties of dates: A review. *Critical Reviews in Food Science and Nutrition, 48*, 877–887.
18. Al-Omari, S. A. B. (2006). Experimental Investigation on the Combustion and Heat Transfer Characteristics in a Furnace with Unconventional Biomass Fuels (Date Stones and Palm Stalks). *Energy Conversion Management*, *47*, 778–790.
19. Al-Omari, S. A. B. (2009). Evaluation of the Biomass "Date Stones" as a Fuel in Furnaces: A Comparison with Coal Combustion. *International Communication in Heat and Mass Transfer*, *36*, 956–961.
20. Sait, H. H., Hussain, A., Salema, A. A., & Ani, F. N. (2012). Pyrolysis and combustion kinetics of date palm biomass using thermogravimetric analysis. *Bioresource Technology, 118*, 382–389.
21. Abed, I., Paraschiv, M., Loubar, K., Zagrouba, F., & Tazerout, M. (2012). Thermogravimetric investigation and thermal conversion kinetics of typical North African and middle eastern lingo cellulosic wastes. *BioResources*, *7*, 1200–1220.
22. Joardder, M. U. H., Uddin, M. S., & Islam, M. N. (2012). The utilization of waste date seed as bio-oil and activated carbon by pyrolysis process. *Advanced Mechanical Engineering*, *2012*, 6 pages.
23. Demirbas, A. (2004). Effects of temperature and particle size on biochar yield from pyrolysis of agricultural residues. *Journal of Analytical Applied Pyrolysis*, *72*, 243–248.
24. Karaosmanoglu, F., Çift, B. D., & Işigigür-Ergüdenler, A. (2001). Determination of reaction kinetics of straw and stalk of rapeseed using thermogravimetric analysis. *Energy Sources*, *23*, 767–774.
25. Şensöz, S., & Can, M. (2002). Pyrolysis of pine (Pinus Brutia Ten.) chips: 1. Effect of pyrolysis temperature and heating rate on the product yields. *Energy Sources, 24*, 347–355.
26. Mansaray, K. G., & Ghaly, A. E. (1997). Physical and Thermochemical Properties of Rice Husk. *Energy Sources, 19*(9), 989–1004. https://doi.org/10.1080/00908319708908904.
27. Kastanaki, E., & Vamvuka, D. (2006). A comparative reactivity and kinetic study on the combustion of coal–biomass char blends. *Fuel, 85*(9), 1186–1193. https://doi.org/10.1016/j.fuel.2005.11.004.
28. Akasha, I. A., Campbell, L., & Euston, S. R. (2012). Extraction and characterisation of protein fraction from date palm fruit seeds. *World Academy of Science, Engineering and Technology*, 70.
29. El Maya, Y., Jeguirim, M., Dorge, S., Trouvé, G., & Said, R. (2012). Study on the thermal behavior of different date palm residues: Characterization and devolatilization kinetics under inert and oxidative atmospheres. *Energy, 44*, 702–709.
30. Sánchez-Zapata, E., Fernández-López, J., Peñaranda, M., Fuentes-Zaragoza, E., Sendra, E., Sayas, E., & Pérez-Alvarez, J. A. (2011). Technological properties of date paste obtained

from date by-products and its effect on the quality of a cooked meat product. *Food Research International, 44*, 2401–2407.

31. Baliga, M. S., Baliga, B. R. V., Kandathil, S. M., Bhat, H. P., & Vayalil, P. K. (2011). A review of the chemistry and pharmacology of the date fruits (*Phoenix dactylifera L.*). *Food Research International, 44*, 1812–1822.
32. Bendahou, A., Dufresne, A., Kaddamai, H., & Habibi, Y. (2007). Isolation and structural characterization of hemicelluloses from palm of *Pheonix dactylifera L. Carbohydrate Polymers, 68*, 601–608.
33. Saadaoui, N., Rouilly, A., Fares, K., & Rigal, L. (2013). Characterization of date palm lignocellulosic by-products and self-bonded composite materials obtained thereof. *Materials and design, 50*, 302–308.
34. Khiari, R., Mhenni, M. F., Belgacem, M. N., & Mauret, E. (2010). Chemical composition and pulping of date palm rachis and *Posidonia oceanica*—A comparison with other wood and non-wood fibre sources. *Bioresource Technology, 101*, 775–780.
35. Mansaray, K. .G., & Ghaly, A. E. (1998). Thermal degradation of rice husks in nitrogen atmosphere. *Bioresource Technology, 65*(1–2), 13–20. https://doi.org/10.1016/S0960-8524(98)00031-5.
36. Várhegyi, G., Antal, M. J., Jakab, E., & Szabó, P. (1997). Kinetic modeling of biomass pyrolysis. *Journal of Analytical Applied Pyrolysis , 42*, 73–87.
37. Al-Zuhair, S., Ahmed, K., Abdulrazak, A., & El-Naas, M. H. (2013). Synergistic effect of pretreatment and hydrolysis enzymes on the production of fermentable sugars from date palm lignocellulosic waste. *Journal of Industrial and Engineering Chemistry, 19*, 413–415.
38. Anuradha, J., Abbasi, T., & Abbasi, S. A. (2015). An eco−friendly method of synthesizing gold nanoparticles using on otherwise worthless weed pistia. *Journal of Advanced Research*, http://doi.org/10.1016/j.jare.2014.03.006.
39. Abbasi, T., & Abbasi, S.A. (2012). Is the use of renewable energy sources an answer to the problems of global warming and pollution? *Critical Reviews in Environmental Science and Technology* (Taylor and Francis), *42*, 99–154.
40. Abbasi, S. A. (2006). Atomic Absorption Spectrometric and Spectrophotometric Trace Analysis of Uranium in Environmental Samples with N-p-MEthoxyphenyl-2-Furylacrylohydroxamic acid and 4-(2-Pyridylazo) Resorcinol. *International Journal of Environmental Analytical Chemistry, 36*(3), 163–172. https://doi.org/10.1080/03067318908026869.
41. Abd-Alla, M. H., Morsy, F. M., & El-Enany, A. W. E. (2011). Hydrogen production from rotten dates by sequential three stages fermentation. *International Journal of Hydrogen Energy, 36*(21), 13518–13527.
42. Mehaia, M. A., & Cheryan, M. (1991). Fermentation of date extracts to ethanol and vinegar in batch and continuous membrane reactors. *Enzyme and Microbial Technology, 13*(3), 257–261.
43. Besbes, S., Drira, L., Blecker, C., Deroanne, C., & Attia, H. (2009). Adding value to hard date (Phoenix dactylifera L.): Compositional, functional and sensory characteristics of date jam. *Food Chemistry, 112*, 406–411.
44. Gupta, N., & Kushwaha, H. (2011). Date Palm as a Source of Bioethanol Producing Microorganisms. In S. M. Jain, J. M. Al-Khayri, & D. V. Johnson (Eds.), *Springer* (pp. 711–727). Date Palm Biotechnology.
45. Gaily, M. H., Sulieman, A. K., Zeinelabdeen, M. A., Al-Zahrani, S. M., Atiyeh, H. K., & Abasaeed, A. E. (2012). The effects of activation time on the production of fructose and bioethanol from date extract. *African Journal of Biotechnology., 11*(32), 8212–8217.
46. Acourene, S., & Ammouche, A. (2012). Optimization of ethanol, citric acid, and α-amylase production from date wastes by strains of Saccharomyces cerevisiae, Aspergillus niger, and Candida guilliermondii. *Journal of Industrial Microbiology and Biotechnology, 39*(5), 759–766.
47. Abd-Alla, M. H., & El-Enany, A. W. E. (2012). Production of acetone-butanol-ethanol from spoilage date palm (Phoenix dactylifera L.) fruits by mixed culture of Clostridium acetobutylicum and Bacillus subtilis. *Biomass and Bioenergy, 42*, 172–178.

48. Khamaiseh, E. I., Kalil, M. S., Dada, O., El-Shawabkeh, I., & Yusoff, W. M. W. (2012). Date fruit as carbon source in RCM-Modified medium to produce biobutanol by Clostridium acetobutylicum NCIMB 13357. *Journal of Applied Sciences, 12*(11), 1160–1165.
49. Ahmadi, F., Rad, A. R., Holtzapple, M. T., & Zamiri, M. J. (2012). Short-term oxidative lime pre-treatment of palm pruning waste for use as animal feedstuff. *Journal of the Science of Food and Agriculture*, Article in press.
50. Bamufleh, H. S., Alhamed, Y. A., & Daous, M. A. (2013). Furfural from midribs of date-palm trees by sulfuric acid hydrolysis. *Industrial Crops and Products, 42*, 421–428.
51. Baroon, Z., El-Nawawy, A. S., & Al-Othman, A. (2004). Ensilage of Cardboard and Date Palm Leaves. *Journal of Environmental Science and Health—Part A Toxic/Hazardous Substances and Environmental Engineering, 39*, 515–533.
52. Chandrasekaran, M., & Bahkali, A. H. (2013). Valorization of date palm (Phoenix dactylifera) fruit processing by-products and wastes using bioprocess technology—Review. *Saudi Journal of Biological Sciences*. Article in press.
53. Devshony, S., Eteshola, A., & Shani, A. (1992). Characterisation and some potential application of date palm (Phoenix dactylifera L.) seeds and seeds oil. *Bioresource Technology, 67*, 291–295.
54. Ghehsareh, A. M., Samadi, N., & Borji, H. (2011). Comparison of date-palm wastes and perlite as growth substrates on some tomato growing indexes. *African Journal of Biotechnology, 10*(24), 4871–4878.
55. Ghosh, S. K., Sengupta, S., & Naskar, M. (2010). Physio-mechanical properties of particle boards from agro-wastes. *Journal of Scientific and Industrial Research, 69*(5), 396–400.
56. Ramzi, K., Nizar, M., Farouk, M., Naceur, B. M., & Evelyne, M. (2011). Sodium carboxyl-methylate cellulose from date palm rachis as a sizing agent for cotton yarn. *Fibers and Polymers, 12*(5), 587–593.
57. Tavakkoli, M., Hamidi-Esfahani, Z., & Azizi, M. H. (2009). Optimization of Corynebacterium glutamicum Glutamic acid production by response surface methodology. *Food and Bioprocess Technology*, 1–8.

Assessment of Groundwater Quality using Pollution Indices with respect to Heavy Metals

Kakarla Jagadeeswara Rao, Kolli Suresh Kumar, Namuduri Srinivas, and Akbar Ziauddin

1 Introduction

The environmental quality of an area depends on extent of industrial and development activities of an area, which causes adverse effect on human health and biota. The rate at which many natural and anthropogenic activities degrade the groundwater quality nowadays is alarming [1]. Industrialization and economic expansion in both developing and developed countries have also contributed heavy metals into groundwater which has become global issue. Groundwater is one of the significant and direct sources of water for many areas used by both rural and urban population for various purposes such as drinking, domestic use, irrigation. Groundwater occurs in widespread and local aquifer layers which can able to move one place to another place through the aquifers. The quality of groundwater is impaired by many factors like climate, soil composition, groundwater movement by rock types, region topography, infiltration of saline water into coastal regions and contaminants due to various man-made activities. Among the contaminants that can impact water quality, heavy metals are given more consideration because of their prevalence and high toxicity even though at low concentrations [2]. Contaminants of heavy metals such as Ni, Zn, Cu, Pb, Cr and Cd are usually more common than organic contaminants [3]. Heavy metals are high-density metalloids and non-biodegradable form that can persist through bioaccumulation in humans as well ecosystems and cause direct and indirect health effects [4]. Few heavy metals such as Zn, Cu, Ni, Fe and Mn are considered as micronutrients required for the growth of microbes, plants and animals. Besides, metals like Cr, Pb and Cd cause health risk beyond the prescribed

K. J. Rao · K. S. Kumar · N. Srinivas (✉)
Department of Environmental Science, GITAM Institute of Science, GITAM Deemed to be University, Visakhapatnam 530045, India
e-mail: snamudur@gitam.edu

A. Ziauddin
Advance Career Training & Consulting, Abu Dhabi, United Arab Emirates

N. A. Siddiqui et al. (eds.), *Advances in Behavioral Based Safety*,
https://doi.org/10.1007/978-981-16-8270-4_4

limit and can easily pass into human body by ingestion (by mouth) and can able to cause cancer, kidney and neurological diseases [5].

Hence, study of groundwater quality assists in establishing methods to identify the source and mitigate groundwater contamination. This has contributed to a growing emphasis on groundwater quality work around the world. Understanding the current condition of groundwater quality in any region forms part of the essentials needed to make wise plans and policies on the safety and management of water quality [6].

The present study was carried out at upcoming industrial area of Visakhapatnam. The objectives of the study are to find out the heavy metal concentration and its contamination level in groundwater along with identification of pollutant sources through principle component analysis (PCA).

2 Materials and Methods

Visakhapatnam is the fastest-growing city in Andhra Pradesh, India. The present identified five locations Baraniakam, Desapatrunipalem, Kondakarla, Mutyalampalem and Devada which are located near to the Parawada sub-urban area of Visakhapatnam (Fig. 1). These five locations are near to various major industrial areas like thermal power plants, steel plants, pharmaceutical and other minor industrials.

2.1 *Geography of Groundwater and Sampling*

The soils of the study area are red sandy loamy in nature, whereas fine sandy soils are confined only near coastal regions. The groundwater aquifers are having both hard and soft formations. In hard formation areas (granite gneisses, charnockites, khondalites, etc.), the groundwater is unconfined to semi-confined state, and in soft rock formations (sandstones and alluvium) areas, the groundwater is unconfined to confined state [7]. The samples were taken from the five boreholes (handle bore) on bi-month frequency for a period of two years. The total number of samples for each borewell is twelve. The water was extracted using handle-based grab sampling. Before sampling the handle, bore was operated for 5–7 min continuously to remove stagnate water. The sampling pretreated containers were cleaned with borewell water. Onsite pH was measured using portable battery-operated pH meter. After pH measurement, one liter of water taken is acidified with one ml HNO_3. The samples were immediately brought to laboratory and stored in refrigerator at 4 °C for further analysis. Before initiation of heavy metal analysis, the water samples were filtered with 0.45 μm millipore size filter; further, it is sent to ICP-MS for heavy metal analysis.

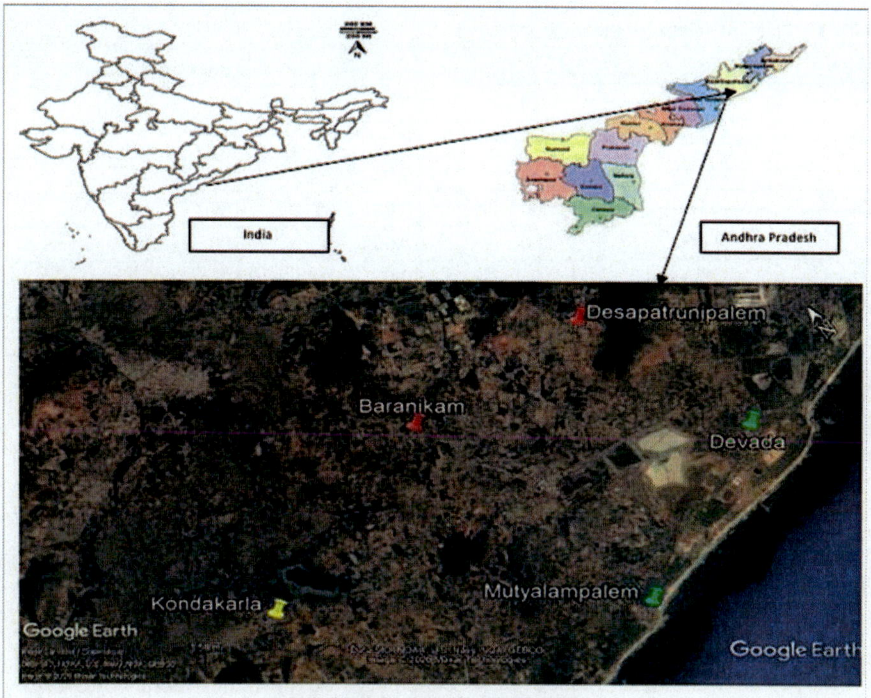

Study locations in Parawada, Visakhapatnam

Fig. 1 Study area and sampling location

Assessment of heavy metal
The groundwater heavy metal is analyzed using ICP-MS model, ELAN DRC-II, Perkin-Elmer Sciex Instrument, USA. The instrument setup, data acquisition and calibration were carried out as [8] recommend (CRM-NIST 1640 used for calibration and CRM SLRS-4 used as unknown to make sure the accuracy and precision of the analysis). The recovery percentage of elements is within the allowable standard (±4%).

To determine the water quality with respect to heavy metals, three indices HPI, HEI and Cd were used which provide overall heavy metal quality in the water.

Heavy metal pollution index (HPI): Heavy metal pollution index assesses the overall quality of water with reference to heavy metals calculated using following Formula [9].

$$\mathrm{HPI} = \frac{\sum_{i=1}^{n} \mathrm{WiQi}}{\sum_{i=1}^{n} \mathrm{Qi}}$$

where Qi = sub-index of ith metal; n = total number of metals; Wi = unit weight of ith metal

Table 1 Standard values for the indices

Metal	Si	Ii	Wi	Hmax or MAC
As	50	10	0.02	50
Cd	5	3	0.3	3
Cr	1	50	0.02	50
Cu	1000	2000	0.001	1000
Fe	300	200	0.005	200
Mn	100	500	0.02	50
Ni	20	20	0.05	100
Pb	100	10	0.7	1.5
Zn	5000	3000	0.0002	5000

$$Qi = \sum_{i=1}^{n} \frac{\{Mi(-)Ii)}{(\text{si} - \text{Ii})}$$

where Mi = measured value of ith heavy metals; Si = standard value (Table 1); Ii = ideal value of ith heavy metal; The (−) specifies the arithmetical variance of the two numbers, discounting the algebraic sign.

The classification of HPI value < 100, consider to be low heavy metal contamination; HPI value = 100, consider to be critical heavy metal contaminated; HPI value > 100, consider to be high heavy metal contamination (not recommended for drinking) [9, 10].

Heavy metal evaluation index (HEI): The HEI is also assessed by considering the total heavy metal content based the following Formula [11].

$$HEI = \sum_{i=1}^{n} \frac{Hc}{Hmac}$$

where Hc = measured metal concentration of corresponding metal; $Hmac$ = maximum admirable concentration of corresponding metal (Table 1).

The classification of HEI value < 10, consider to be low heavy metal contamination; HEI value between 10 and 20, consider to be moderate contaminated by heavy metal; HEI value > 20, consider to be highly contaminated by heavy metal [12].

Degree of contamination (Cd): Cd is the sum of the contamination factor (Cfi) of individual metals. The Cd is calculation using following Formula [13].

$$Cd = \sum_{i=1}^{n} \frac{CAi}{CNi} - 1 = \sum_{i=1}^{n} Cfi$$

where CAi = measured metal conc. of ith metal; CNi = maximum permissible concentration of ith metal (Table 1); Cfi = contamination factor of ith metal.

The classification of Cd value < 1, consider to be low heavy metal contamination; Cd value between 1 and 3, consider to be medium heavy metal contamination; Cd value > 3, consider to be high heavy metal contamination.

The results were compiled and compared with Bureau of Indian Standards 10,500 and WHO [14, 15] for analyzing heavy metals. SPSS statistical package (Version 20) is used for correlation and principal component analysis (PCA).

3 Results and Discussion

The mean and standard deviations values of results obtained for groundwater samples of the study areas are presented in Figs. 2 and 3, with its BIS and WHO standard. The pH value ranges from 6.9 to 7.5 and within the permissible limits in all the five sites. The maximum mean pH was found in Baranikam (7.5) followed by Kondakarla (7.3) Mutyalampalem (7.3), Devada (6.9) and Desapatrunipalem (6.9).

Heavy metal distribution in the study area:

The concentration of nine metals was analyzed at five study areas, and the results were discussed metal-wise below:

Arsenic: Arsenic is present in the environment in organic and inorganic forms. As_4O_6 is emitted from combusted fossil fuels and condensed and finally transferred to water reservoirs. The non-biodegradable waste can produce inorganic arsenic which is highly toxic and causes cancer to the humans even though at less concentration [16].

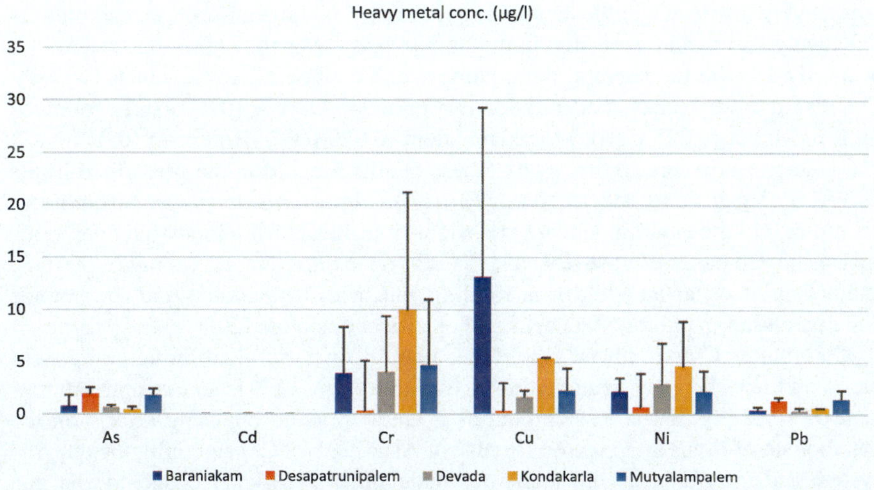

Fig. 2 Heavy metal concentration (As, Cd, Cr, Cu, Ni and Pb) in groundwater of study areas

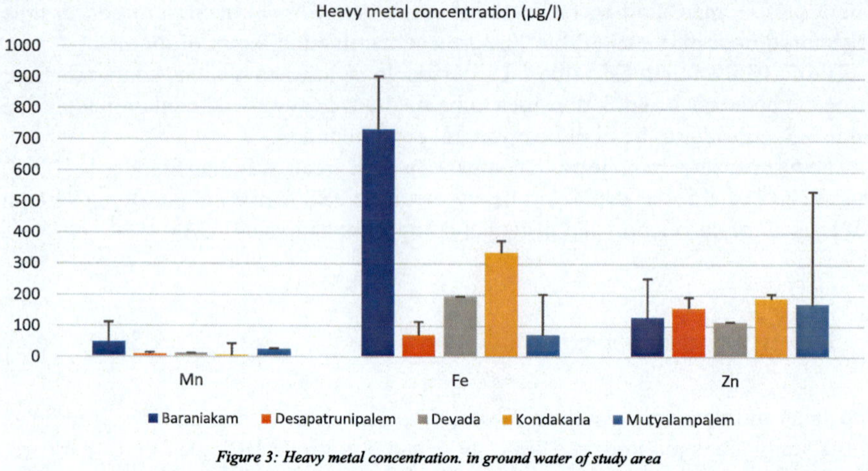

Fig. 3 Heavy metal concentration (Mn, Fe and Zn) in groundwater of study area

The maximum mean concentration of arsenic (As) is reported in Desapatrunipalem (1.919 µg/l) followed by Mutyalampalem (1.838 µg/l), Baranikam (0.761 µg/l), Devada (0.616 µg/l) and Kondakarla (0.389 µg/l); however, these results are within the prescribed limits of WHO (10 µg/l) and BIS standards (50 µg/l) (Fig. 2). The natural origin from weathering of rocks and man-made sources of arsenic in groundwater is runoff from agricultural fields and contains remains of pesticides and fertilizers and sewage run off from urban areas.

Cadmium: In nature, cadmium is distributed uniformly throughout the soil crust. It exists as inorganic materials such as hydroxides, carbonates, sulfates or chlorides at relatively low levels in the aquatic ecosystem [17]. Upon absorption, cadmium is effectively stored and deposited in the human body over the entire life, where it is primarily toxic to the nerve system, kidneys and demineralizes the bones [18, 19]. The maximum Cd mean concentration is reported in Devada (0.031 µg/l), followed by Kondakarla (0.025 µg/l), Mutyalampalem (0.020 µg/l), Branikam (0.017 µg/l) and Desapatrunipalem (0.016 µg/l). These results are within the prescribed limits of WHO (3 µg/l) and BIS standards (3 µg/l) (Fig. 2). The major contributors of cadmium into the environment are emissions or effluents from industrial operations such as Ni–Cd batteries, rust-resistant coatings on metals, dyes on ceramics, plastics, enamels, glasses, as an additive in welding, electrical connections and compounds which are used in photovoltaic cells and electrical detectors [20].

Chromium: Chromium may be either helpful or harmful to biotic component based on their chemical nature and bio-available form. Cr^{3+} is an essential component of a healthy diet at less amount as it tends to avoid opposing effects on the metabolism of lipids and glucose (Piyush & Asha, 2016). Cr can inhibit the enzyme system and interference with numerous metabolisms at higher concentration due to bonding nature with some organic compounds. Specific industries, like electro-forming, processing of paints and pigments, fabric, pesticide and leather tanning, Cr discharged in two varieties, such as Cr^{3+} and Cr^{6+} in waste. Cr^{6+} is potentially lethal, carcinogenic and epigenetic modification due to its solubility and mobility [21]. The maximum Cr mean concentration is reported in Kondakarla (9.958 μg/l) followed by Mutyalampalem (4.697 μg/l), Devada (4.014 μg/l), Branikam (3.903 μg/l) and Desapatrunipalem (0.300 μg/l). These results are within the prescribed limits of WHO (50 μg/l) and BIS standards (50 μg/l) (Fig. 2).

Copper: Copper is abundance element in earth surface and widely used in daily activity in the human for electrical, electronic appliances. Apart from this, it plays an important role in metabolic activities, protein synthesis and catalyst in living being [22]. However, at higher concentration, it will interfere in biological pathways; therefore, it is considered as hazard element in ecosystem [23]. The maximum Cu mean concentration is reported in Branikam (13.160 μg/l), followed by Kondakarla (5.305 μg/l), Mutyalampalem (2.268 μg/l), Devada (1.639 μg/l) and Desapatrunipalem (0.294 μg/l), and these results are within the prescribed limits of WHO (2000 μg/l) and BIS standards (50 μg/l) (Fig. 2). The major sources of copper into the environment are electrical appliances, copper smelting operations, as well as copper-based pesticides.

Iron: Iron is most abounded element in the earth crust as well groundwater in the form of $Fe^{2+}$and Fe^{3+}. It is the vital element for human and involves formation of blood, cytochrome, and metallo-enzymes. The excessive consumption of iron causes hemochromatosis which adversely effects on regular metabolism [24]. The maximum mean iron concentration is reported in Baranikamsite (732 μg/l) followed by Kondakarla (336 μg/l), Devada (194 μg/l), Desapatrunipalem (71 μg/l) and Mutyalampalem (72 μg/l) (Fig. 3). Baranikam and Kondakarla exceeded the BIS (300 μg/l) and WHO (300 μg/l) prescribed standards, and remaining three sites are well within the prescribed limit. The exceeded Fe in groundwater may influence by both natural and anthropogenic activities, including sewage wastewater disposal.

Manganese: Manganese is abundance in nature and essential element for biota. It acts as co-factor in many metabolic reactions, cholesterol, fatty acid synthesis and is bio-available form in the water in certain conditions [25]. At higher concentration, it accumulates into the body cells and causes postural dysfunction, mood disturbances and other shifts in psychiatry which are called manganese madness (characteristic neurotoxicity-linked disorder) [26]. The maximum Mn mean concentration is reported in Branikam (47.806 μg/l), followed by Mutyalampalem (24.480 μg/l), Devada (11.100 μg/l), Desapatrunipalem (9.383 μg/l) and Kondakarla (4.663 μg/l). However, these results are within the prescribed limits of WHO (50 μg/l) and BIS standards (100 μg/l) (Fig. 3). The key man-made sources of this element are sewage waste water and bio-solids.

Nickel: Nickel is present in less amount in both soil and water. The higher concentration exposure of human causes metabolic dysfunction, reducing body mass, allergy, cardiovascular, hepatic damage and carcinogenesis [27]. The maximum Ni mean concentration is reported in Kondakarla (4.590 μg/l), followed by Devada

(2.877 μg/l), Branikam (2.205 μg/l), Mutyalampalem (2.197 μg/l) and Desapatrunipalem (0.659 μg/l), and these results are within the prescribed limits of WHO (100 μg/l) and BIS standards (20 μg/l) (Fig. 2). The major sources of nickel to environment are various man-made activities such as metallic smelter, sewage treatment, pesticides and heavy metal-contaminated soils.

Lead: Lead is available in very minute quantity in the nature. The inorganic form of Pb could be highly toxic which causes lethal health effect on kidney, hemoglobin, digestive system, nerve system and carcinogenic [28]. Usually, it is non-biodegradable form emitted from fossil fuel, vehicular emission to the atmosphere, and later, it deposits into the soil which finally reached to groundwater [29]. The maximum Pb mean concentration is reported in Mutyalampalem (1.464 μg/l), followed by Desapatrunipalem (1.261 μg/l), Kondakarla (0.556 μg/l), Branikam (0.411 μg/l) and Devada (0.327 μg/l), and these results are within the prescribed limits of WHO (10 μg/l) and BIS standards (10 μg/l) (Fig. 2). The common man-made sources such as manufacture batteries, industrial reaction tanks, metal goods, plumbing paints, PVC pipes, waste land fill leachates, house hold items, alloys and electrical fuse cables as well process industrial like energy and automobile operations.

Zinc: Zinc is a most abundance earth crust element and available in atmospheric emission which can be able to adsorb water-borne suspended particulates threatened to the water ecosystem [30]. Zinc serves as a co-factor for many biomolecules (enzymes and protein) required to reproduce and convert genetic material in many organisms [31]. The excessive Zn can induce system disorders like yellowing of the mucous membranes, kidney and liver damage and effect on growth and reproductive system [30]. The maximum Zn mean concentration is reported in Kondakarla (189.347 μg/l) followed by Mutyalampalem (172.076 μg/l), Desapatrunipalem (160.038 μg/l), Branikam (129.444 μg/l) and Devada (113.851 μg/l). These results are within the prescribed limits of WHO (5000 μg/l) and BIS standards (5000 μg/l) (Fig. 3). Untreated residential and industrial sewage waste, land fill leachates and agricultural runoff are the major man-made sources of Zn into the groundwater.

Pollution Indices

Heavy metal pollution index (HPI): HPI score gives an idea on overall heavy metal contamination in the study areas. The sum of HPI results is present in Table 2 for the five sites. Among the five sites, Baranikam (55.976) has shown the maximum HPI followed by Kondakarla (51.490), Desapatrunipalem (51.345), Devada (51.161) and

Table 2 HPI results of groundwater in the study area

Study area	ΣQi.Wi	ΣWi	HPI	Pollution status
Baranikam	62.481	1.116	55.976	Low
Desapatrunipalem	57.312	1.116	51.345	
Devada	57.106	1.116	51.161	
Kondakarla	57.473	1.116	51.490	
Mutyalampalem	56.842	1.116	50.924	

Table 3 HEI results of groundwater in the study area

Study area	HEI	Pollution status
Baranikam	7.841	Low
Desapatrunipalem	1.476	
Devada	1.568	
Kondakarla	2.452	
Mutyalampalem	2.022	

Mutyalampalem (50.924) (Table 3). According to Prasad and Bose (2001), HPI value less than 100, consider to be low in heavy metal contamination, in all the five study sites is scored less than 100; hence, all the five sites are low heavy metal contamination. The HPI has given an indication on heavy metal quality in the groundwater of five sites that are well within the range. However, tremendously increasing heavy metal content in the water may cause adverse health effect on living organisms.

Heavy metal evaluation index (HEI): HEI is another index for the heavy metal quality assessment. It gives an overall heavy metal contamination approach based on maximum admissible value (MAC). Sum of HEI values is present in Table 3 for the five sites. Among the five sites, Baranikam (7.841) has shown the maximum HEI followed by Kondakarla (2.452), Mutyalampalem (2.022), Devada (1.568) and Desapatrunipalem (1.476). Due to the higher Fe concentration, Baranikam scored the highest HEI value. HEI value less than 10, consider to be low heavy metal contamination. In all the five study sites are scored less than 10, indicates low heavy metal contamination. The HEI has given an indication on heavy metal quality in the groundwater of five sites that are well within the range. Rapid increasing heavy metal concentration in the water may cause adverse health effect on biota.

Degree of contamination (Cd): Cd score is a cumulative heavy metal contamination in the study areas. The sum of Cd values is present in Table 3 for the five sites (Table 4). Among the five sites, Baranikam (−1.159) has shown the maximum Cd followed by Kondakarla (−6.548), Mutyalampalem (−6.978), Desapatrunipalem (−7.432) and Devada (−7.524). Cd value less than 1, consider to be low heavy metal contamination; in the study area, all the five sites are scored less than 1; hence, all the five sites are low heavy metal contamination. The Cd has given an indication on heavy metal quality in the groundwater of five sites that are well within the range.

Table 4 Cd results of groundwaters in the study area

Study area	Cd	Pollution status
Baranikam	−1.159	Low
Desapatrunipalem	−7.524	
Devada	−7.432	
Kondakarla	−6.548	
Mutyalampalem	−6.978	

Table 5 Correlation analysis among the parameters

	pH	As	Cd	Cr	Cu	Fe	Mn	Ni	Pb	Zn
pH	1.00									
As	−0.26	1.00								
Cd	−0.33	−0.64	1.00							
Cr	0.48	−0.68	0.46	1.00						
Cu	0.83	−0.49	−0.27	0.25	1.00					
Fe	0.70	−0.45	−0.28	0.08	0.98	1.00				
Mn	0.70	0.01	−0.47	−0.21	0.80	0.82	1.00			
Ni	0.35	−0.81	0.65	0.97	0.22	0.08	−0.26	1.00		
Pb	−0.13	0.93	−0.57	−0.37	−0.52	−0.56	−0.13	−0.55	1.00	
Zn	0.24	0.24	−0.28	0.48	−0.21	−0.37	−0.40	0.28	0.54	1.00

4 Statistical analysis

Correlation analysis: The correlation matrix was performed between the pH and metals. The strong correlation ($r > 0.60$) among the pH has shown significant correlation with Cu–Fe–Mn. Similarly, relation between As–Pb, Cd–Ni, Cu–Mn–Fe metals is observed (Table 5).

Principle component analysis: PCA was plotted for the five location among the nine variants to recognize the heavy metal source. PC with eigen values greater than one considered to be significant and the loading value greater than forty were taken into consideration for the data interpretation [19]. The screen plot total variance about 96.8%. PC-1 with 4.15 eigen and 46.21% of total cumulative variants have shown the highest loading for Cu, Fe and Pb. PC-2 with 2.97 eigen and 33.03% of total cumulative variants have shown loading for As, Ni and Zn. PC-3 with 1.58 eigen and 17.61% of total cumulative variants have shown considerable loading for Cr and Mn (Fig. 4) indicating significant contribution of metals from mixed source of natural soil crust (red soil) and common man-made source.

5 Conclusion

The finding from the present study is heavy metals concentration in the groundwater in the five study areas that are well within the prescribed standards of WHO and BIS. Fe in Baranikam and Kondakarla areas was exceeded due to mixed sources of natural and man-made. The three pollution indices HPI, HEI and Cd have shown low heavy metal content. PCA has given three principal components with a total variance of 96.8%. PC-1 grouped with Cu–Fe–Pb; PC-2 grouped with As–Ni–Zn; and PC-3 grouped with Cr and Mn. Cd is not fit in the PCA due to its very low concentration. Even though the present values were well within the permissible limits,

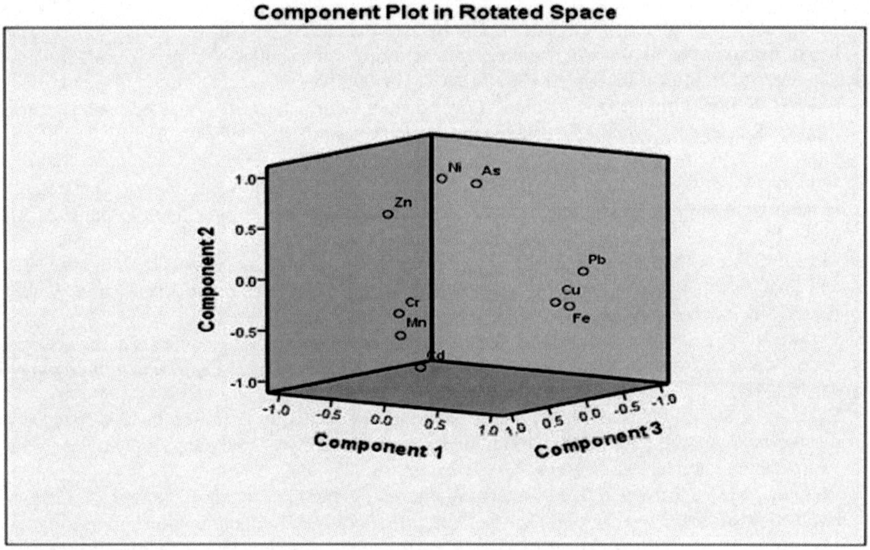

Fig. 4 Principle component analysis for heavy metals of groundwater

the PCA gives an idea to categorize the possible sources of metals based on their groupings. The results are useful for the future pollution source management to avoid groundwater contamination particularly in areas where residential and industrial activities commingled.

Acknowledgements Authors are thankful to DST-FIST, New Delhi, for providing infrastructural support. Special thanks to NGRI, Hyderabad, to carryout heavy metal analysis.

References

1. Guo, G., Wu, F., Xie, F., & Zhang, R. (2012). Spatial distribution, and pollution assessment of heavy metals in urban soils from southwest China. *Journal of Environmental Sciences, 24*(3), 410–418.
2. Pazand, K., Khosravi, D., Ghaderi, M. R., & Rezvanianzadeh M. R. (2018) Hydrogeochemistry and lead contamination of groundwater in the north part of Esfahan province, Iran. *Journal of Water & Health, 16* (4), 622–634.
3. Lou, S., Liu, S., Dai, C., Tao, A., Tan, B., Ma, G., Chalov, R. S., Chalov, S. R. (2017). Heavy metal distribution and groundwater quality assessment for a coastal area on a Chinese Island. *Polish Journal of Environmental Studies, 26*(2), 733–745.
4. Duffus, J. H. (2002). "Heavy metals" a meaningless term (IUPAC Technical Report). *Pure and Applied Chemistry, 74*(5), 793–807.
5. Karavoltsos, S., Sakellari, A., Mihopoulos, N., Dassenakis, M., & Scoullos, M. J. (2008). 2: Evaluation of quality of groundwater in regions of Greece. *Desalination, 224*(1–3), 317–329.

6. Mgbenu, C. N., Egbueri, J. C. (2019). The hydrogeochemical signatures, quality indices and health risk assessment of water resources in Umunya district, southeast Nigeria. *Applied Water Science, 9*(22). https://doi.org/10.1007/s13201-019-0900-5.
7. CGWB (Central Ground Water Board): Groundwater Information Booklet; Ludhiana District; Punjab. Retrieved Nov 28, 2018, from http://cgwb.gov.in/District_Profile/Punjab/Ludhiana.pdf.
8. Balaram, V., & Gnaneshwar Rao, T.: Rapid determination of REEs and other trace elements in geological samples by microwave acid digestion and ICP-MS. *Atomic Spectroscopy, 24*(6), 206–212.
9. Mohan, S. V., Nithila, P., & Reddy, S. J. (1996). Estimation of heavy metals in drinking water and development of heavy metal pollution index. *Journal of Environmental Science and Health, Part A Environmental Science, 31*, 283–289.
10. Prasad, B., & Bose, J. M. (2001). Evaluation of heavy metal pollution index for surface and spring water near a limestone mining area of the lower Himalayas. *Environmental Geology, 41*, 183–188.
11. Edet, A.E., & Offiong, O.E. (2002). Evaluation of water quality pollution indices for heavy metal contamination monitoring. A study case from Akpabuyo–Odukpani area, Lower Cross River Basin (southeastern Nigeria). *GeoJournal, 57*, 295–304.
12. Prasanna, M. V., Praveena, S. M., Chidambaram, S., Nagarajan, R., & Elayaraja, A. (2012). Evaluation of water quality pollution indices for heavy metal contamination monitoring: A case study from Curtin Lake, Miri City, East Malaysia. *Environmental Earth Science, 67*, 1987–2001.
13. Backman, B., Bodiš, D., Lahermo, P., Rapant, S., & Tarvainen, T. (1998). Application of a groundwater contamination index in Finland and Slovakia. *Environmental Geology, 36*, 55–64.
14. BIS (Bureau of Indian Standards—10500). (2012). Specification for drinking water, Report Indian Standards Institution, New Delhi, pp. 1–5.
15. WHO (World Health Organization). (2011). Guidelines for drinking water quality, 4th edn. WHO, Geneva
16. Shankar, S., & Uma Shanker, S. (2014). Arsenic contamination of groundwater: A review of sources, prevalence, health risks, and strategies for mitigation. *Scientific World Journal*. https://doi.org/10.1155/2014/304524.
17. Namieśnik, J., & Rabajczyk, A. (2010). The speciation and physicochemical forms of metals in surface waters and sediments. *Chemical Speciation & Bioavailability, 22*(1), 1–24.
18. Rodríguez, J., & Mandalunis, P. M. (2018). A review of metal exposure and its effects on bone health. Journal of Toxicology. https://doi.org/10.1155/2018/4854152.
19. Shrestha, S., & Kazama, F. (2007). Assessment of surface water quality using multivariate statistical techniques: A case study of the Fuji river basin, Japan. *Environmental Modelling Software, 22*, 464–475.
20. Yuan, Z., Luo, T., Liu, X., Hua, H., Zhuang, Y., Zhang, X., Zhang, L., Zhang, Y., Xu, W., & Rend, J. (2019). Tracing anthropogenic cadmium emissions: From sources to pollution. *Science of the Total Environment., 676*, 87–96.
21. Zhitkovich, A. (2011). Chromium in drinking water: Sources, metabolism, and cancer risks. *Chemical Research in Toxicology, 24*(10), 1617–1629.
22. Chellan, P., & Sadler, P. J. (2015). The elements of life and medicines. *Philosophical Transactions of the Royal Society A Mathematical Physical and Engineering Sciences, 373*(2037). https://doi.org/10.1098/rsta.2014.0182.
23. Tchounwou, P. B., Yedjou, C. G., Patlolla, A. K., & Sutton, D. J. (2012). Heavy metals toxicity and the environment. EXS, *101*, 133–164.
24. Mesías, M., Seiquer, I., & Pilar Navarro, M. (2013). Iron nutrition in adolescence. *Critical Reviews in Food Science and Nutrition, 53*(11), 1226–1237.
25. Chen, P., Bornhorst, J., & Aschner, M. (2018). Manganese metabolism in humans. *Front Biosci (Landmark Ed)., 23*, 1655–1679.
26. O'Neal, S. L., & Zheng, W. (2015). Manganese toxicity upon overexposure: A decade in review. *Current Environmental Health Reports, 2*(3), 315–328.

27. Monisha, J., Tenzin., T., Naresh., A. Blessy., N, Mathew Krishnamurthy., & Beeregowda (2014). Toxicity mechanism and health effects of some heavy metals. *Interdisciplinary Toxicology, 7*(2), 60–72.https://doi.org/10.2478/intox-2014-0009.
28. Genchi, G., Carocci, A., Lauria, G., Sinicropi, M. S., & Catalano, A. (2020). Nickel: Human health and environmental toxicology. *International Journal of Environmental Research and Public Health, 17*(3), 679.
29. Chatterjee, M., Roy, K., Janarthan, M., Das, S., & Chatterjee, M. (2012). Biological activity of carotenoids: Its implications in cancer risk and prevention. *Current Pharmaceutical Biotechnology, 13*(1), 180–190.
30. Hose, G. C., Symington, K., Lott, M. J., & Lategan, M. J. (2016). The toxicity of arsenic(III), chromium(VI) and zinc to groundwater copepods. *Environmental Science and Pollution Research, 23*(18), 18704–18713.
31. Maret, W. (2013). Zinc biochemistry: From a single zinc enzyme to a key element of life. *Advances in Nutrition, 4*(1), 82–91.

Intuitionistic Fuzzy-Based ETA for Risk Assessment of DHDS Unit

Arya Dinesh and V. R. Renjith

1 Introduction

The relevance of petroleum industries and its activities have always been a boon to humankind, as well as they have a significant influence on the economies of oil producing as well as oil importing countries. This accounts for the petroleum countries to have a far-reaching influence on the politics of the producing and exporting countries and the world at large. With the production of petroleum products and its by-products, there had been a staggering advancement to industrial growth worldwide, and it has created an ease of transportation at both international and local levels.

The Diesel Hydrodesulfurization Unit (DHDS) of a petroleum industry is of prime importance as it has become an inevitable part of petroleum industry. To meet the revised diesel specification petroleum industry, DHDS Unit of a standard capacity is installed in all petroleum industries. The main purpose of the unit is to hydro treat a blend of straight run, cracked gas oils, heavy naphtha and breaker gas oil in order to produce a low sulfur content diesel oil at the required capacity in order to achieve a sulfur content of prescribed parts per million (ppm) weight in the hydro-treated diesel.

It is essential to measure the likelihood of mishaps by dissecting and anticipating every possible danger associated with the selected area. In this study, event tree analysis (ETA) is utilized as a novel strategy for possible danger investigation and has been applied at the for the purpose of assessing the risk associated with DHDS Unit. An event tree analysis (ETA) is an inductive procedure that shows all possible outcomes resulting from an accidental (initiating) event, taking into account whether installed safety barriers are functioning or not, considering all the additional events and factors.

A. Dinesh · V. R. Renjith (✉)
Faculty of Division of Safety and Fire Engineering, School of Engineering, CUSAT, Kochi, Kerala, India

N. A. Siddiqui et al. (eds.), *Advances in Behavioral Based Safety*,
https://doi.org/10.1007/978-981-16-8270-4_5

An initiating event with high potential has been selected, the safety barriers associated with it are identified, and their probability of failure on demand (PFD) is assessed quantitatively. For obtaining the probability of failure on demand (PFD) of safety barriers, linguistic opinion of experts was taken, and intuitionistic fuzzy theory is applied [1].

2 DHDS Unit

High speed dDiesel (HSD) contains pollutants such as sulfur, metallic compounds, nitrogen and other contaminants that contribute to increased air pollution, reduced efficiency, mechanical corrosion, etc. So, it is of high importance that the sulfur content is maintained in a controlled limit, and the recommended level of sulfur in HSD is less than 100 ppm. In Diesel Hydrodesulfurization (DHDS) Unit, the sulfur in HSD is reduced to a level less than 100 ppm in the presence of catalyst, hydrogen (H_2). The unit uses a fixed bed preparation process to improve the quality of the petroleum distillate particles by decomposing impurities with minimal impact on the boiling point of the feed.

The steam naphtha reforming process is carried out for the purpose of producing of hydrogen in the unit. The hydrogen unit uses light naphtha or high aromatic naphtha as feedstock. In the final desulfurization unit (FDS), to maintain the sulfur level of less than 0.05 ppm, the raw feedstock is used to meet the hydrogen requirement. Sweet feed (non-sulfur) is recycled to convert high hydrocarbon into methane. Thus, a methane-rich feed is generated. The hydrogen obtained by conversion is 70% pure. Again, on processing this in the pressure swing adsorption (PSA) unit also improves hydrogen purity to 99.99% volume.

Amine is used to treat H_2S gas from DHDS Unit to enrich and to obtain sulfur as a by-product, and the H_2S-rich gas again is treated in the sulfur recovery unit (SRU). Clauss process is carried out for the production of sulfur. The acid gas from the Clauss reactor is treated in maximum Clauss recovery concept (MCRC) unit to recover sulfur to the maximum possible extend (Fig. 1).

2.1 *Process Design*

In the Diesel Hydrodesulfurization Unit (DHDS), hydrogen is compressed and fed to the reactor system along with feed at an operating temperature of 350 °C. In the presence of catalysts, the hydrogen reacts with sulfur content in diesel and gets converted into hydrogen sulfide (H_2S). This hydrogen sulfide will be removed on later stage as the process continues. Hydrogen is being supplied by a makeup gas compressor, where it takes suction from a hydrogen header which is having a pressure of 20 bar. This hydrogen is compressed to a higher pressure of 40 bar for feeding to the DHDS Unit. So, in case of any abnormalities or any unprecedented shutdown

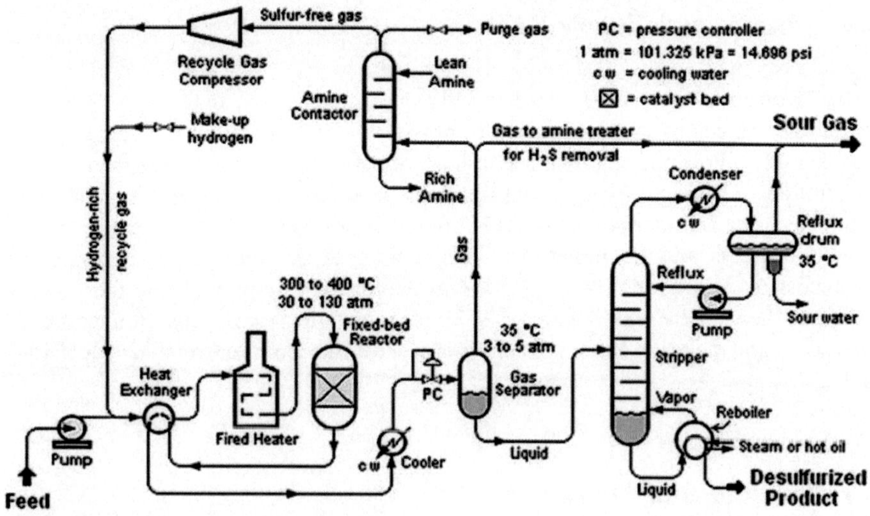

Fig. 1 DHDS unit

of the system, the compressor discharge valve will be automatically closed and will safeguard the mug compressor. This is an undesired event (which is considered as an initiating event here) which has the potential to cause release of hydrogen from the compressor outlet if all the provided safety barriers are not functioning as intended.

3 Materials and Methods

3.1 Event Tree Analysis

An event tree analysis (ETA) illustrates the conceptual integration of the various events that may follow from an initiation event (e.g., an accident event such as an LPG release). The first event of the tree uses dichotomous conditions, i.e., success/failure (true or false or yes/no) to spread the effect of an event on different branches of a tree. Each individual path is followed by different branches which ultimately identifies the occurrence of potential outcome event or consequences.

Thus, in this risk analysis, the outcome events are sequential and used for pre-incident determination and post-incident examination. Thereby, the path followed for each unwanted events is identified to spot the failure events. The potential outcome events or the consequent events are identified quantitatively in an ETA, and frequency or probability of these events is also estimated quantitatively.

3.2 Intuitionistic Fuzzy ETA

ETA—event tree analysis is a traditionally used quantitative analysis, where crisp probability values of events are used to obtain the consequent event probability or frequency [2]. Pointing to a real scenario, in most of the cases, the estimates taken are limited to a single expert's opinion, and it is both difficult and expensive to obtain accurate estimates of event probabilities because in many cases, the result are limited in access, and incomplete knowledge of the expert selected also accounts for vagueness in estimation. Thus, the poor or incomplete interpretation of the failures increases the vagueness of the analysis. These inevitable issues transmit uncertainty to the ETA and make the entire risk analysis process to become unreliable in decision-making.

3.3 Intuitionistic Fuzzy

In 1986, Zadeh's paradoxical theory [3] was added to the intuitionistic fuzzy concept by Atanassov [1]. The intuitionistic fuzzy sets consist of three parts: membership, non-membership and skepticism or hesitation. Combining these three components, intuitionistic fuzzy sets (IFSs) succeeded in describing a more fragile set for different applications than the conventionally used traditional and non-traditional sets. In order to point out the skepticism in the process of quantification of data failure uncertainty, Atanassov used this concept of intuitionistic fuzzy set (IFS) to get a better result. Thus, IFS theory gained a broader scope of application than a preconceived notion of conventional fuzzy set. It can portray a more delicate ambiguity of the objective world. Therefore, this theory has been widely used. The IFS concept has been used in a variety of areas such as decision-making problems, medical diagnosis, pattern recognition and logic programming.

Basic Preliminaries—Intuitionistic Fuzzy Number

Considering X as universal set, the intuitionistic fuzzy set A in X can be considered as an ordered triple set as follows [1]:

$$A = \{\langle x, \mu_A(x), \upsilon_A(x)\rangle | x \in X\}$$

In which $\mu_A : X \to [0, 1]$, $\upsilon_A : X \to [0, 1]$ represent the membership function, that is, the degree of acceptance and non-membership function and the degree of rejection of an element such that $x \in A \subset X$ and also for every $x \in X$ $0 \leq \mu_A(x) + \upsilon_A(x) \leq 1$.

So, the intuitionistic fuzzy index of x in A can be defined as follows (Fig. 2):

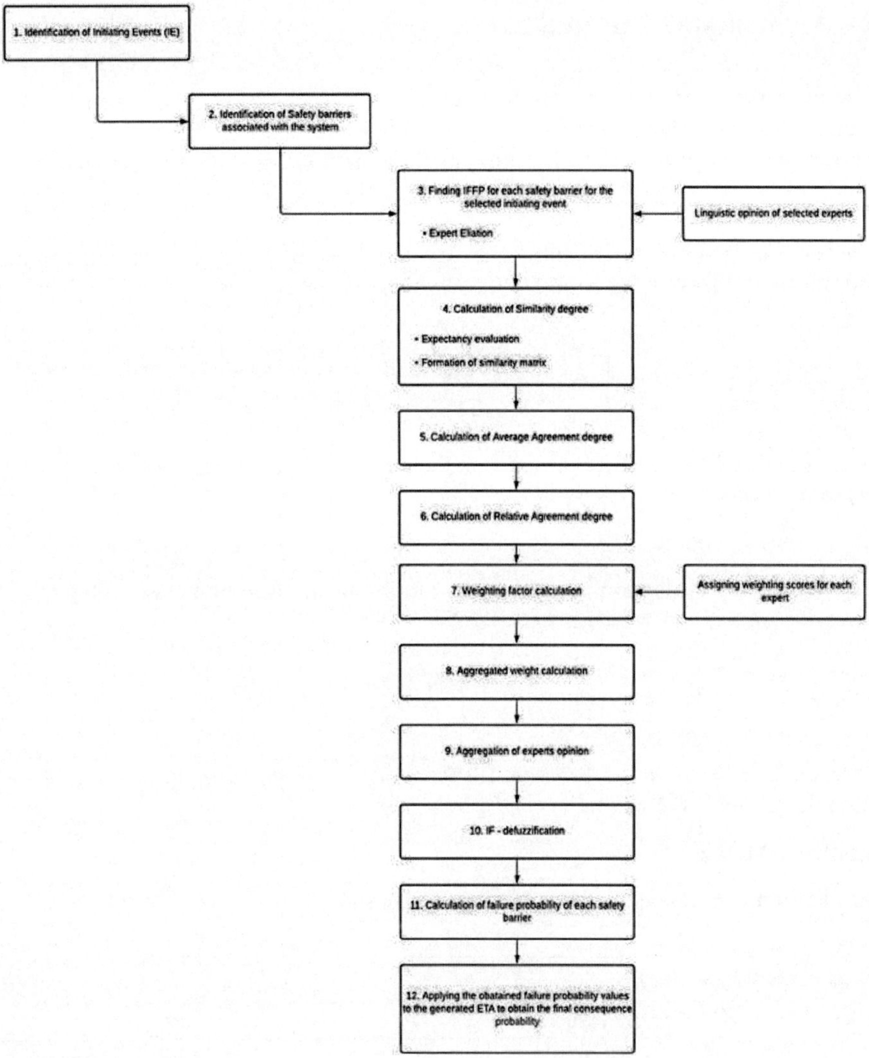

Fig. 2 Flow diagram

$$\pi A(x) = 1 - \mu_A(x) - v_A(x)$$

4 Application of IF to ETA

The steps for incorporating intuitionistic fuzzy approach to ETA are given in this section. The steps given are to be sequentially done in order to get the crisp probability of each safety barriers. The calculation methods are explained in the given steps:

Similarity Degree

Considering the opinions P_i and P_j of experts E_i and E_j, it can be calculated using the similarity measure function $S(P_i, P_j)$ which is expressed as:

$$S(P_i, P_j) = \begin{cases} {EV(P_i)}/{EV(P_j)};\ for EV(P_i) \leq EV(P_j) \\ {EV(P_j)}/{EV(P_i)};\ for EV(P_i) \geq EV(P_j) \end{cases}$$

Here, $EV\ (P_i)$ and $EV\ (P_j)$ represent the expectancy evaluation for IFNs P_i and P_j, respectively.

Expectancy Evaluation

The expectancy evaluation of a triangular intuitionistic fuzzy number (TIFN).
$P_{ij} = (p, q, r; p', q', r')$ is expressed as:

$$EE(P_i) = \frac{(p_i + p'_i) + 4q_i + (r_i + r'_i)}{8}$$

Here, $p, q, r; p', q', r'$ are the IFFP values corresponding to the linguistic variables as given below (Table. 1).

Similarity Matrix

Similarity matrix calculated from experts is as follows:

Table 1 IFFP values of various opinions

Opinion	IFFP values of opinions					
	P	Q	R	p’	q’	r’
VL	0	0.04	0.08	0	0.04	0.08
L	0.07	0.13	0.19	0.06	0.13	0.2
RL	0.17	0.27	0.37	0.15	0.27	0.39
M	0.35	0.5	0.65	0.32	0.5	0.68
RH	0.63	0.73	0.83	0.61	0.73	0.85
H	0.81	0.87	0.93	0.79	0.87	0.95
VH	0.92	0.96	1	0.92	0.96	1

$$SM = \begin{bmatrix} 1 & S_{(12)} & S_{(13)} & . & S_{(1m)} \\ S_{(21)} & 1 & . & . & S_{(2m)} \\ . & . & 1 & . & . \\ . & . & . & 1 & . \\ S_{(m1)} & S_{(m2)} & . & . & 1 \end{bmatrix}$$

Here, it is to be noted that $s_{ij} = S(P_i, P_j)$, and if $i = j$, then $s_{ij} = 1$.

Average Agreement Degree

It is the linguistic agreement between a group of employees or an individual regarding a particular event. Considering m experts, the average agreement degree can be obtained as:

$$AA(E_i) = \frac{1}{m-1}\sum_{j=1}^{m} S(P_i, P_j); i = (1, 2, \ldots, m), j \neq i$$

Relative Agreement Degree

It indicates the relative linguistic agreement between the experts for the failure of a component.

The relative agreement degree (RAD) considering 'm' experts can be found out using:

$$RAD(E_i) = \frac{AA(E_i)}{\sum_{i=1}^{m} AA(E_i)} i = (1, 2, \ldots, m)$$

Weighting Factor

Weighting scores of experts are selected on the basis of educational qualifications, professional positions and years of working experience. Table 2 showing the weighting scores of different experts is shown below.

On selecting m experts, weighting scores for any expert are defined as follows:

$$WF(E_i) = \frac{WS(E_i)}{\sum_{i=1}^{m} WS(E_i)}; i = (1, 2, \ldots, m)$$

Here, WS is weighting score, and WF is weighting factor of the considered expert Ei. Since each expert is having different educational qualifications, professional positions and years of working experience the weighing scores.

Aggregated Weight

It is the sum of relative agreement degree (RAD) and weighting factor $WF(E_i)$ of the selected expert in which a relaxation factor is considered.

Table 2 Weighting scores

Condition	Classification	Scores
Professional position	Professor, GM/DGM, chief engineer, director	5
	Assistant Professor, manager, factory inspector	4
	Engineer, supervisors	3
	Foreman, technician, graduate apprentice	2
	Operator	1
Job experience (years)	≥20	5
	15–19	4
	10–14	3
	5–9	2
	<5	1
Education	Ph.D. or M.Tech	5
	MSc or B.Tech	4
	Diploma or B.Sc.	3
	ITI	2
	Secondary school	1

$$w_i = \beta \cdot RAD(E_i) + (1 - \beta).WF(E_i); i = 1, 2, \ldots, m)$$

Here, β is the relaxation factor. β denotes the dominance of $WF(E_i)$ over $RAD(E_i)$ or vice-versa.

Aggregation of Expert Opinions

For obtaining the aggregated IFFP (pj) of any safety barrier X_j, $j = 1, 2, ...,$ the following equation can be used:

$$P_j = \sum_{i=1}^{m} w_i \otimes p_{ij};\ j = 1, 2, .., n$$

IF Defuzzification

The obtained IFFP of any safety barrier, say (a; b; c; a0; b; c0), needs to be defuzzified and converted to crisp possibility score. This is accomplished using the given equation:

$$S = \frac{1}{3}\left[\frac{(c' - a')(b - 2c' - 2a') + (c - a)(a + b + c) + 3(c'^2 - a'^2)}{(c' - a' + c - a}\right]$$

Failure Probability of Each Safety Barrier

For obtaining the failure probability of each safety barrier, the obtained crisp values are transformed using failure probability of each safety barrier by using Onisawa's logarithmic function [4] as shown:

$$P = \begin{cases} \dfrac{1}{10^{\left(\left[\frac{1-S}{S}\right]^{\frac{1}{3}} \times 2.301\right)}}; & S \neq 0 \\ 0; & S = 0 \end{cases}$$

Here, 's' represents the crisp probability score.

5 Proposed IFETA Approach

IFETA's proposed approach uses intuitionistic fuzzy to obtain details of the quantitative failure data of each security barrier to assess the probability of system failure and to evaluate the probability of different outcomes.

5.1 *Selection of Initiating Event*

The makeup gas flow rate to the reaction section in the DHDS Unit is controlled by means of a compressor spillback system. A reciprocating compressor is used in the plant. As per the normal process, hydrogen is being supplied by this makeup gas compressor, where it takes suction from a hydrogen header which is having a pressure of 20 bar. This hydrogen is compressed to a higher pressure of 40 bar for feeding to the DHDS Unit. So, in case of any abnormalities or any unprecedented shutdown of the system, the compressor discharge valve will be automatically closed and will safeguard the mug compressor. This is an undesired event (which is considered as an initiating event here) which has the potential to cause release of hydrogen from the compressor outlet if all the provided safety barriers are not functioning as intended.

5.2 *Construction of ETA*

The most important step in the construction of ETA is to identify the initiating event and its corresponding probability. The initiating event is the any undesired event that is having the potential to cause a deviation from the normal process. The next step after identification of the initiating event is to identify the safety barriers that are installed in the system to provide protection or safe operating systems that are provided to bring back to normal functioning in case of any deviation or mishappening in the system. After the identification of all safety barriers associated with the system, next

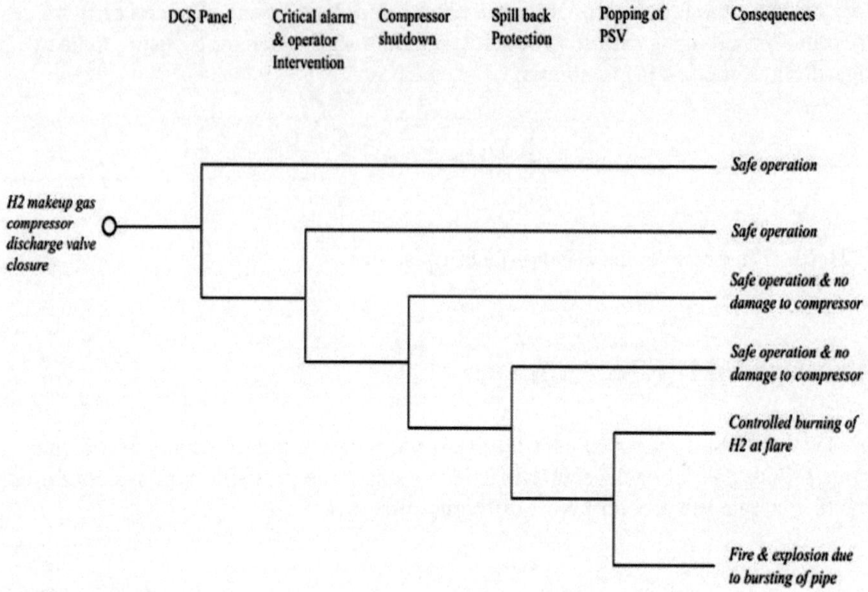

Fig. 3 Constructed ETA

is to plot the event tree. Construction of event tree starts with the initiating event at the left extreme end, proceeding through failures of the safety functions toward right (Fig. 3).

Explanation of ETA

The initiating event is taken as 'hydrogen makeup gas compressor discharge valve closure.'

Different safety barriers and their safety functions are given below:

Distributed control system (DCS) panel—It is considered as BPCS, and on detection of DCS fault, the panel operator can act in the situation in the desired manner and make the system safe.

Critical alarm and human intervention—The system can be brought to safe operating condition by human intervention.

Compressor shutdown—It acts as a safety instrumented system (SIS). If the pressure increases beyond the tolerable limit, the compressor shuts down so that the pressure will not buildup on the compressor outlet.

Spillback protection—This is also a safety instrumented system (SIS), and if the compressor fails to shutdown, the spillback mechanism works and the hydrogen is sent back to the hydrogen header.

Pressure safety valve (PSV)—In case, if the spillback protection is not functioning, PSV will be popped and the hydrogen gas is released to flare where it is allowed for controlled burning.

If all the safety barriers fail, there will be an over pressure at the compressor outlet pipe. When the pressure tolerable limit of the pipe line is reached, the bursting of the pipe line happens and hydrogen gas will be released. On availability of any ignition source, burning of hydrogen happens which will lead to a high consequence scenario.

5.3 Failure Probability of Different Safety Barriers

For the obtaining the failure probability of different safety barriers and for obtaining the linguistic opinions, the following experts were selected as given in Table. 3.

5.4 Conversion of Linguistic terms to IF numbers

For obtaining the probability of failure of each safety barrier, three experts handling different designation, having different professional qualification and different experience, were selected, and their opinion about the failure of each safety barrier was recorded linguistically. The experts were given seven different linguistic terms to give their opinions which include very low, low, reasonably low, moderate, reasonably high, high and very high.

Linguistic Opinions of Experts Given for Each Safety Barrier

See Table. 4.

Table 3 Details of experts selected

Sl. no	Professional position	Job Experience (years)	Education
1	Manager	5–10	B.Tech
2	Assistant Manager	<5	B.Tech
3	Operator	10–14	Diploma

Table 4 Linguistic opinions of selected experts

Safety barriers	Expert (E1)	Expert (E2)	Expert (E3)
DCS panel	VL	VL	L
Critical alarm	RL	VL	VL
Compressor shutdown	L	L	R
Spillback protection	VL	RL	L
Popping of PSV	VL	VL	VL

Table 5 Crisp values of different safety barriers

Sl. no	Safety barriers	IF defuzzified values	Crisp values
1	DCS panel	0.047617098	5.66566E−07
2	Critical alarm	0.064616072	2.46665E−06
3	Compressor shutdown	0.130666667	4.70219E−05
4	Spillback protection	0.105190299	2.00837E−05
5	Popping of PSV	0.0311	5.75052E−08

5.5 *Intuitionistic Fuzzy Values for Each Safety Barrier*

For obtaining the failure probability of the initiating event, values from offshore and onshore reliability data (OREDA) were used. Thus, the failure probability of the reciprocating compressor was obtained as 2.65×10^{-2}(Table. 5).

6 Results

On the successful completion of construction of ETA and incorporating the intuitionistic fuzzy to the ETA, the consequence probabilities are obtained as given below:

- Safe operation is 1.89×10^{-14}
- Safe operation and no damage to compressor are 1.71×10^{-18}
- Controlled burning of H_2 at flare is 1.89×10^{-30}
- Fire and explosion due to bursting of pipe are 1.97×10^{-3}

Resulting ETA is shown below (Fig. 4).

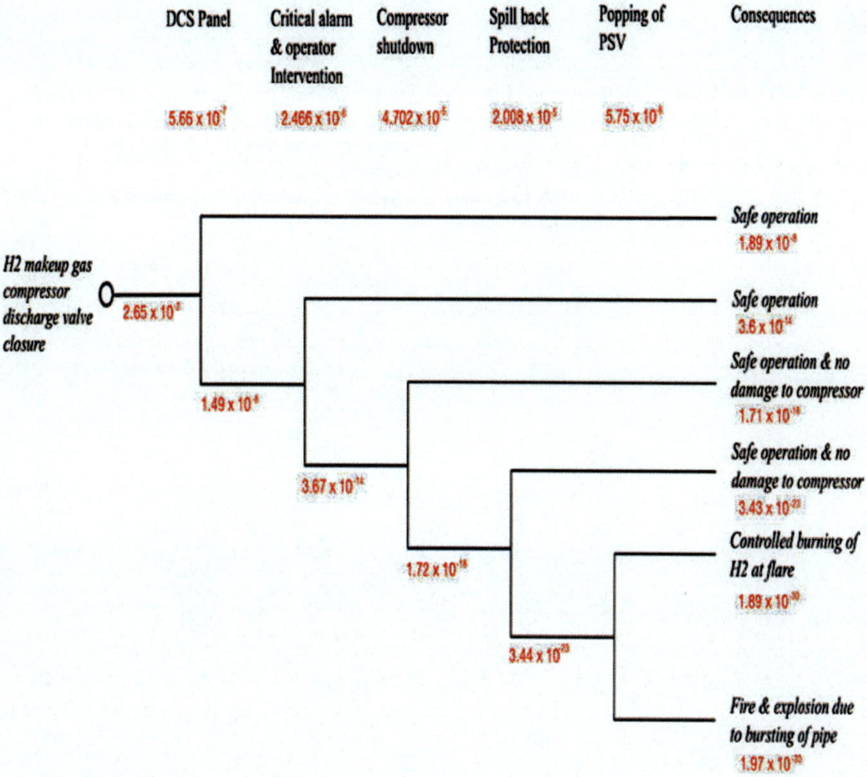

Fig. 4 Resulting ETA

7 Conclusions

Construction of an event tree based on the selected initiating event (IE) was done, and the consequence probabilities of different events were calculated using the intuitionistic fuzzy set theory. Based on the probability of occurrence of different consequent events, suitable engineering controls and administrative controls can be brought in the process for an uninterrupted and continuous working of the plant.

With the application of intuitionistic fuzzy theory, the quantitative risk assessment procedure has been lightened and easy application of the theory has been enabled. It also proved that the use of intuitionistic fuzzy has a wide range of application over the conventional fuzzy when the probability distributions of different components are not readily available. However, the proposed IFETA has to be further investigated for application in different systems or fields. Thus, more effective results can be obtained, and its significance in quantitative risk assessment can be confirmed.

References

1. Atanassov, K. T. (1999). Intuitionistic fuzzy sets. In *Intuitionistic fuzzy sets* (pp. 1–137). Physica.
2. Ferdous, R., Khan, F., Sadiq, R., Amyotte, P., & Veitch, B. (2009). Handling data uncertainties in event tree analysis. *Process Safety and Environmental Protection, 87*(5), 283–292.
3. Zadeh, L. A. (1965). Fuzzy sets. *Information and Control, 8*(3), 338–353.
4. Onisawa, T. (1988). An approach to human reliability in man-machine systems using error possibility. *Fuzzy sets and Systems, 27*(2), 87–103.

Termigradation of Un-Compostible Parts of Major Weeds Prosopis *(Prosopis Juliflora)* and Ipomoea *(Ipomoea Carnea)*

Pratiksha Patnaik, S. A. Abbasi, and Tabassum-Abbasi

1 Introduction

As brought out earlier [1], quote—'anthropogenic processes for the treatment of biodegradable solid waste revolve round the use of aerobic, anaerobic, and facultative bacteria [2, 3]. Be it a sanitary landfill, a composting system, a solid-feed anaerobic digester, or a bioprocess of some other kind, bacterial digestion has been central to the treatment of biodegradable solid wastes [4–8]. The only exception to this general rule has been vermicomposting wherein the action of bacteria and enzymes on solid waste is mediated (and controlled) by earthworms. The animal ingests solid waste along with soil and deposits the digested material in the form of seed-like vermicast. During the passage through the worm gut, the feed is acted upon by the gut microflora and gets significantly stabilized. The resulting vermicast is a good soil conditioner and fertiliser [9–14].

But neither vermicomposting nor direct bacterial action during any of the economically viable solid waste degradation processes can handle lignin [5, 6]. 'Hard' biowastes such as coconut shells and woody biomass also defy swift biodegradation.

In an attempt to find a quicker and more widely applicable way to dispose large volumes of biowaste, especially the type of biowaste—mentioned above—which resists treatment methods currently in general use, we have begun exploring a new frontier: termigradation,"—unquote.

As further noted earlier [1], quote—' termites are among the nature's most powerful scavengers and earth movers, alongside earthworms and ants [15–17]. But

P. Patnaik · S. A. Abbasi
Centre for Pollution Control and Environmental Engineering, Pondicherry University, Kalapet, Puducherry 605014, India
e-mail: abbasi.cpee@gmail.com

Tabassum-Abbasi (✉)
School of Engineering, University of Petroleum and Energy Studies, Bidholi, Dehradun 248007, India

N. A. Siddiqui et al. (eds.), *Advances in Behavioral Based Safety*,
https://doi.org/10.1007/978-981-16-8270-4_6

unlike the other two, termites harbour in their midst microflora which have the special ability not possessed by other animals: ability to digest lignin [18]. In case of lower termites, ligneous material is masticated and ingested which is then digested by microflora present in certain species of protozoa living symbiotically in the termite gut [19, 20]. In case of higher termites, the microflora capable of digesting lignin is present directly in the animal gut [19, 20].

Other characteristics of termites which make them potential candidates for bioprocessing of solid waste are [21]: (a) their voracious appetite; (b) their ability to consume a wide variety of wastes; (c) diversity of their habitat preference which makes it possible to always find one or other species suitable for a given geo-climatic situation [22–24]; (d) their very fast rate of population growth; and (e) good quality of protein represented by termite bodies, making them ideal as poultry feed or source of chemicals such as biofuel [21, 25],'—unquote.

But,as noted earlier [1], quote—"any endeavour to develop bioreactors based on termites has to overcome a unique challenge. It lies in the fact that termites are 'eu-social' animals with well-defined and uncompromising social hierarchy [26]. Unlike earthworms, of which every single individual has the potential to reproduce sexually while it is feeding upon the waste in a bioreactor, the worker termites cannot breed. It is, therefore, not possible to inoculate a pile of waste with worker termites and expect that the workers would feed and breed till the entire waste is consumed. For any termireactor to function sustainably, it has to be ensured that the workers keep coming from termite nests where the workers are being born and reared continuously (along with, of course, other termites of higher caste viz soldiers),'—unquote.

To achieve this objective, we have developed two types of termite-based reactors:

(a) Developing life-size but captive termite colonies indoors as a kind of 'bioangines'. In them the termites are made to feed upon the waste that is supplied in specially designed reactors. These termigradation systems can be termed *ex-situ* processes.
(b) *In situ* systems wherein termireactors containing the waste are placed near pre-existing termite mounds. The reactors are designed to facilitate termite entry, feeding and exit while also protecting the waste from being disturbed by wind and other animals. Chambers have sufficient openings to allow access to termites but are otherwise closed from all sides. The reactors are provided with trails so that the scout termites who keep looking for food (so as to call others to the food source once it is located) are led to the reactors.

Termireactors of three sizes as were fashioned from 3mm thick aluminium sheets. The dimensions of the larger sized reactors of 432 L volume were as in Fig. 1. The smaller reactors of dimensions 63(L), 40(B), 28(H) cm and 48(L), 30(B), 23(H) cm, having 71 and 33 L volume, respectively, had similar shape and design.

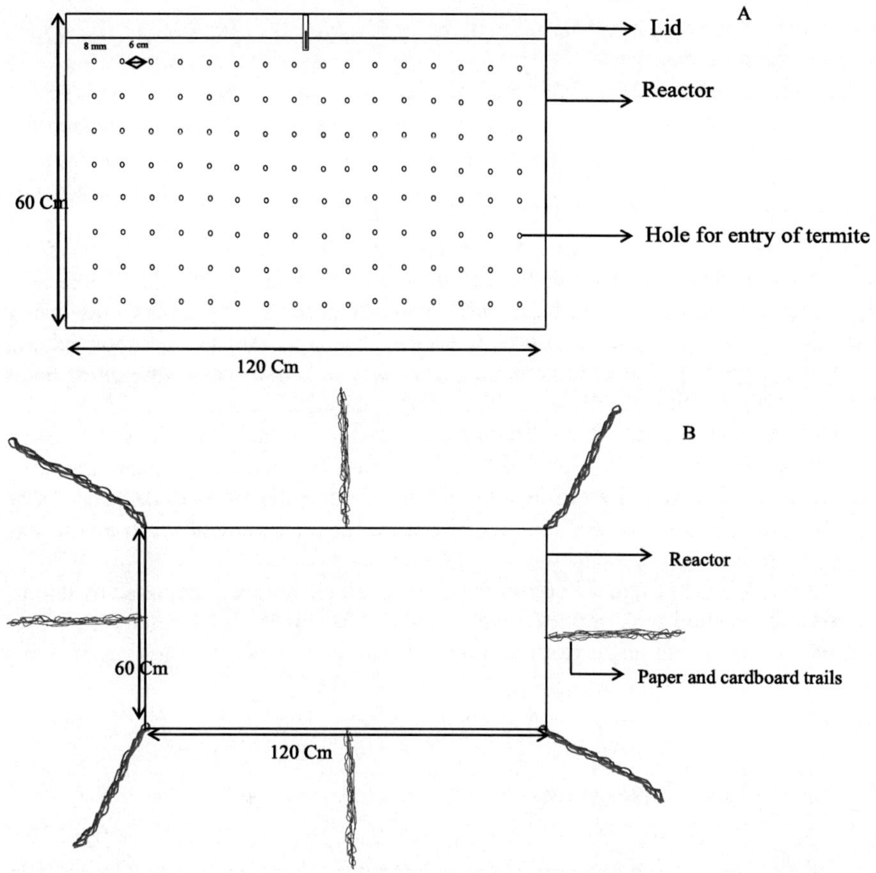

Fig. 1 (**a**) Sectional view of the termireactor and (**b**) top view of the reactor with trails made up of squeezed waste paper (to lead the termite food scouts to the reactor).The dimensions belong to the 71 L reactor; the other reactors deployed in the study have identical design but smaller sizes

2 Materials and Method

The substrates—'hard' branches of prosopis and ipomoea (parts too woody to be amenable to composting or vermicomposting yet not woody enough for use as fuel)—were collected from in and around Pondicherry University. After removing the debris and leaves of other species, dry weights of three randomly picked and pooled samples of either substrate were estimated by oven drying weighed samples at 105 °C to constant weight.

For *in-situ* experiments, active mounds were identified within the University campus and termireactors were placed inside holes dug near these mounds. To protect the reactors from rain and sunlight, they were covered with soil from all the four sides.

For *ex-situ* studies a captive termite colony of *Hypotermes obscuriceps,* developed in the laboratory, was used.

The experiments in the in situ mode were started with the two largest sized (71 L) reactors, each fed with 35 kg prosopis branches, and two 71 L reactors, each fed with 2 kg of ipomoea stem. Simultaneously *ex-situ* studies were commenced with two 33 L reactors, fed with 1 kg of prosopis and 500 g of ipomoea, respectively. The quantities differed because of difference in the density of the substrates.

Unlike what happens in vermicomposting, during which the substrate is converted to vermicast and the vermicast becomes the quantifiable and reproducible outcome of the vermicomposting process, there is no 'termicast' generated by the termites during termigradation. Most of the substrate consumed by the termites is metabolised and what little excreta termites do generate, they carry it to their nests wherein 'fungal gardens' are organised by them. In the 'fungal gardens' fungus is 'cultivated' by termites as a food source. The excreta is deposited in the fungal gardens to fertilize it. Hence almost nothing is left behind when termites feed upon prosopis or ipomoea (or any other substrate) in a termireactor. For this reason, the extent of degradation of the substrate has to be worked out on the basis of the unconsumed substrate present in the termireacotrs.

Accordingly the progress of termigradation in all reactors was assessed by taking out the unconsumed substrate once every 30 days, freeing it from termites and a few soil-like particles that result from termite foraging and movement, and quantifying it.

3 Results and Discussion

As explained earlier, termites tend to assimilate most of what they consume and take away the little excreta they might have generated. What remains in the termireactor is still-to-be-consumed substrate of which some part is finely fragmented by the termites in the course of their foraging activity. The reddish brown deposits over pieces of prosopis branches, seen in Fig. 2, represents that component. The situation in the ipomoea fed termireactor is as seen in Fig. 3. Earlier, in trial experiments, we had allowed termites to act on the substrate till no unconsumed substrate remained. In such termireactors only traces of the reddish-brown particulates were found. This indicates that by-and-by, the termites consume these particulates as well.

A close-up of a partly consumed prosopis branch is shown in Fig. 4 while Fig. 5 displays individuals of *H.obscuricaps* (tiny, milk-white animals) active near their mound.

The results of in-situ experiments with prosopis and ipomoea are presented in Tables 1 and 2, respectively. Despite the reactor contents being heterogeneous, and dependent upon termite behaviour, there is remarkable agreement between the duplicates; the relative error being well below 10% in all the assessments. We have encountered similar reproducibility in the past [26, 27, 28, 29] for different substrates and in different situations.

Fig. 2 Termite-worked branches of prosopis

Fig. 3 Termite-worked pieces of ipomoea stem

Fig. 4 Close-up of a partially consumed piece of prosopis

Fig. 5 Individuals of *H. obscuriceps* near their mound (colony)

In case of prosopis (Table 1), the maximum reactor utilization —21.1%—occurred during the first 30 days. In subsequent months the rate gradually declined. The pattern was similar with ipomoea (Table 2): 30.2% of all the substrate was consumed in the first 25% of the experiment duration.

Table 1 Extent of termigradation (%) of prosopis (35 kg) at 30-days intervals in in-situ reactors

Days	Reactors		Termigradation %	
	A	B	During each run	Cumulative
30	20.8	21.3	21.1 ± 0.4	21.1 ± 0.4
60	18.4	19.8	19.1 ± 1.0	40.2 ± 1.4
90	14.6	13.9	14.3 ± 0.5	54.4 ± 1.9
120	11.2	12.4	11.8 ± 0.8	66.2 ± 2.7

Table 2 Extent of termigradation (%) of ipomoea (2 kg) at 30-days intervals in in-situ reactors

Days	Reactors		Termigradation %	
	D	E	During each run	Cumulative
30	30.5	29.9	30.2 ± 0.4	30.2 ± 0.4
60	27.4	26.7	27.1 ± 0.5	57.3 ± 0.9
90	21.6	20.5	21.1 ± 0.8	78.4 ± 1.7
120	17.4	16.8	17.1 ± 0.4	95.5 ± 2.1

To understand these findings we have to consider the termite behaviour. The termite scouts, who keep looking for food source, are the ones who launch, and in a way control, the termigradation process. It is these scouts who, upon locating the food contained in the termireactors, send signals to their colonies. From there, and apparently based on the content of the singles, appropriate number of worker termites come out to the food source and consume it. Apparently, as the quantity of the source goes down the termites are able to sense it and give signals due to which lesser number of termites forage upon the remains. It also appears that for some reason, that we have not been able to fathom, termites never depute such a large number of foragers to any termireactor that they can finish off all the substrate content in a few hours or a few days. There is always a pattern whereby utilization of a food source is slowly rolled out with time. Perhaps they tend to go slower as the source of the food dwindles so as to enhance the duration of its availability—as a form of food security measure.

To benefit from this termite behaviour it appears advisable that instead of running the termireactors in batch mode—as we have done—they should be operated in a semi-continuous mode. The quantity of substrate consumed in the first 30 days should be augmented with equivalent quantity of fresh substrate.

The results of the *ex-situ* reactors operated with prosopis and ipomoea are shown in Tables 3 and 4, respectively. Due to the past experience of the authors, which had shown the high degree of reproducibility in termireactor performance, supported by similar reproducibility achieved in the present study on *in-situ* termigradation described above, we have deployed only one reactor each for the two weeds. It may be seen that the performance of the *ex-situ* reactors has been very similar to the performance of the *in-situ* reactors. Whereas in the *in-situ* reactors 21.1, 19.1,

Table 3 Extent of termigradation (%) of prosopis (1 kg) at 30-days intervals in ex-situ reactors

Days	Termigradation % in reactor –C	
	During each run	Cumulative
30	20.7	20.7
60	17.5	38.2
90	13.9	52.1
120	10.5	62.6

Table 4 Extent of termigradation (%) of ipomoea (500 gm) at 30-days intervals in ex-situ reactors

Days	Termigradation % in reactor -F	
	During each run	Cumulative
30	31.1	31.1
60	25.6	56.7
90	20.2	76.9
120	12.3	89.2

14.3 and 11.8% of prosopis was consumed by 30th, 60th, 90th and 120th day, the respective figures for the *ex-situ* reactors are 20.7, 17.5, 13.9 and 10.5%. Whereas the average cumulative prosopis consumption was 66.2% in the *in-situ* reactors, it was very similar—62.6%—in the *ex-situ* reactors. In a like fashion, the consumption by the 30th, 60th, 90th and 120th day of ipomoea, which was 30.2, 57.3, 78.4, and 95.5% in the *in-situ* reactors matched closely with the pattern in the ex-situ reactors—31.1, 56.7, 76.9 and 89.2%. Figure 6 brings this out graphically. And this level of match in performance was seen even though the capacities of the reactors were different while the reactor contents were heterogeneous.

The rate of termidegradation of ipomoea was over 20% greater than of prosopis perhaps because of the lesser content of lignin in ipomoea.

4 Summary and Conclusion

The paper recounts the concepts of 'termigradation'—which represents termite-induced biodegradation—and associated technology, introduced and patented earlier by S. A. Abbasi and co-workers. The technology enables assimilative and eco-friendly disposal of such ligninous biowaste which defies conventional biodegradation processes such as anaerobic digestion, composting and vermicomposting. The paper also provides a gist of special challenges associated with the use of termites as bioagents. They stem from the highly eu-social character of termites. Similar to wasps, bees and ants termits have rigid hierarchy and social order wherein the worker termites do most of the work but have no ability to breed (except in rare and special cases). Also termites cannot survive for long if isolated from the colony to which they

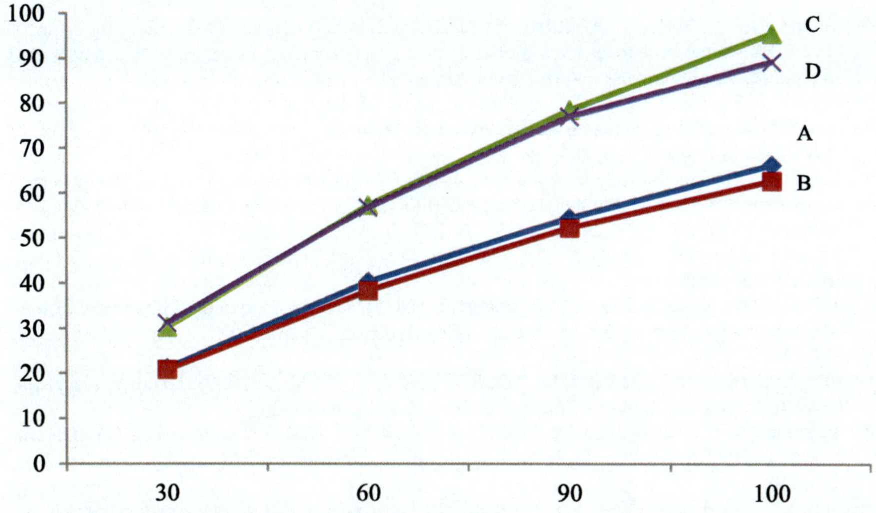

Fig. 6 Relative performance of the *in-situ* and the *ex-situ* reactors. Curves A and C pertain to *in-situ* termigradation of prosopis and ipomoea, respectively, while curves B and D pertain to *ex-situ* termigradation these weeds.

had belonged. The author's termigradation technology overcomes all these impediments while harnessing termites for the final disposal of those ligninous parts of harmful weeds like prosopis and ipomoea, which defy conventional bioprocesses.

The paper then describes experiments on the degradation of those ligninous constituents of prosopis and ipomoea which cannot be otherwise biodegraded. Utilization of *in-situ* and *ex-situ* termireactors has been described in experiments lasting 4 months on the trot. The results show that the most rapid termidegradation occurs during the first 30 days. The mechanism of the process and the use of the findings in process design, optimization and control have been described.

Acknowledgements SAA thanks Indian National Science Academy (INSA), New Delhi, and PP thanks Council for Scientific and Industrial Research (CSIR), New Delhi, for support.

References

1. Abbasi, S. A., & Gajalakshmi, S. (2015). Disposal of municipal solid waste with in situ termireactors: Proof-of-concept. *Bioresources and Bioprocessing, 2*(1), 1–5.
2. Abbasi, S. A. (1985). Occurrence, toxicity and treatment of lignin in pulp and paper effluents-state of the art. *Journal of the South African Institution of Civil Engineers, 65*, 1–8.
3. Abbasi, T., & Abbasi, S. A. (2010). Biomass energy and the environmental impacts associated with its production and utilization. *Renewable and Sustainable Energy Reviews, 14*(3), 919–937.

4. Abbasi, T., & Abbasi, S. A. (2010). Production of clean energy by anaerobic digestion of phytomass—New prospects for a global warming amelioration technology. *Renewable and Sustainable Energy Reviews, 14*(6), 1653–1659.
5. Abbasi, S. A., Bhatia, K. K. S., Nandeshwar, M. D., & Nipaney, P. C. (1991). Environmental impact of pulpmill effluents on an estuary on the Malabar coast. *Journal of the Institution of Public Health Engineers, 4*, 28–37.
6. Abbasi, T., Gajalakshmi, S., & Abbasi, S. A. (2009). Towards modeling and design of vermicomposting systems: Mechanisms of composting/vermicomposting and their implications.
7. Abbasi, T., Tauseef, S. M., & Abbasi, S. A. (2012). Anaerobic digestion for global warming control and energy generation—An overview. *Renewable and Sustainable Energy Reviews, 16*(5), 3228–3242.
8. Abbasi, S. A., Nayeem-Shah, M., & Abbasi, T. (2015). Vermicomposting of phytomass: Limitations of the past approaches and the emerging directions. *Journal of Cleaner Production, 93*, 103–114.
9. Abe, T., Bignell, D. E., Higashi, M., Higashi, T., & Abe, Y. (Eds.). (2000). *Termites: evolution, sociality, symbioses, ecology*. Springer Science & Business Media.
10. Anantharaju, T., Kaur, G., Gajalakshmi, S., & Abbasi, S. A. (2014). Sampling and identification of termites in Northeastern Puducherry. *Journal of Entomology and Zoology Studies, 2*(3), 225–230.
11. Bignell, D. E., Roisin, Y., & Lo, N. (Eds.). (2010). *Biology of termites: a modern synthesis*. Springer Science & Business Media.
12. Breznak, J. A., & Brune, A. (1994). Role of microorganisms in the digestion of lignocellulose by termites. *Annual Review of Entomology, 39*(1), 453–487.
13. Butler, J. H. A., & Buckerfield, J. C. (1979). Digestion of lignin by termites. *Soil Biology and Biochemistry, 11*(5), 507–513.
14. Casatelli, L. M. (2007). Vermicomposting: Turning paper sludge into 'black gold.' *Pap Package, 48*(2), 36–37.
15. Dhamija, U. (2006). *Sustainable solid waste management: Issues, policies, and structures*. Academic Foundation.
16. Domínguez, J., Edwards, C. A., & Webster, M. (2000). Vermicomposting of sewage sludge: Effect of bulking materials on the growth and reproduction of the earthworm Eiseniaandrei. *Pedobiologia, 44*(1), 24–32.
17. Eggleton, P. (2010). An introduction to termites: biology, taxonomy and functional morphology. In *Biology of termites: A modern synthesis* (pp. 1–26). Springer.
18. Gajalakshmi, S., & Abbasi, S. A. (2008). Solid waste management by composting: State of the art. *Critical Reviews in Environmental Science and Technology, 38*(5), 311–400.
19. Ganeshkumar, T., Premalatha, M., Gajalakshmi, S., & Abbasi, S. A. (2014). A new process for the rapid and direct vermicomposting of the aquatic weed salvinia (Salviniamolesta). *Bioresources and Bioprocessing, 1*(1), 1–5.
20. Grube, S. (2001). Soil modification by the harvester termite Hodotermesmossambicus (Isoptera: Hodotermitidae) in a semi-arid savanna glass land of Namibia. *Sociobiology, 37*, 757–767.
21. Gupta, R., Mehta, G., Khasa, Y. P., & Kuhad, R. C. (2011). Fungal delignification of lignocellulosic biomass improves the saccharification of cellulosics. *Biodegradation, 22*(4), 797–804.
22. Harit, A., Gajalakshmi, S., & Abbasi, S. A. (2011). Species composition and duration of termite swarms at a wooded location in Puducherry-II: The pattern in the year 2010, ECOCORPS-2011, National Seminar on Environmental Challenges in conservation of Resources of 'ONE'Planet for Sustainable Development. Department of Environmental Sciences, Bishop Heber College, Tiruchirappalli, India.
23. Harit, A. K., Gajalakshmi, S., & Abbasi, S. A. (2014). Swarming of the termite Coptotermes-gestroi in north-eastern Puducherry. *Zoology and Ecology, 24*(1), 62–69.
24. Karthikeyan, M., Gajalakshmi, S., & Abbasi, S. A. (2014). Effect of storage on the properties of vermicompost generated from paper waste: With focus on pre-drying and extent of sealing. *International Journal of Energy and Environmental Engineering, 5*(4), 291–301.

25. Karthikeyan, M., Hussain, N., Gajalakshmi, S., & Abbasi, S. A. (2014). Effect of 326 vermicast generation form an allelopathic weed lantana (Lantana camara) 327 on seed germination, plant growth, and yield of cluster bean (Cyamopsis 328 tetragonoloba). *Environmental Science and Pollution Research.* https://doi.org/10.1007/s11356-014-3103-5
26. Karthikeyan, M., Gajalakshmi, S., & Abbasi, S. A. (2014). Ingestion of sand and soil by phytophagous earthworm Eudriluseugeniae: A finding of relevance to earthworm ecology as well as vermitechnology. *Archives of Agronomy and Soil Science, 60*(12), 1795–1805.
27. Anuradha, J., Abbasi, T., & Abbasi, S. A. (2015). An eco–friendly method of synthesizing gold nanoparticles using on otherwise worthless weed pistia. Journal of Advanced Research, http://doi.org/10.1016/j.jare.2014.03.006.
28. Abbasi, T., & Abbasi, S. A. (2012). Is the use of renewable energy sources an answer to the problems of global warming and pollution? Critical Reviews in Environmental Science and Technology (Taylor and Francis), 42, 99–154.
29. Abbasi, S. A., (1989). Atomic absorption spectrometric and spectrophotometric trace analysis of uranium in environmental samples with Np-methoxyphenyl-2-furylacrylohydroxamic acid and 4-(2-pyridylazo) resorcinol. International journal of environmental analytical chemistry, 36(3), 163–172.

Roadmap of Uttarakhand Power Sector: Major Policies and Current Scenario

Deepali Yadav and Madhu Sharma

1 Introduction

Electrical Power Sector for any country is a vital parameter in shaping its progress and future development. Power forms a crucial aspect of any sector's growth and thus reliable and quality power supply becomes integral to the country's economic growth. For India, this is evident from the power sector (generation, transmission and distribution) being an integral part of the twelfth five-year plan (year 2012–17) laid by the Planning Commission [1]. India is among the fastest growing economies in the world with average 8% annual GDP growth rate and to sustain this it requires an annual grid power growth at approximately 6% [1, 2]. Globally, India stands in top three and top two nations for largest power generation and power consumption, respectively, with 373.43 GW installed capacity as on October 2020 [3].

The Ministry of Power (MoP) is the central governing body of the Indian power sector—generation, transmission and distribution. The generation is further divided into central, state and private units [4]. The Power Grid Corporation of India Ltd (PGCIL) is the apex body for national grid development and inter-state power transmission. More than fifty distribution companies (Discoms), divided into north, south, east, west and northeast zones include both government and private ownership [5]. Over the past few decades, India's electricity reforms have been stimuli in bringing its power sector out of the botched state in 1980s. The Electricity Laws Act 1991, introduction of privatization in generation in 1991, opening the transmission sector to private companies in 1998, the Electricity Regulatory Commissions Act 1998 and the Electricity Act 2003 (open access, captive generation and unbundling of state electricity boards) are a few measures that played a pivotal role in reforming the electricity sector in India.

D. Yadav · M. Sharma (✉)
University of Petroleum and Energy Studies, Bidholi, Dehradun, India
e-mail: madhusharma@ddn.upes.ac.in

N. A. Siddiqui et al. (eds.), *Advances in Behavioral Based Safety*,
https://doi.org/10.1007/978-981-16-8270-4_7

Apart from the reforms taken at central level, the states also have taken initiative in having their own reform act like Orissa (1995), Haryana, Andhra Pradesh, Uttar Pradesh, Karnataka, Rajasthan, New Delhi, Madhya Pradesh, Gujarat and Maharashtra. The reforms here primarily focussed on unbundling SEBs into generation, transmission and distribution, however, Delhi, Orissa and Maharashtra implemented privatization of the DISCOMs as well [2, 6].

In this paper, the power sector roadmap of the state of Uttarakhand since its formation in year 2000 is discussed. This paper throws light on the state's current power scenario in context of abundant hydropower, central and state policies, load mix, tariff, structure and challenges. The authors have chosen the state of Uttarakhand as it is one of the newest formed states of India and has an abundant hydropower and renewable energy potential, which can prove to be effective in supporting its growth and development and overcoming the challenges posed by state's demography as well.

2 Uttarakhand State Power Sector—Background and Scenario

Uttarakhand, formerly known as Uttaranchal, was formed as the 27th Indian state in November 2000, bifurcated from the north region of Uttar Pradesh. On its formation the state inherited a fragile power sector and has since then seen major developments in almost all sectors and thriving against the some challenges. The state's power sector, aided by its abundant hydropower electric potential estimated approximately at 25 GW, has witnessed major policies and developments in the last two decades [7]. The vast hydro resources play a pivotal role in Uttarakhand being developed as an "energy state". The 2022 vision of the state focusses on global leader in green energy, green economy and 100% state electrification, all of which are direct or indirect outcomes of harnessing hydro potential [7].

2.1 Uttarakhand Power Sector Structure

The state implemented the three utilities model—Uttarakhand Power Corporation Ltd. (UPCL) set up in 2001, Uttarakhand Jal Vidyut Nigam Ltd (UJVNL) set up in 2001 and Power Transmission Corporation of Uttarakhand Ltd. (PTCUL) set up in 2004, mainly governing the state's power sector, taking care of distribution, generation and transmission, respectively. Uttarakhand Electricity Regulatory Commission (UERC) set up in 2002 under the Electricity Act of 2003, regulates power purchase and tariffs. Uttarakhand Renewable Energy Development Agency (UREDA), set up in year 2008 under state government, leads in RE sector growth and development in the state. UREDA works in line with the Ministry of New and Renewable Energy (MNRE) and as state nodal point for Bureau of Energy Efficiency (BEE).

2.2 *Hydropower Generation*

The total installed power capacity of Uttarakhand as on November 2020 is 4.868 GW including 3.756 GW from hydel, 662.5 MW from other renewable energy sources (RES) and 450 MW from gas-based thermal [8]. Hence, Uttarakhand is among the very few states in India with more than 90% installed power capacity derived from RES. There are 15 operational hydropower stations in the state, three hydroelectric stations (Lakhwar Dam, Kishau Dam and Tapovan Vishnugad Hydropower Plant) either are under construction or proposed and Loharinag-Pala Hydropower Project is discontinued [9–11]. (Table 1, Fig. 1)

2.3 *Other RES Generation*

Other RE generation sources of the state include micro/mini/small hydro, solar photovoltaic (PV), solar thermal and bio energy generation (through biomass or agro

Table 1 List of Hydropower Stations (Operational) in Uttarakhand

Name	Capacity	Location
Chibro Power Plant	4 × 60 MW	Dakpathar
Khodri Power Plant	4 × 30 MW	Dakpathar
Dhakrani Power Plant	3 × 11.25 MW	Dakpathar
Dhalipur Power Plant	3 × 17 MW	Dakpathar
Dharasu Power Station	4 × 76 MW	Uttarkashi
Dhauliganga Power Station	4 × 70 MW	Dharchula
Koteshwar Dam	4 × 100 MW	Tehri
Maneri Dam	3 × 30 MW	Uttarkashi
Pathri Power Plant	3 × 6.8 MW	Haridwar
Mohammadpur Power Plant	3 × 3.1 MW	Haridwar
Ramganga Power Plant	3 × 66 MW	Pauri Garhwal
Srinagar Hydro-electric Power Project	4 × 82.5 MW	Pauri Garhwal
Tehri Dam	1000 MW	Tehri
Tanakpur Power Station	3 × 31.4 MW	Tanakpur
Chilla Power Plant	4 × 36 MW	Chilla, Haridwar

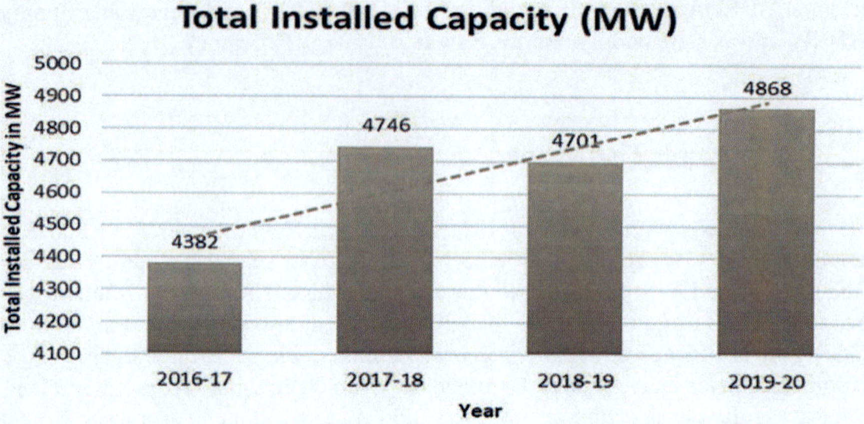

Fig. 1 Total installed power capacity in MW

remains). It is estimated that the state also has a satisfactory wind energy and cogeneration potential and by year 2020, approximately 1 kW power could be harnessed from these other RES [12]. With this objective, the state government also enacted a policy for harnessing RES in Uttarakhand with private sector/community participation in year 2008 [12]. The geography and demography of the state make it a rich reserve of non-conventional energy resources, which are, yet to be tapped (Table 2) [12].

The grid connected Bachelikhal Wind Power Project of 2.4 MW (year 2005) is under development [13]. Solar energy harnessing is executed by UREDA under various state government programs, namely, solar water heating, solar cooking, solar street and home light, solar lantern, etc. (Table 3) [14].

Table 2 Estimated RES potential that can be harnessed by year 2020

Renewable energy resource	Estimated potential
Micro/Mini/Small hydropower	600 MW
Co-generation	220 MW
Biomass/Agro/Waste	300 MW
Energy conservation	25 kW

Table 3 Solar power plants in Uttarakhand

Solar power project	Capacity
Grid connected	221 MW
Rooftop solar	6.075 MW
Off-grid	689 kW
Canal bank to canal top	20 MW

2.4 Thermal Power Generation

Uttarakhand has a single gas fired combined cycle thermal power plant—Kashipur Sravanthi Power Station with four units of 75 MW each located in Kashipur. The phase 1 of this project was commissioned in year 2016 (225 MW) and phase 2 completion is expected by March 2021 [15].

2.5 Load Mix and AT & C Losses

Power consumers in Uttarakhand, although a mix of all types, but majority power is consumed by industrial load. Since the formation in 2000, several factories and industries have been set up here because of centre's hill policy implementation (Table 4). Domestic, commercial, agricultural and others consume less than fifty per cent of the total state consumption making a favourable load mix for the state [16]. Hence, higher industry tariffs giving huge revenue from industrial consumers and abundant hydroelectric power are key drivers for the state's power sector in balancing the lower domestic and agriculture tariffs and the distribution losses.

Figure 2 shows variation in the aggregate transmission and commercial (AT&C) losses from 2012 to 2018 [16]. In the early years, these losses amounted to about 50%, however the industry hill policy and hydro potential has worked in favour of the state's distribution utility performance as can be seen in Fig. 2. In year 2018, the MoP awarded UPCL with "A" grade for its performance [17]. The state losses from year 2016 to 2019 are much less than the national average of AT&C losses. However, this trend of losses cannot overshadow the fact that apart from industrial consumers, the losses incurred by other consumers amount to more than 30%.

2.6 Policies and Schemes

Since its formation in year 2000, Uttarakhand's power sector has seen a transformation with major improvements. Government policies and initiatives have played an important role in achieving this. Table 5 presents a brief review of major policies, projects and regulations that have played a vital role in the state's power sector transformation since its formation in year 2000.

Table 4 Consumer wise percentage power consumption

Year	Domestic (%)	Commercial (%)	Agriculture (%)	Industrial (%)	Others (%)	Total (in MU)
2016–17	23.39	11.08	3.30	54.65	7.58	10,629
2017–18	23.59	10.63	2.34	53.01	10.43	11,621
2018–19	22.54	10.27	2.73	52.73	11.73	12,640

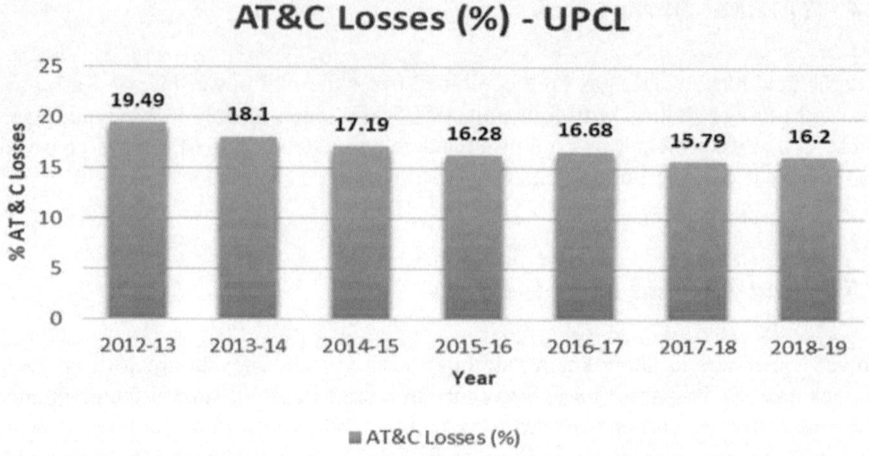

Fig. 2 AT&C Losses in % from year 2012 to 2019

Table 5 Policy and Schemes implemented for Uttarakhand Power Sector

Policy/Scheme	Aim and objective
The Hill Industrial Policy implemented in year 2003	Attract industry establishments in the state giving them excise and tax benefits. This policy has played a pivotal role in providing a favourable load mix to the state's power sector with more than fifty percent power consumed by industrial customers. Although the policy expired in year 2018, but it was further extended until 2025 [18].
24x7 Power for All (PFA), year 2015	Joint venture of GoI and Uttarakhand state government to provide electricity to all households, factories, industry units, commercial units and other entities consuming power [19].
Uttarakhand Transmission Strengthening and Distribution Project	Joint venture of GoI, PTCUL and UPCL to provide reliable and efficient power supply to all consumers with financial sustainability. This project proposes to widen the power transmission network capacity, cultivate resilient distribution network and tap all available state resources for community aligned power generation [20].
Uttarakhand state government policy for harnessing Renewable Energy Sources, year 2008	Provide encouraging conditions to private companies and community for renewable energy power projects [21].
UERC regulations from year 2002 to 2020	Provide regulation supply code, new connections, metering, tariff, grievance redressal and related renewable power obligations [22].

2.7 *Drivers and Challenges*

Hydropower potential of more than 20 GW remains one of key drivers for the Uttarakhand power sector. Also, the estimated RE potential if harnessed can be a boon in making Uttarakhand a clean energy state. The state currently enjoys a perfect mix of low cost hydropower and high revenue making industrial load.

High AT&C losses from non-industrial loads, electricity theft from plain region and need for resilient system owing to harsh climate are some challenges that the state faces. In addition, the hill policy stands expired in year 2025. The cheap hydropower and high revenue industry mix is the key driver in the agriculture, domestic and other consumers enjoying very low tariffs and negligible power cuts. New hydropower projects face environmental challenges and hence are either cancelled or stalled. With introduction of open access (OA) the share of industrial load has fallen since year 2008.

3 Recommendations

Uttarakhand's reliability on low cost hydropower and high tariff industry load does not seem to go a long way as the other issues and challenges remain unattended by the utilities. With high AT&C losses from non-industrial consumers, the state stands not far away from other financially struggling discoms. In addition, poor metering, electricity theft, pending bill payments and poor load factor are other concerns that add to the core problem. Although UERC has repeatedly recognized and emphasized these issues the problem persists even after a decade with very little efforts to resolve the same.

There is a need to see beyond the favourable hydropower-industry combination, as the hill policy benefits to industries will soon be expired. There is a need to promote RE generation, increase industry participation in discoms, and most importantly to resolve the decade long pending problems related metering, billing and collection.

4 Conclusion

This paper discusses the roadmap and current scenario of the power sector of newly formed state of Uttarakhand. Its transformation from year 2000 until date has been relatively good. However, the same can be attributed to only one factor—low cost hydropower and high industrial revenue. In addition, there is an immediate need to lower the dependency on same and explore other options as in the past few year the state has witnessed slower industrialization and limited scope of expansion in terms of land availability.

The state has enormous RE potential which remains untapped. Majority challenges faced by the state power sector is owing to two factors—major dependency on favourable load mix and decade long unresolved discom utility issues.

References

1. Final Approach Paper Cover.cdr (niti.gov.in).
2. Totare, N. P., & Pandit S. (2010). Power sector reform in Maharashtra, India. *Energy Policy, 38*(11), 7082–7092. ISSN 0301-4215. https://doi.org/10.1016/j.enpol.2010.07.027.
3. Power Sector in India: Market Size, Industry Analysis, Govt Initiatives I IBEF (Govt. Trust).
4. Overview of Indian Power Sector I Power Sector in India I INDIAN POWER SECTOR.
5. List of beneficiary DISCOMs under CAP-DISCOMs programme of BEE (1).pdf (beeindia.gov.in).
6. Power Sector Reforms in India - EnergySector.in.
7. "Uttarakhand", Indian Brand Equity Foundation (IBEF) Report, May 2018, https://www.ibef.org/download/Uttarakhand-May-20181.pdf.
8. States of India by installed power capacity—Wikipedia.
9. Category:Hydroelectric power stations in Uttarakhand - Wikipedia, https://en.wikipedia.org/wiki/Category:Hydroelectric_power_stations_in_Uttarakhand.
10. Chilla, Welcome to Uttarakhand Irrigation Department, https://uttarakhandirrigation.com/chilla.
11. NHPC Limited : Projects : Power Stations : Tanakpur (nhpcindia.com) .
12. UREDA Booklet (cbip.org) .
13. Wind Energy Projects I wind power projects I Wind Resource Assessment I Wind Project Implementation I Wind Farm I Renewable Energy I Planning I Developing Site I Wind Energy Estimation I Wind energy experts I Wind Maps (cecl.in) .
14. Uttarakhand Renewable Energy Development Agency (uk.gov.in).
15. Sravanthi Group, Sravanthi Industries, Sravanthi Energy, Sravanthi Hydro.
16. Uttarakhand Power Corporation Ltd (UPCL) website, https://www.upcl.org/.
17. Ranking Discoms—Power Line Magazine.
18. Uttarakhand to set up 11 new industrial hubs I Business Standard News (business-standard.com).
19. 24x7 power for all (powermin.nic.in).
20. 51308–008: Uttarakhand Transmission Strengthening and Distribution Project (adb.org).
21. state-government-policies-issued-by-IREDA17.pdf (windpro.org).
22. UERC-Current_Regulations, http://www.uerc.gov.in/Current_Regulations.html.

A Study on the Influence of Human Factors in Safety Performance

Amal S. George and V. R. Renjith

1 Introduction

In almost all industries and workplaces, there is a common misconception that productivity is contrary to good health and safety practices, which to a certain limit is true, and the management systems usually value productivity more than safety [1]. An accident arising due to below par safety performance can adversely impact the volume and cost of production in industries. Neither safety nor productivity will work well without each other. A balance between safety and productivity can be achieved only with the enforcement of a proper safety culture in the organization.

For most of the safety mishaps which are related to the human factors, the role of emotional intelligence (EI) of workers had some potential influence liable to these accidents [2]. Study on how these emotional intelligence factors could help reason out the ability to rule own individual emotions to facilitate thinking and dealing with it precisely was found to have a correlation with safety performance of that particular individual. Also, it is ascertained that the emotional intelligence factors of workers can distress their health and safety performance in a workplace.

A hybrid model for human factor analysis based on the Human Factor Analysis and Classification Systems (HFACS) could tackle issues related to the conventional methods of human factors assessment which were highly uncertain and static. This model uses soft computing methods like intuitionistic fuzzy set theory and Bayesian network, and it also incorporates the concepts of behavior-based safety [3]. Application of this particular model on various accident scenarios in an industry or workplaces could accurately predict the contributing factors leading to the accident and could also suggest potential risk-based safety actions for preventing the same.

A recent area of research in the field of human factors affecting safety performance is the management of human factors by the application of augmented reality. Augmented reality is a next generation virtual reality which visualizes the reality of

A. S. George (✉) · V. R. Renjith
School of Engineering, Cochin University of Science and Technology, 682022 Kochi, India

N. A. Siddiqui et al. (eds.), *Advances in Behavioral Based Safety*,
https://doi.org/10.1007/978-981-16-8270-4_8

an object better. Augmented reality technology has the scope to bring in changes and new values on the industrial point of view to reduce human errors and improving the safety performance [4]. The research focuses on identifying the various human factors related to an industry and checks whether augmented reality can be applied to address those concerns.

Human factors associated with improper decision and behavior can have an obstructive impact on safety, effectiveness and safety performance. Although there is an increasing shift in automating the manual work, still the humans are closely associated with design of the engineered system of the workplace or industry, and this will ultimately affect the safety performance throughout its life cycle. Many of the industries and workplaces do require a managerial decision support for human factors affecting safety performance; however, they do not consider the same and their impact on the safety performance and the system [5]. This could in turn lead to underperforming systems and results, increased health hazards and poor safety performance. Here, we will be reviewing few recent methods associated with human factors affecting safety performances in a workplace.

2 Emotional Intelligence

Regardless of improvements in technology and equipment design, accidents continue to occur at workplaces due to the role of human element. Despite the implementation of various safety regulations and policies to reduce accident causation and loss of life, accidents still occur at almost all industries and workplaces. It is found that human behavior is influenced by various factors such as emotion, attitude, culture, environment, health, motivation, training and experience [6]. If any of these factors that are compromised at a workplace results in abnormal human action thereby creating a human hazard and leading to accidents. Out of all these factors, emotion or emotional intelligence is found to have a greater influence on human behavior.

Human emotion or emotional intelligence has been defined as the capability to understand, observe and manage emotions. It can also be a direct or unintended derivative of other factors resulting in unusual and unsafe human behavior [2]. A better emotion makes better people and aids them to take smart decisions and also enables them to act and respond to various critical situations around them in a safe manner.

There are various models of emotional intelligence such as the [6, 7], Goleman model (2012) which give an idea about how a human emotion can influence the causation of an incident or even the action of others associated with the person. The contemporary models of emotional intelligence are the Goleman model which deals with four fundamental domains such as self-awareness, social awareness, self-management and relationship management. It also consists of 19 confidences under the four broad domains. Figure 1 shows the Goleman model2.

A recent research method conducted by Ifelebuegu et al. [2] deals with emotional intelligence of professionals in an oil and gas industry. They identified key success

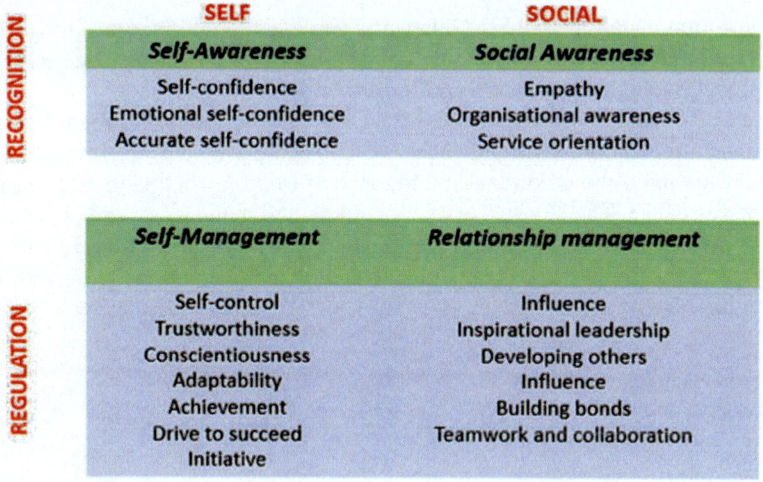

Fig. 1 Four domains of emotional intelligence [2]

factors of human emotion using questionnaires and other data based on nature of job and experience. This method helped the authors to ascertain the strength of emotional disposition of the person at the workplace. The primary and secondary data sources were collected regarding the human emotional intelligence at the workplace, and these data were used as the basis for design of the study questionnaire which consists of a combination of closed-ended and open-ended questions. The researcher had used both inductive and deductive methods of data analysis to define the key success factors of emotional intelligence [1, 2]. The factors were also determined with the support of other existing literatures. The statistical package for social sciences (SPSS) was used to analyze the relationship between the participants' emotional intelligence and how these could affect the derived key success factors. To ascertain the strength of relationship between emotional intelligence and safety performance, regression analysis was conducted.

A thorough review of the literatures related to human factors showed that safety performance will be compromised without a better understanding of the involvement of the worker's emotional intelligence. The study identified 14 success indicators as the key to effective occupational health and safety in a workplace. The key factor 'effective communication skills' was determined as the fundamental competencies, and it deals with the capacity to deliver emotions from a person to another. Also, in the study, it was found that after reviewing the various models of emotional intelligence, some commonalities were observed, and these fall under the four key areas of emotional intelligence. Being conscious of one's own emotions, being able to rule one's own emotion to facilitate thinking, being able to recognize the emotions of others at any given moment in time and being able to deal with the emotions of others were the common competencies. Also, being able to discuss one's own emotions accurately found out to be the major key success factor associated with effective communication skill [2]. This was a gap missed out in most of the models. Finally, the strength of worker's emotional disposition based on key success factors of emotional intelligence was analyzed on a five-point scale.

The results from the study found out well-built connection between the safety performance and emotional intelligence, and they could evolve a model which shows the extent to which the emotional intelligence affects the occupational health and safety performance. The key success factors identified in the study can be enhanced by the workers' emotional intelligence. The key factors, 'being able to rule own emotion to ease thinking' and 'being able to deal with the emotions of others', were found to have a strong relationship.

3 Human Factor Analysis and Classification Systems

Among the various methods that are broadly used to deal with human factors element in accident causation, Human Factor Analysis and Classification System (HFACS) model is the most popular and is particularly crafted for analyzing the contribution of human factors in an accident. HFACS method is more reliable than the other methods such as the Reason's Swiss cheese model, classification of socio-technical systems, STAMP, systems-theoretic accident model and process because of its taxonomic nature. It was found that developing the existing qualitative model into a quantitative HFACS can improve the procedure of accident investigation. A new hybrid model was developed incorporating Bayesian network and fuzzy set theory to compensate ambiguity and uncertainty connected with the existing model. Zarei et al. [8] improvised the existing accident analysis model by integrating these approaches such as HFACS framework, fuzzy set theory and expert elicitation, Bayesian network and fuzzy analytical hierarchy process (FAHP).

Human Factors Analysis and Classification System (HFACS) is based upon Reason's Swiss cheese model of human error. This model looks into active and latent failures within an industry or a workplace that led up to an accident at four levels such as:

1. Organizational influences
2. Unsafe supervision
3. Preconditions for unsafe acts
4. Unsafe acts

This is a detailed human error framework which includes 19 casual categories and 69 subcategories. The first step in applying this HFACS framework to an accident is by defining the accident scenario clearly. The scope of the study is defined, and the concerned data and information are collected in this phase. The second step is the qualitative analysis where HFACS is established for a defined scenario. Here, the human factors are identified in all the four levels, and finally, a hierarchy of the human and organizational factors (HOFs) model is constructed which acts as a casual representation model. The human factors identification for four levels can be done by holding safety meetings in the workplace or by interviewing the experts. The third and final step includes a quantitative and inference analysis where fuzzy set theory, expert elicitation and Bayesian networks are applied [8]. The HFACS model framework is shown in Fig. 2.

In another method, the relationship between various human factors of the four levels of the HFACS model in a workplace can also be found out by preparing the accident database with the help of expert elicitation and soft computing analysis methods. The accident causational routes and corresponding probability could be found out, and recommendations (direct and indirect measures) can be suggested to improve the safety performance. The accidental causational routes are found out using the leading human factors obtained from the analysis. Relationship between various levels of human factors can be identified by odds ratio and chi-square test using the SPSS software.

Fig. 2 HFACS model framework [3]

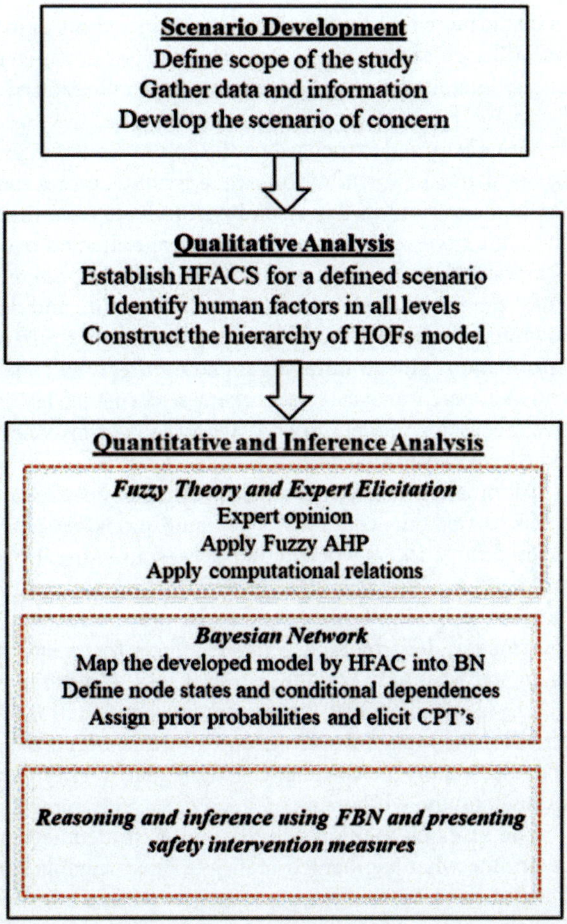

The HFACS model based on fuzzy set theory, Bayesian network and fuzzy analytical hierarchy process (FAHP) will help in identifying, distinguishing and ranking human factors leading to accidents [3]. This method could comprehensively identify the failures of the workers and also deals with uncertainty and ambiguity associated with accident analysis and safety performance in a workplace. The second method deals with the statistical analysis of accidents which results in obtaining the relationship between the human factors at various levels of the HFACS framework. Both the methods also help to identify specific safety measures based on the results and analysis to improve the overall safety performance in the workplace and to prevent recurring of similar accidents.

4 Managing Human Factors Using Augmented Reality

With the increase in automation in various sectors of industries and workplaces, there are still accidents resulting due to human errors which influence productivity, quality and ultimately the safety performance. Application of augmented reality is found to be an appropriate tool to address such issues.

Qeshmy et al. [4] focused on the application of augmented reality to ensure safety performance in assembly lines in a manufacturing industry. The major root causes for human errors in the assembly line were identified and characterized based on literature review. Preliminary data were collected based on interviews and reports; also, interviews with the experts in the field of augmented reality were conducted. From the initial analysis of the causing factors and data collected, the main causes of human error were found to be the amount of thinking, deciding and searching for information. This in turn was found to affect the cognitive capacity of the worker's and their performance. The workers performance levels are divided into three cognitive processing models, knowledge-based performance, rule-based performance and skill-based performance.

Automation in any workplace is efficient only if the worker receives feedback for a task carried out accurately [9]. A multimodal feedback was found to be an efficient method for workers working in the assembly line. Haptic feedback, visual feedback and auditory feedback were found to support assemblers, and among these, visual and haptic feedback are found to be more efficient. The amount of searching for information done by an assembler affects his mental workload, and deciding what to do and whether it is done correctly also affects the cognitive load of the worker.

The mental workload framework gives the relationship between the various factors affecting the mental workload and how this affects the consciousness of the workers and their safety performance. Augmented reality can be introduced as a way of automating the workers mental activities and giving a digital feedback (Fig. 3).

The research methodology begins with detailed pre-study and investigation to determine what augmented reality is most suitable for the assembly or production line. It also includes certain practical works and informal interviews with the workers and supervisors. After preliminary investigation problem formulation, purpose and

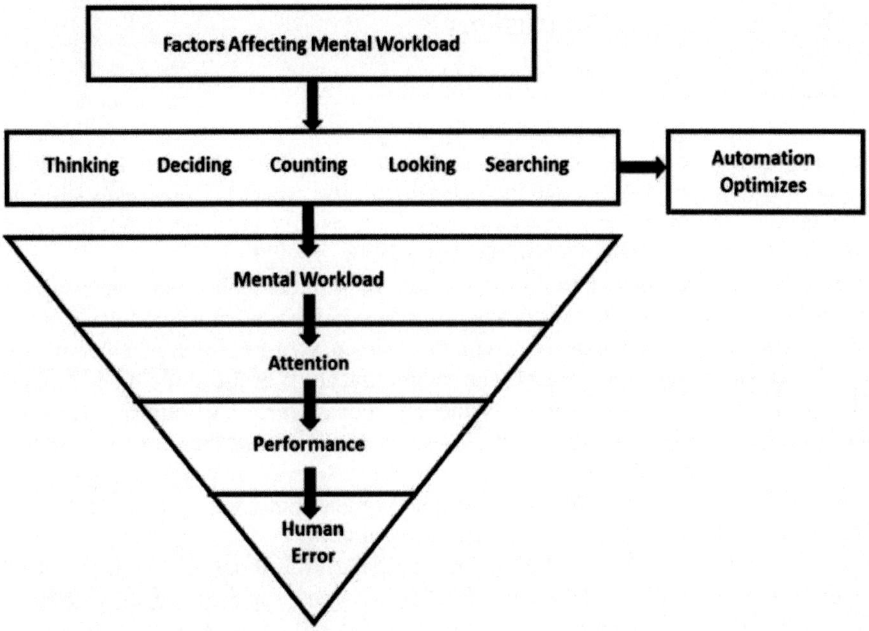

Fig. 3 Mental workload framework [4]

research questions were developed along with literature review on augmented reality. Data were collected from workers using augmented reality concepts and video samples. The collected data were analyzed, and conclusions were made on the research question.

It was found that the workers work under highly automated work; hence, they are working under skill-based performance level with slip and lapses. Another observation concluded that thinking, deciding and searching for information affected mental workload of the workers. All these were found to affect the safety performance and production. Therefore, if the automation system has to be efficient, then the workers have to face these skill-based errors and the factors affecting cognitive workload [4]. An ideal augmented reality system should those address these concerns in an assembly line in an industry. It was analyzed in the research that the augmented reality with a digital information system to trace the error that has occurred and to visualize the same is required. Also, a need for independent indirect interaction with the system is required to effectively improve the performance.

5 Discussions and Conclusions

Although the different control systems in an industry or a workplace are modernized and have a high degree of automation, the safety performance associated with this is quite low because of the presence of human factors, which has the overall immediate responsibility for safe and economic operation of the process. The worker still has the vital function of running a process in order to avoid shut down conditions in a workplace, which makes the job of the worker more crucial.

The role of emotional intelligence factors in workers safety performance and occupational health was found to have a greater influence on the human factors. The emotional intelligence factors had a direct impact for improving the health and safety performance of the workers. The major advantage of the study on emotional intelligence was that it could identify the key success factors of human emotions. On contrary, the model developed in this study also revealed that not every safety performance factor is affected by the emotional intelligence. In order to improve the safety performance, it was necessary to incorporate emotional intelligence factors into regular human resource development programs.

The second model discussed in this review paper was about the Human Factors Analysis and Classification Systems (HFACS), a broad human error analysis framework which can examine the mechanism of human failures and analyze the safety performance in a workplace. From the accident causation perspective, this hybrid model could identify, characterize and rank the human factors with the application of fuzzy set theory, Bayesian network and fuzzy analytical hierarchy process. But, in order to apply this in day-to-day operations and process, there is a requirement to further detailed validation and testing by a third party. Also, human factors out of the enterprise could also be taken into consideration to obtain a better result.

The third method reviewed about the qualitative analysis of managing human factors using augmented reality in manufacturing. The augmented reality was found to be an appropriate tool when the system was functioning according to the skill-based performance and the causing factors of cognitive workload. If the manufacturing systems are too complex, they need complex augmented reality systems relying on artificial intelligence. The need for using such an augmented reality system may not be effective as there are existing technologies that are more or less working in the similar manner. Also, the amount of augmented reality that is required can make the major function of the system more redundant.

References

1. Lees P. F. (2002). Loss prevention in the process industries. *Journal of Hazardous Materials*https://doi.org/10.1016/S0304-3894(02)00033-X
2. Ifelebuegu, A. O., Martins, O. A., Theophilus, S. C., & Arewa, A. O. (2019). The role of emotional intelligence factors inworkers' occupational health and safety performance—A case study of the petroleum industry. *Safety, 5*. https://doi.org/10.3390/safety5020030.

3. Zarei, E., Yazdi, M., Abbassi, R., & Khan, F. (2019). A hybrid model for human factor analysis in process accidents: FBN-HFACS. *Journal of Loss Prevention in the Process Industries, 57*, 142–155. https://doi.org/10.1016/j.jlp.2018.11.015
4. Qeshmy, D. E., Makdisi, J., Da Silva, E. H. D. R., & Angelisa, J. (2019). Managing human errors: Augmented reality systems as a tool in the quality journey. In *Procedia Manufacturing*. https://doi.org/10.1016/j.promfg.2018.12.005.
5. Shi, W., Jiang, F., Zheng, Q., & Cui, J. (2011). Analysis and control of human error. *Procedia Engineering, 26*, 2126–2132. https://doi.org/10.1016/j.proeng.2011.11.2415
6. Mayer, J. D., & Salovey, P. C. (2004). Emotional intelligence: Theory, findings, and implications. *Psychological Inquiry, 15*, 197–215.
7. Petrides, K. V., & Furnham, A. (2001). Trait Emotional Intelligence: Psychometric Investigation with Reference to Established Trait Taxonomies. *European Journal of Personality*, 15, 425–448. https://doi.org/10.1002/per.416
8. Zarei, E., Khakzad, N., Cozzani, V., & Reniers, G. (2019). Safety analysis of process systems using Fuzzy Bayesian Network (FBN). *Journal of Loss Prevention in the Process Industries*. https://doi.org/10.1016/j.jlp.2018.10.011
9. Tang, A., Owen, C., Biocca, F., Mou, W. (2003). Comparative effectiveness of augmented reality in object assembly. *Proceedings of the SIGCHI Conference on Human factors in computing systems, 3*, 73

Preliminary Study on Seawater Intrusion in Coastal Aquifers of Visakhapatnam

K. Mamatha, K. Suresh Kumar, N. Srinivas, D. Gopamma, and Akbar Ziauddin

Abbreviations

APHA	American Public Health Association
BIS	Bureau of Indian Standards
Ca	Calcium
CaH	Calcium Hardness
Cl	Chlorides
EC	Electrical Conductivity
EEC	European Economic Community
MgH	Magnesium Hardness
Na	Sodium
TA	Total Alkalinity
TDS	Total Dissolved Solids
TH	Total Hardness
WHO	World Health Organisation

1 Introduction

Water is essential to people's lives and livelihoods. It causes the demand for clean water to rise in lockstep with global population growth. Without groundwater, the

K. Mamatha · K. S. Kumar (✉) · N. Srinivas · D. Gopamma
Department of Environmental Science, Institute of Science, GITAM Deemed to be University, 530045 Visakhapatnam, India
e-mail: skolli@gitam.edu

A. Ziauddin
Advance Career Training & Consulting, Abu Dhabi, United Arab Emirates

N. A. Siddiqui et al. (eds.), *Advances in Behavioral Based Safety*,
https://doi.org/10.1007/978-981-16-8270-4_9

largest and most reliable of all freshwater resources, it would be impossible to maintain secure water supplies for drinking, industry, and agriculture. Groundwater, unlike other natural resources or raw materials, is found all over the world. Because of rainfall patterns and aquifer distribution, the possibilities for its extraction vary greatly from place to place (rocks, sand layers, and so on, in whose pore spaces the groundwater sits). Because of the hydraulic gradient of groundwater above mean sea level, freshwater is also widespread in the coastal aquifer [1, 2]. The mass transport of saline waters into previously occupied zones by fresher waters is known as saltwater intrusion. The movement of saltwater with high total dissolved solids (TDS) content into freshwaters is characterised as saltwater intrusion. Coastal aquifers are important sources of freshwater supply in many countries around the world, particularly in arid and semi-arid zones [3]. Many coastal areas in India are heavily urbanised, exacerbating the need for freshwater. Aquifers along the coast are vulnerable to disturbances. Inadequate management of a coastal aquifer may result in its demise as a source of freshwater much sooner than other aquifers not connected to the sea. The threat of seawater intrusion is the reason [4, 5].

The chemical composition of groundwater is determined primarily by three processes: atmospheric inputs, water interaction with soil and rock, and anthropogenic activities. Precipitation, climate change, and natural hazards all contribute to atmospheric inputs, while water interaction with soil and rock causes weathering and erosion of crustal materials. Over-exploitation of groundwater and the location of industries above sensitive zones have a significant impact on groundwater quality [6]. The presence of highly brackish waters in certain localised pockets of the coastal aquifer, such as Kaniti-Paravada near Visakhapatnam, is caused by hydrogeological processes in the area, not by sea influence [7]. The purpose of this paper is to investigate the controlling mechanism of seawater intrusion in Visakhapatnam's coastal aquifers.

- MeghadriGedda Reservoir
- Raiwada Link Canal
- Yeleru Left Bank Canal

2 Materials and Methods

2.1 Study Area

Greater Visakhapatnam, once a small fishing village, has grown into a major port city over the years. It also serves as the administrative centre for the Visakhapatnam district. Visakhapatnam is located on India's east coast, at 17o42' North latitude and 82o02' East range of hills. The city and its surroundings can be divided into four categories based on topographical conditions: Hilly region, Upland tracks, Rolling plains, and Plains.

The main surface sources of water for the city of Visakhapatnam are

Table 1 Details of sampling sites with codes in Visakhapatnam

Sample code	Sample site	Source	Use of water
S1	Bheemunipatnam	Hand pump	Drinking, domestic
S2	Chepaluppada	Open well	Domestic, laundry
S3	Mangamaripeta	Bore well	Drinking, domestic
S4	Timmapuram	Bore well	Drinking, domestic
S5	Rushikonda	Bore well	Drinking, domestic
S6	SagarNagar Colony	Hand pump	Drinking, domestic
S7	Jodugullapalem	Hand pump	Domestic, laundry
S8	Appugarh	Hand pump	Drinking, domestic
S9	MVP Colony	Hand pump	Drinking, public usage
S10	PeddaWaltair	Hand pump	Drinking, domestic
S111	EastPoint Colony	Hand pump	Drinking, domestic
S12	Babuji Nagar	Hand pump	Drinking, domestic
S13	Pandurangapuram	Hand pump	Drinking, domestic
S14	MaharaniPeta	Hand pump	Drinking, domestic
S15	Opp. KGH Hospital	Hand pump	Drinking, public usage
S16	Burujupeta	Hand pump	Drinking, domestic

- Mudasarlova Reservoir
- GambheeramGedda Reservoir
- Gosthani River
- Thatipudi Reservoir

Ground water samples (from open wells, bore wells, and hand pumps) were collected from sixteen locations along Visakhapatnam's coastline (from Burujupeta to Bheemunipatnam). The sampling sites are located 5–8 kms from the coast, and the details of the sampling sites are provided in Table 1. Water samples were collected in 2-L polythene bottles (soaked overnight in 2% nitric acid and washed thoroughly in distilled water) from each location and analysed for pH, Conductance, Total Dissolved Solids, Total Hardness, Calcium Hardness, Magnesium Hardness, Alkalinity, Chloride, Sulphate, Nitrate, Phosphate, Iron, Sodium, and Potassium as per standard methods [8].

3 Results and Discussion

Table 2 shows the results of the various quality parameters for all samples. A brief description of the essential physicochemical properties of groundwater is provided below.

Table 2 Physico-Chemical characteristics of ground gater samples collected from different areas of Visakhapatnam city

Sample Code	pH	Electrical Conductivity (mmhos)	Total Dissolved Solids	Chlorides	Alkalinity	Total Hardness	Calcium Hardness	Magnesium hardness	Phosphates	Sulphates	Nitrates	Iron	Sodium	Potassium
S1	7.49	0.78	429	24.99	160	170	70	100	0.5	7.5	0.04	0.153	60.3	6.5
S2	7.85	2.08	1144	149.99	380	340	160	180	0.62	31.5	0.069	0.22	132.8	7.3
S3	7.1	0.31	170.5	24	110	120	90	30	0.64	3.5	0.027	0.04	32.2	1.8
S4	7.38	1.06	583	74	160	280	150	130	0.54	11	0.042	0.15	49.0	9.6
S5	8.12	1.41	775.5	99	290	320	180	140	0.96	11	0.034	0.2	71.9	7.7
S6	8.01	2.02	1111	124	300	360	200	160	0.3	16	0.049	BDL	170.6	25.8
S7	7.21	1.25	687.5	99	160	290	160	130	0.6	10	0.037	BDL	91.2	4.3
S8	8.2	2.36	1298	199	440	380	220	160	0.72	33	0.067	BDL	200.1	11.3
S9	7.91	1.81	995.5	149	370	320	170	150	0.89	10	0.04	0.01	127.6	2.5
S10	7.84	1.67	918.5	149	260	340	180	160	0.68	17.5	0.059	0.06	66.2	3.7
S11	7.72	1.96	1078	174	340	280	170	110	0.74	20	0.066	BDL	87.1	11.4
S12	7.02	2.18	1199	199	300	280	130	150	0.2	33.5	0.078	0.02	161.5	11.7
S13	7.5	1.49	819.5	174	250	240	140	100	0.2	18.5	0.064	BDL	92.6	4.3
S14	7.73	1.39	764.5	99.99	280	270	130	140	0.48	3.5	0.028	0.09	93.0	3.7
S15	7.69	1.4	770	99.99	280	240	130	110	0.62	3.5	0.022	0.15	93.0	2.8
S16	8.21	2.11	1160.5	149.99	360	340	160	180	0.8	23.5	0.072	0.08	150.7	68.4

* pH: No measurements, EC: mmhos; Cl, Alkalinity, CaH, MgH, TH, phosphates, Sulphates, Nitrates, Iron, Sodium, Potassium measures in mg/l
BDL = Below Detection Level

3.1 pH

The pH in the study areas ranged from 8.21 at (S-16) in Burujupeta to 7.02 at (S-12) PeddaWaltair. All of the pH values in the samples are well within the ranges for various water uses, including drinking and other domestic supplies [8, 9].

3.2 Conductance

The conductance ranged from a maximum of 2.36 mmhos/cm at (S-8) Appugarh to a minimum of 0.31 mmhos/cm at (S-3). Due to the high concentration of Ionic constituents present in the water bodies under study, almost all of the water samples under study exceeded the permissible limit (0.3 mmhos/cm), reflecting the contributions from salinity intrusion and pollution by domestic wastes. The dissolved ions such as bicarbonates, chlorides, sodium, potassium, magnesium, and sulphate are primarily responsible for electrical conductivity [11].

3.3 Total Dissolved Solids (TDS)

The values of Total Dissolved Solids (TDS) in groundwater in the current study range from 170.5 to 1298 mg/l, indicating high mineralization in the study area. Furthermore, organic substances commonly found in polluted water may contribute to the Dissolved Solids. Water with more than 500 mg/l TDS is not considered desirable for drinking water supplies, though more highly mineralized water is used in places where better water is unavailable. As a result, for drinking water, 500 mg/l has been proposed as the desirable limit and 2000 mg/l as the maximum permissible limit [8, 9]. TDS levels greater than 500 mg/l in water cause gastrointestinal irritation [8, 9].

3.4 Chlorides

Ranged from 24 to 199 mg/l in the study areas. Chloride limits have been established primarily based on taste considerations. A desirable limit of 250 mg/l chlorides has been recommended for drinking water [8, 9]. However, no adverse health effects on humans have been reported from drinking water with a higher chloride content [10, 11].

3.5 Alkalinity

Alkalinity in the natural system is primarily composed of carbonates, bicarbonates, and hydroxides. These constituents are formed as a result of mineral dissolution in the soil and atmosphere [12]. Carbonate and bicarbonates can also be produced by microbial decomposition of organic matter. The alkalinity ranged from 440 mg/l at S-8 in Appugarh to 110 mg/l at S-3 in Mangamaripeta. The alkalinity of almost all of the areas studied was found to be higher than the allowable limit (120 mg/l). This could be due to the percolation of the area's domestic sewage [13].

3.6 Total Hardness

The carbonates, sulphates, and chlorides of calcium and magnesium, as well as their carbonates, sulphates, and chlorides, harden the water. Total Hardness ranged from a maximum of 380.0 mg/l at S-8, Appugarh, to a minimum of 120.0 mg/l at S-3, Mangamaripeta. For drinking water, a desirable limit of 300 mg/l and a permissible limit of 600 mg/l have been recommended [8]. Hardness has no known negative effects on health; however, some evidence suggests that it may play a role in heart disease [14].

3.7 Calcium Hardness

Calcium concentrations in the study area ranged from 70 mg/l in Bheemili to a maximum of 220 mg/l in Appugarh. Calcium in drinking water has a maximum desirable limit of 75 mg/l and a maximum permissible limit of 200 mg/l. The calcium ions in the majority of the study samples exceeded the desirable limit, but remained within allowable limits. Calcium is a necessary component of the human body. Rickets and tooth decay can be caused by low calcium levels in drinking water. It is required for nervous system function, cardiac function, and blood coagulation [15].

3.8 Magnesium Hardness

Magnesium concentrations in the current study area ranged from 30 mg/l in Mangamaripeta to a maximum of 180 mg/l in Burujupeta. Magnesium in drinking water has a maximum desirable limit of 30 mg/l and a maximum permissible limit of 100 mg/l. Magnesium ions exceeded the allowable limit in 90 per cent of the study samples. Magnesium concentration in water is lower than calcium concentration, possibly due to the lower occurrence of Magnesium Minerals. Magnesium is a moderately toxic element when present in high concentrations in drinking water [15].

3.9 Phosphates

Phosphate concentrations in the study area are low in all locations. Phosphorus is a necessary plant nutrient that is widely used in fertilisers. In acidic soils, phosphate is adsorbed or fixed as aluminium or iron phosphate, while in alkaline or neutral soils, it is fixed as calcium phosphate. As a result, phosphate concentrations in groundwater are typically low, but phosphate mobility in subsoil and groundwater may be induced by various chemical processes in soil strata.

3.10 Sulphates

The soluble salts of calcium, magnesium, and sodium make up the majority of the sulphate content in groundwater. The sulphate content of rainfall infiltration and groundwater recharge, primarily from stagnant water pools and surface runoff water collected in low-lying areas, changes significantly over time. Sulphate levels in the study area ranged from 3.5 mg/l in Maharanipeta to 33.5 mg/l in Babuji-nagar. Sulphate levels are within the prescribed standard (200 mg/l) at all sampling stations. Sulphate is poorly absorbed from the human intestine and slowly penetrates into the cellular membranes of mammals before being rapidly eliminated through the kidneys. In humans, sodium and magnesium sulphates have a cathartic effect. It has also been linked to respiratory diseases [16].

3.11 Nitrates

The presence of nitrates in drinking water is thought to be necessary for its negative health effects. The presence of high nitrate levels in groundwater is a major issue in many parts of the country. The Nitrate content in the study area is low in all sampling locations. Nitrate is a plant nutrient that is both effective and moderately toxic. [8, 9] have set a limit of 45 mg/l for drinking water supplies. Its concentrations above 45 mg/l may be hazardous to human health. At higher concentrations, nitrate can cause Methaemoglobinaemia (blue baby syndrome) [17], a disease that primarily affects bottle-fed infants. Ingestion of high doses of nitrates on a regular basis may also result in carcinogenic diseases.

3.12 Iron

The highest iron content was found in the current study areas at Chepaluppada, at 0.22 mg/l. Iron content was found to be low in the other regions, and in Sagarnagar colony, Jodugullapalem, Appugarh, MVP Colony, and EastPoint Colony, iron content was found to be below the detectable level of 0.3 mg/l, as prescribed by BIS [9] for drinking water supplies. Iron concentrations above 0.3 mg/l stain clothing and utensils. The water is also unsuitable for processing food, beverages, ice, dyeing, bleaching, and other similar processes. The limit for iron in water is determined by aesthetic and taste considerations rather than physiological effects [18].

3.13 Sodium and Potassium

In the study area, sodium concentrations range from 32 mg/l in Mangamaripeta to 200 mg/l in Appugarh. Sodium has not been included in drinking water standards set by the Bureau of Indian Standards. The high sodium levels could be due to base-exchange phenomena. The potassium concentration in groundwater in the study area ranged from 1.8 to 68.4 mg/l. Potassium is an essential element for humans, plants, and animals, and it is derived primarily from vegetation and soil in the food chain. Rainwater, potash silicate minerals, the use of potash fertilisers, and surface water for irrigation are the primary sources of potassium in groundwater. It is more common in sedimentary rocks and is found in feldspar, mica, and other clay minerals. Potassium has not been included in drinking water standards set by the Bureau of Indian Standards. However, the European Economic Community [19] has set a guideline level of potassium in drinking water at 10 mg/l. Potassium levels were found to be high in Sagarnagar Colony (25.8 mg/l) and Burujupeta (68.4 mg/l).

3.14 Correlation Analysis

The correlation analysis of 13 physicochemical parameters of groundwater from 16 different sites in Visakhapatnam city was performed, and the correlation coefficients (r) are shown in Table 3. Between EC and TDS (r = 1), Sulphates and Nitrates (r = 0.94), Total Hardness and Calcium Hardness (r = 0.91), Total Hardness and Magnesium Hardness (r = 0.9), TDS and Chlorides (r = 0.90), and TDS and Alkalinity (r = 0.90), there is a strong correlation (r = 0.9). A very high positive correlation also exist between TDS and Sodium (r = 0.86), EC and Total Hardness (r = 0.84), EC and Magnesium Hardness (r = 0.80), Chlorides and Nitrates (r = 0.78), Alkalinity and Sodium (r = 0.78), Chlorides and Sulphates (r = 0.77), TDS and Sulphates (r = 0.77), EC and Nitrates (r = 0.73), pH and Alkalinity (r = 0.72), EC and Calcium Hardness (r = 0.72), Chlorides and Sodium (r = 0.71), pH and Total Hardness (r = 0.70).

Table 3 Correlation coefficient (r) among various water quality parameters

	pH	EC	TDS	Chlorides	Alkalinity	TH	CaH	MgH	Phosphates	Sulphates	Nitrates	Sodium	Potassium
Ph	1	0.553	0.553	0.331	0.729	0.700	0.666	0.611	0.530	0.142	0.171	0.455	0.439
EC		1	1	0.902	0.902	0.842	0.728	0.809	0.010	0.770	0.738	0.865	0.408
TDS			1	0.902	0.902	0.842	0.728	0.809	0.010	0.770	0.738	0.865	0.408
Chlorides				1	0.788	0.681	0.652	0.591	−0.093	0.770	0.785	0.716	0.211
Alkalinity					1	0.753	0.673	0.702	0.273	0.582	0.550	0.789	0.321
TH						1	0.913	0.911	0.234	0.535	0.483	0.691	0.360
CaH							1	0.664	0.296	0.393	0.371	0.588	0.217
MgH								1	0.131	0.584	0.510	0.674	0.441
Phos									1	−0.204	−0.212	−0.133	0.113
Sulphates										1	0.949	0.633	0.360
Nitrates											1	0.580	0.436
Sodium												1	0.435
Potassium													1

It can thus be assumed that the concurrent increase/decrease in ion composition in these waters is primarily due to dissolution/precipitation reactions and concentration effects.

3.15 Sea Water Contamination Using Chemical Ratio

The ratio of specific chemical parameters such as Ca^{2+}/Mg^{2+}, $Cl/(CO_3 + HCO_3)$, and TA/TH can be used to assess saltwater intrusion into coastal aquifers [20]. Because magnesium is much more abundant in seawater than calcium, the Ca/Mg ratio is recommended as a parameter for determining saltwater contamination. A low Ca/Mg ratio could also indicate saltwater contamination [21].

. If the ratio is less than one, the area is thought to be heavily impacted by saltwater intrusion. Similarly, one of the criteria used to assess the presence of seawater intrusion is the $Cl/(CO_3 + HCO_3)$ ratio. Table 4 shows the $Cl/(CO_3 + HCO_3)$, TA/TH, and Ca/Mg ratios.

A low Ca/Mg ratio may also indicate saltwater contamination. If the ratio is less than one, the area is considered to be heavily impacted by saltwater intrusion. Similarly, the $Cl/(CO_3 + HCO_3)$ ratio is one of the criteria used to determine the

Table 4 Verification of seawater contamination in different location in Visakhapatnam

Sample site	Conductivity mmhos	$Cl/CO_3 + HCO_3$	TA/TH	Ca/Mg
Bheemunipatnam	0.78	0.119227	0.941	0.7
Chepaluppada	2.08	0.301306	1.117	0.88
Mangamaripeta	0.31	0.173421	0.916	3
Timmapuram	1.06	0.357777	0.571	1.15
Rushikonda	1.41	0.263201	0.906	1.28
SagarNagar Colony	2.02	0.318041	0.833	1.25
Jodugullapalem	1.25	0.477052	0.551	1.23
Appugarh	2.36	0.346964	1.157	1.37
MVP Colony	1.81	0.309449	1.156	1.13
PeddaWaltair	1.67	0.44037	0.764	1.125
EastPoint Colony	1.96	0.392883	1.214	1.545
Babuji Nagar	2.18	0.50888	1.071	0.866
Pandurangapuram	1.49	0.534321	1.041	1.4
MaharaniPeta	1.39	0.272601	1.037	0.928
Opp.KGH Hospital	1.4	0.272601	1.166	1.181
Burujupeta	2.11	0.318045	1.058	0.88

* measures in mg/l

presence of seawater intrusion. Table 4 shows the ratios of $Cl/(CO_3 + HCO_3)$, TA/TH, and Ca/Mg.

3.16 GIBB'S Diagram

The Gibb's ratio for the ions (Na/Na + Ca and $Cl/(Cl + HCO_3)$ of groundwater samples was plotted against the TDS values. The plot shows that all of the groundwater samples belong to the rock dominant category. The interaction between aquifer rocks and groundwater causes rock dominance in the samples (Fig. 1).

$$\text{TDS vs Na}/(\text{Na} + \text{Ca})$$

$$\text{TDS vs Cl}/(\text{Cl} + \text{HCO}_3)$$

Fig. 1 Gibb's Diagram showing the controlling mechanism of groundwater

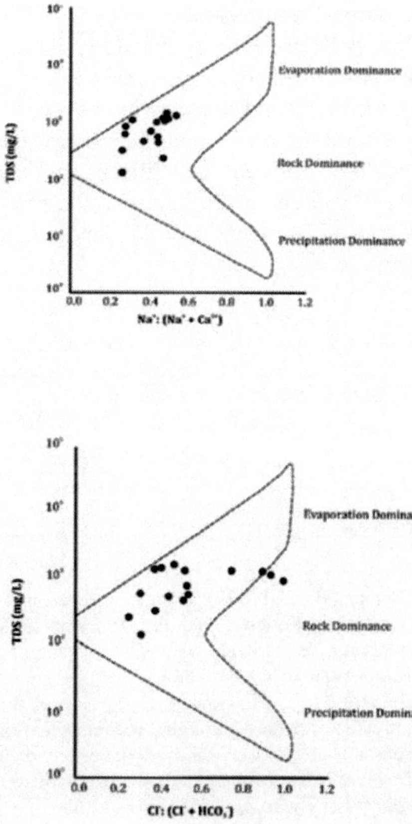

4 Summary and Conclusion

Water samples were collected and analysed to assess the quality (in terms of physical and chemical content) of Visakhapatnam City's coastal aquifers. The study revealed the following findings. Groundwater in the study area is moderately complex and slightly alkaline. According to the current study, seawater intrusion is the cause of increased ion concentrations in the coastal aquifer. According to the findings of this study, the coastal aquifer is slightly contaminated by seawater intrusion, which is most likely due to increased over-exploitation of groundwater sources. More research on seawater intrusion into the coastal aquifer and geological weathering is needed to accurately quantify groundwater contamination.

5 Suggestions

- To protect coastal ground water, the sources of saline water intrusion and mobility mechanisms must be identified and investigated for sustainable groundwater resource development.
- Data collected during this study should be expanded to include all seasons of the year, as changes in water quality that occur as a result of normal seasonal variations are not reflected in this report due to the limited period of performance.
- The 2002 National Water Policy emphasises that "overexploitation of groundwater should be avoided, particularly near the coast, to prevent ingress of seawater into sweet water aquifers". As a result, determining the extent of seawater intrusion is critical for planning and managing groundwater resource protection in coastal aquifers.

Acknowledgements The authors are thankful to GITAM Management to carry out the work. The authors also thanking Prof. Sabu Joseph, Department of Environmental Sciences, University of Kerala for GIBBS software assistance.

References

1. Ghyben, B. W. (1889). Nota in verband met de voorgenomen put boring put boring nabij, Amsterdam Kononkl. *Inst. Ing. Tijdschr., 1888–1889*, 8–22.
2. Herzberg, B. (1901). Die Wasserversorgung einiger Nordeebader. *Jour Gasbeleuchtung Wasserversorg, 44*, 815–844.
3. Behera, A. K., Chakrapani, G. J., Kumar, S., & Rai, N. (2019). Identification of sea water intrusion signature through geochemical evolution of groundwater. A case study based on coastal region of the Mahanadi delta, Bay of Bengal, India. *Natural Hazards, 97*, 1209–1230.
4. Davood, M., & Mohammad, K. (2019). Seawater intrusion and aquifer under flooding events. *Journal of Hydrology, 568*, 1118–1130.

5. Korrai, S., Gangu, K. K., Rao, P. P., & Jonnalagadda, S. B. A study on assemment of vulnerability of seawater intrusion to groundwater in coastal areas of Visakhapatnam, India. *Environment, Development and Sustainability, 23*, 5937–5955.
6. Kannan, N. (2010). A study on the growth of Indian sector. *International Journal of Management, 1*(1), 17–22.
7. Jagannadha , V. V., & Krishnaiah, N. (1979). Quality of groundwater in the costal aquifer near Visakhapatnam. *India Ground water, 14*(5), 290–295.
8. APHA. (2005). Standard methods for the examination of water and waste water (21st ed.). American Public Health Association.
9. Bureau of Indian Standards (BIS). (2012). Amendment no.1 June 2015 to IS 10500: 2012 Drinking water-specification. Bureau of Indian Standards (BIS), Manak Bhavan,9 Bahadur Shah Zafar Marg, New Delhi.
10. World Health Organization (WHO). (1996). Water quality monitoring—A practical guide to the design and implementation of freshwater quality studies and monitoring programmers. In J. Bartram & R. Balance (Eds.), Published on behalf of United Nation Environment Programme and the World Health Organization (UNEP).
11. Peavy, H. S., Rowe, D. R., & Tchobanoglous, G. (1985). Donnan dialysis removal of nitrate from water: Effects of process parameters. Environmental Engineering (p. 696). McGraw-Hill, Book Company.
12. Mittal, S. K., & Verma, N. (1997). Critical analysis of ground water quality parameters. *Indian Journal of Environmental Protection, 17*(6), 426–429.
13. Naik, S. (2005). Studies on pollution status of Bondamunda area of Rourkela industrial complex, Doctor of Philosophy, Department of Chemistry, National Institute of Technology, Rourkela, Orissa.
14. Sarala, C., & Ravi Babu, P. (2012). Assessment of ground water quality parameters in and around Jawaharnagar, Hyderabad. *International Journal of Scientific and Research Publications, 2*(10), 2250–3153.
15. Dwivedi, A., Dwivedi, H. P., Tiwari, S. (2017). Quality analysis of Hydrosphear water of Deosar Block in District Singrauli, Madhya Pradesh, India. *International Journal of Advanced Research in Chemicals Science (IJARCS), $*(1), 22–26.
16. Sarda, P., & Sadgir, P. (2015). Assessment of multi parameters of water quality in surface water bodies—A review. *International Journal for Research in Applied Sciences & Engineering Technology, 3*(VIII), 2321–9653.
17. Soni, N., & Bhatia, A. (2015). Analysis of quality of drinking water of private bore well and piped water supply in Jaipur city, Rajasthan, India. *Research Journal pf Recent Science, 4*, 313–316.
18. Kumari, K. (2014). Statistical and Correlation studies of ground water parameters of Kamleshwarpur area at Mainpat, Chhattisgarh. *IOSR Journal of Engineering, 04*(I), 2278–8719.
19. European Economic Community (EEC). (1998). Directive of the European Parliament and of the council relating to the quality petrol and diesel fuels and Amending Council Directive 93/12/EEC.
20. Rengaraj, S., Elampooranan, T., Elango, L., & Ramalingam, V. (1996). Groundwater quality in suburban regions of Madras City, India. *Pollution Research, 15*(4), 325–328.
21. Ravi Prakash, S., & Krishna Rao, G. (1996). Delineation of salt water contamination zones by chemical parameters in Parvada Area, Vishakapattnam, AP. *Journal of Indian Water Research Society, 2*, 63–38.

COVID-19 Impact on Air Quality of Twenty-Three Most Polluted Indian Cities and Lessons to Implement Post-lockdown

Ankit Dasgotra, Vishal Kumar Singh, Gurpreet Singh, S. M. Tauseef, N. A. Siddiqui, and Suvendu Manna

1 Introduction

China, the United States and India are always considered as the main three air polluting countries in the world contributing 30%, 15% and 7% of world air pollution, respectively [1]. Also, China and India are the world's highest populated countries having combine 38 % of world's population. Coronavirus affected almost every part of the world resulted in implementing lockdown in major impacted cities. The first case was reported by china at the ending of December 2019 having its main epicentre Wuhan. COVID-19 lockdown restricted the human and vehicle moment due to sealed borders across states; industries go on shut-down stage and production got impacted majorly. These factors help nature to rejuvenate itself by reducing the air pollution to a greater extent. Air pollution is majorly measured by six factors i.e., Air Quality Index (AQI) and five air pollutants (SO_2, PM2.5, PM10, NO_2, and CO) [2]. PM2.5 and PM10 are atmospheric particulate matter having diameter equivalent or less than 2.5 μm and 10 μm, respectively. These two factors along with AQI are the main parameters to calculate air pollution quality and in general, PM2.5 level is monitored to observe the air pollution [3, 4, 5] AQI is calculated as Eq. 1 [6] (Fig. 1):

$$AQI = \frac{(PM_O - PM_L) \times (AQI_H - AQI_L)}{PM_H - PM_L} + AQI_L \tag{1}$$

A. Dasgotra
Research and Development, School of Engineering, University of Petroleum & Energy Studies, 248007 Bidholi, Dehradun, India

G. Singh
HyprSci Private Limited, Dwarka, India

V. K. Singh · S. M. Tauseef · N. A. Siddiqui · S. Manna (✉)
Department of Health Safety, Environment and Civil Engineering, School of Engineering, University of Petroleum & Energy Studies, 248007 Bidholi, Dehradun, India
e-mail: smanna@ddn.upes.ac.in

N. A. Siddiqui et al. (eds.), *Advances in Behavioral Based Safety*,
https://doi.org/10.1007/978-981-16-8270-4_10

Fig. 1 Graphical abstract representation

Here,

AQI	Air Quality Index
PM_0	Observed twenty-four-hour average concentration in $\frac{\mu g}{m^3}$
PM_H	maximum concentration of AQI colour category that contains PM_0
PM_L	minimum concentration of AQI colour category that contains PM_0
AQI_H	maximum air quality index value for colour that corresponds to PM_0
AQI_L	minimum air quality index value for colour category that corresponds to PM_0

Colour category corresponds to AQI sections of health concerns as there are six categories for specific numeric value i.e., green (0–50), yellow (51–100), orange (101–150), red (151–200), purple (201–300) and maroon (301–500) [6]. Generally, red, purple and maroon colour categories are considered dangerous resulting in some serious health issues in living beings. PM maximum and minimum concentration can be categorized as mentioned in Table 1.

Recently, researchers focused on many aspects of COVID-19 pandemic related to many environmental issues and collect data for analysing the impact. Some studies are being published on-air quality-related topics covering COVID-19 epicentre city Wuhan and major COVID-19 impacted countries [2, 7, 8]. Some articles reflect air pollution problems that occurred in past in top three polluting countries, i.e., China [9, 10] and India [11]. Also, some of the recent publications covered air quality

Table 1 Air quality index breakpoint table

Category	Good (AQI: 0–50)	Moderate (AQI: 51–100)	Unhealthy for sensitive groups (AQI: 101–150)	Unhealthy (AQI: 151–200)	Very unhealthy (AQI: 201–300)	Hazardous (AQI: 301–500)
PM_2 (μg/m^3) 24 h avg	0–12	12.1–35.4	35.5–55.4	55.5–150.4	150.5–250.4	250.5–500.4
PM_{10} (μg/m^3) 24 h avg	0–54	55–154	155–254	255–354	355–424	425–604

studies in India and the impact of lockdown on major COVID-19 affected cities [3, 4, 12]. According to data aggregated from over 60,000 data points considering the main pollutant factor PM2.5, it has been noticed that 26 Indian cities are included in the top 50 polluted cities in the world in 2019. In our study, PM2.5, PM10 level of 23 most polluted Indian cities were compared during the lockdown and before the lockdown period. Lockdown related to COVID-19 has been initiated in India on 22 March 2020 imposing 14-h lockdown followed by Lockdown 1.0 from 24 March 2020 to 14 April 2020. This has been extended further as Lockdown 2.0 till 3 May 2020. In this article, the data between 22 March 2019 to 3 May 2019 and 22 March 2020 to 3 May 2020 were compared to monitor the AQI.

2 Sources of PM2.5 and PM10 Emissions in India

Transportation, industrial pollution and burning waste are the major sources of PM2.5 emission, and Brick Kilns contributes to 8–9% of industrial pollution in India. References reported that major PM10 emission existed from dust and burning waste having 45% and 17% source, respectively. Detailed data are shown in Fig. 2 [13–15].

Figures 3 and 4 show PM2.5 and PM10 emission comparison of 23 Indian polluted cities for Lockdown 1.0 and Lockdown 2.0 within the same period of last year's data. Average data have been compared for 23 most polluted Indian cities where some cities showing more than 50 % reduction in emissions. Ghaziabad has been reported as world's most polluted city [16] having 94.56 μg/m^3 of PM2.5 emission and 284.62 μg/m^3 of PM10 emission as average data collected for fixed period, i.e., 22 March 2019 to 3 May 2019. Surprisingly, it was noticed that for the same period in 2020 the PM2.5 and PM10 level reduced drastically. This data set pointing to the fact that lockdown could be a solution to control air pollution. Control of traffic and air pollutants from industries during the lockdown period and their impact on the air quality is evident from the comparison presented in Figures 2 and 3. Lockdown 1.0 and Lockdown 2.0 restricted Ghaziabad's PM2.5 emission to 58.43 μg/m^3 and PM10 emission to 123.65 μg/m^3. These data give us a rough estimation that approximately

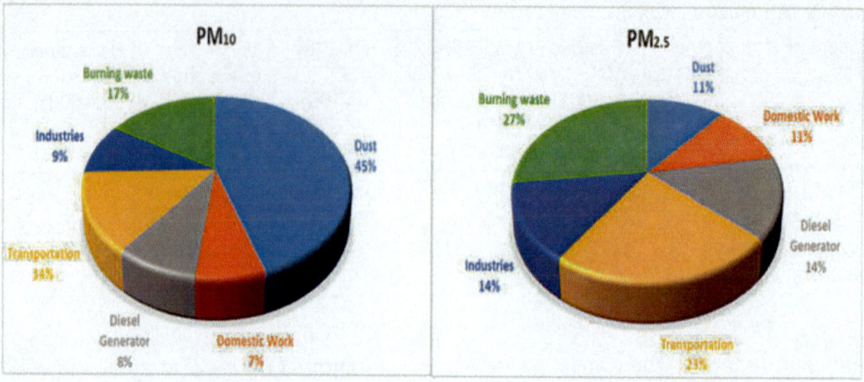

Fig. 2 PM10 and PM2.5 sources in India with their respective stats [13–15]

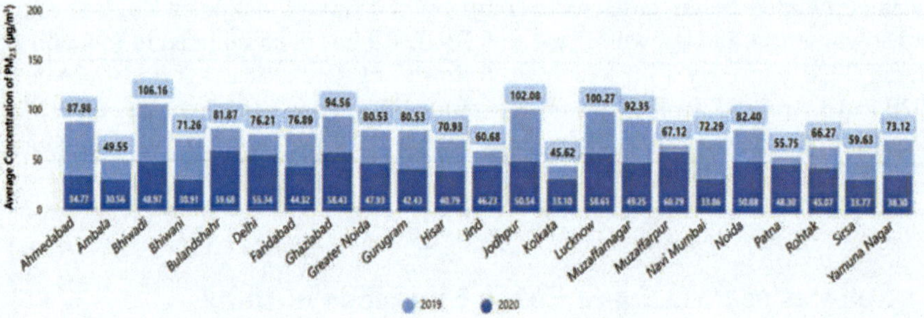

Fig. 3 Comparison of PM2.5 emission data between 22 March 2019 to 3 May 2019 and 22 March 2020 to 3 May 2020 in Indian major polluted cities [17]

38 and 57% reduction in PM2.5 and PM10 emissions took place in Ghaziabad, respectively.

Referring AQI equation and Table 1, the colour category was estimated of the city's air quality index. Based on the emission of the PM2.5 and PM10 the AQI for

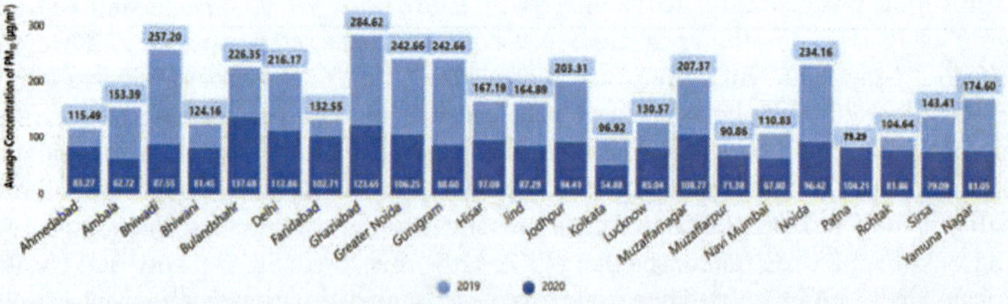

Fig. 4 Comparison of PM10 emission data between 22 March 2019 to 3 May 2019 and 22 March 2020 to 3 May 2020 in Indian major polluted cities [17]

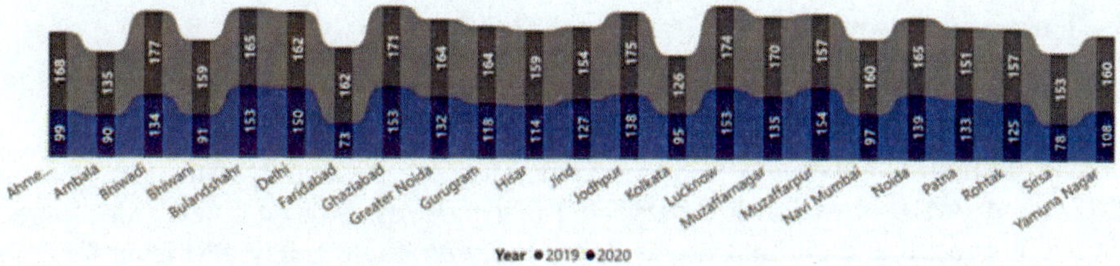

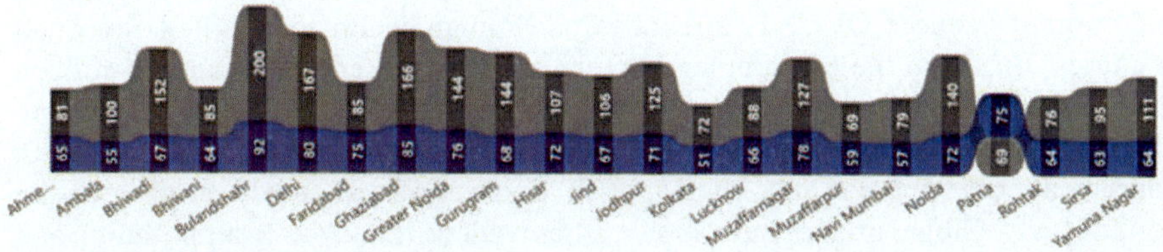

Fig. 5 Comparison of AQI of PM2.5 (above) and PM10 (below) for major polluted cities between 22 March 2019 to 3 May 2019 and 22 March 2020 to 3 May 2020 [17]

the polluted cities are compared and shown in Fig. 5. The data indicated the significant improvement in the air quality of all the cities posts lockdown due to restriction in almost all types of anthropogenic activities responsible for air pollution. While in the case of Patna city there has been variation observed in PM10 during the lockdown period which predominantly differs from the other cities. This is possibly indicating that in Patna the lockdown could not be implemented efficiently. As per the referenced Table 1, it could be said that the AQI of most of the cities shifted to unhealthy (red) to moderate (yellow) post-lockdown.

3 City Wise Data Comparison

Day-wise average PM emissions in major polluted Indian cities during 22 March–3 May 2019 and Lockdown 1.0, 2.0 are presented with Figs. 6, 7, 8, 9, 10, 11, 12, 13, 14, 15, 16, 17, 18, 19, 20, 21, 22, 23, 24, 25, 26, 27, 28 (See supplementary information presented at end). The data indicated that PM emissions reduced in the range of 25% and 65% due to lockdown in all the 23 Indian cities. These figures also indicated that some cities like Kolkata, Lucknow, Muzaffarpur and Bulandshahr are among the few cities which showed high PM level for the initial periods of lockdown whereas others showed high PM level at last few days. These pointing out how efficiently the lockdown was maintained day wise throughout the periods in different parts of India. These also indicate that the lockdown is not completely successful and more stringently the lockdown should be implemented in Indian cities to tackle the COVID-19 pandemic.

4 Lessons Learned and Proposal for Post-Lockdown Strategies

Green colour air quality index is categorized as a good and healthy zone. Last year's PM2.5 and PM10 emission level revealed that most of these polluted cities were marked in the red and purple zone as it referred with Table 1. The PM level for the same period this year showed that those cities air quality is mostly green, yellow and/or orange zone. COVID-19 affect badly to human health and GDP of almost all countries. But from the environmental point of view, this could be our ray of hope to clean our environment. Lockdown restricted traffic, industries and manufacturing units as people got more concerned about coronavirus transmission. Social distancing and stay at home policies have been applied all-over India as cases are still growing continuously. Global images have shown a tremendous reduction in air pollution and greenhouse gas emissions got minimal to a greater extent [2, 7, 8, 18]. Through this data analysis and comparison, we have shown our concern for global environmental conditions taking India as an example. Following points can be proposed for a fixed period every year once COVID-19 lockdown gets over.

a. 15-day vacation type lockdown can be implemented every six months imposing restrictions on personal vehicles and main polluting industries like brick kilns, etc. People will be advised to stay at their homes and to avoid unusual travel. This can help in emission reduction and nature can heel meanwhile. Effective planning will be needed here as the economy cannot be affected too much.
b. As an alternative personal vehicle, adopt cycling for short travels and public vehicles for long routes if needed. This will minimize the PM2.5 emission mainly and PM10 emission to some extent. Dust is the main source of PM10 emission, so unwanted crowd gathering and movement restriction during a fixed period will minimize its existence. Also, waste material dumping should be encouraged instead of burning it.
c. Vocational lockdown for a proposed period will state crowd/vehicle movement on hold and this will give time to municipal organizations for sanitizing or cleaning the area. Maintenance work can also be proposed more effectively.

However, implementation of these could create an economic crisis. Thus, proper legislation, a rigorous discussion is necessary to develop a sustainable economic crisis management strategy. Global fraternity and environmental policymaker should formulate strategies to minimize the economic crisis due to the propose lockdown.

5 Conclusion

Current pandemic crisis not only risk human health but also put economic growth on hold globally. However, the other side of the situation was found to be satisfying and way beyond our imagination. The comparison presented in this article indicated a significant improvement in the air quality of the previously known most polluted 23 Indian cities. Also, analysis of the day-by-day data pointed out that not all the cities implemented the lockdown in the same way. Nevertheless, this lockdown mother nature gets the much-needed time to heal itself. As it is a temporarily influence on the environment, but international/national fraternity and policymakers should take lessons from this lockdown situation to plan future strategies to reduce pollution.

Acknowledgements All the authors would like to thank the efforts of all the anonymous researchers and environmental professionals who collected and made available the data on air quality.

Supplementary Information

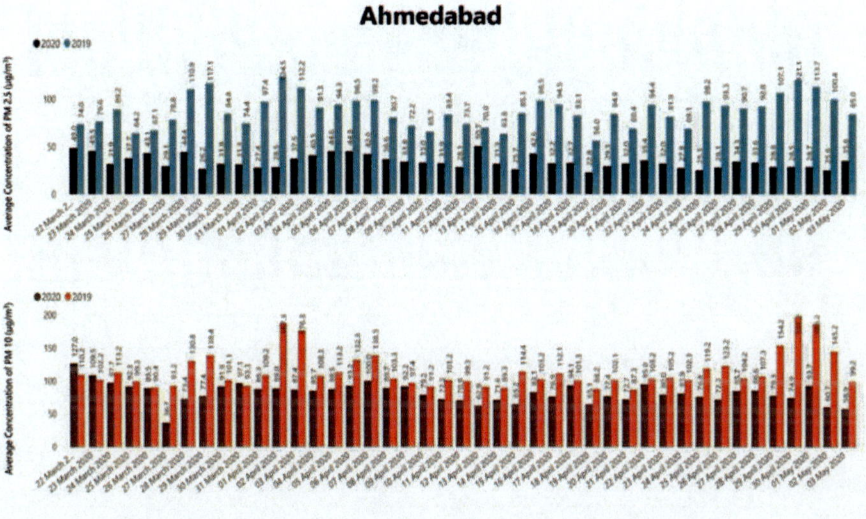

Fig. 6 $PM_{2.5}$ and PM_{10} emissions in Ahmedabad, India before and after lockdown

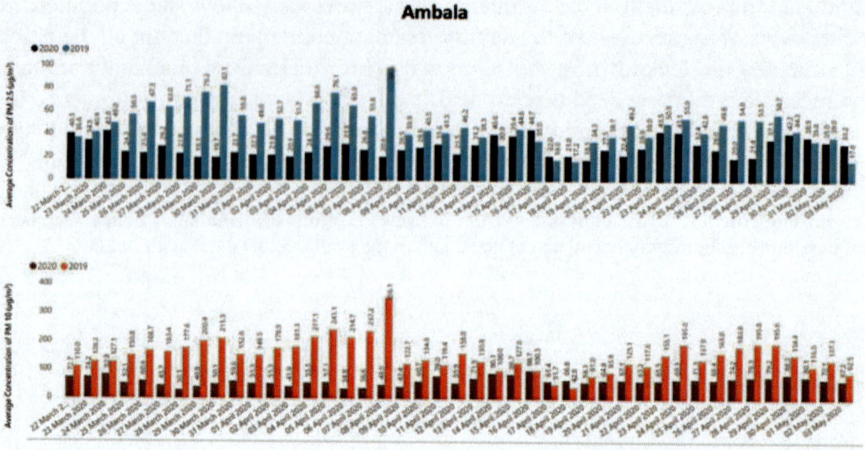

Fig. 7 $PM_{2.5}$ and PM_{10} emissions in Ambala, India before and after lockdown

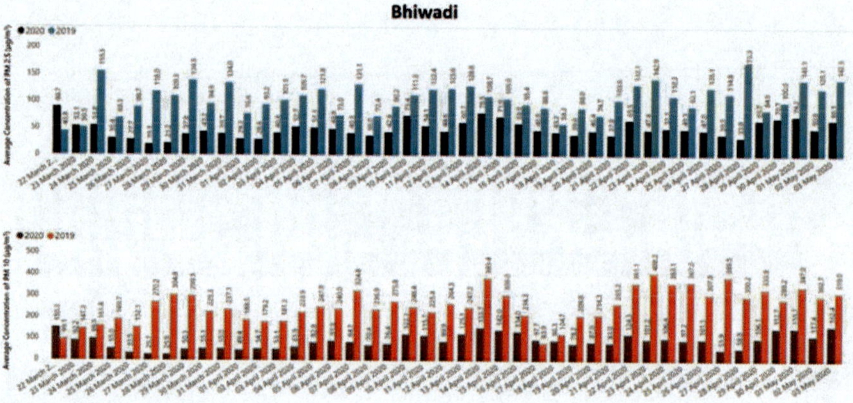

Fig. 8 $PM_{2.5}$ and PM_{10} emissions in Bhiwadi, India before and after lockdown

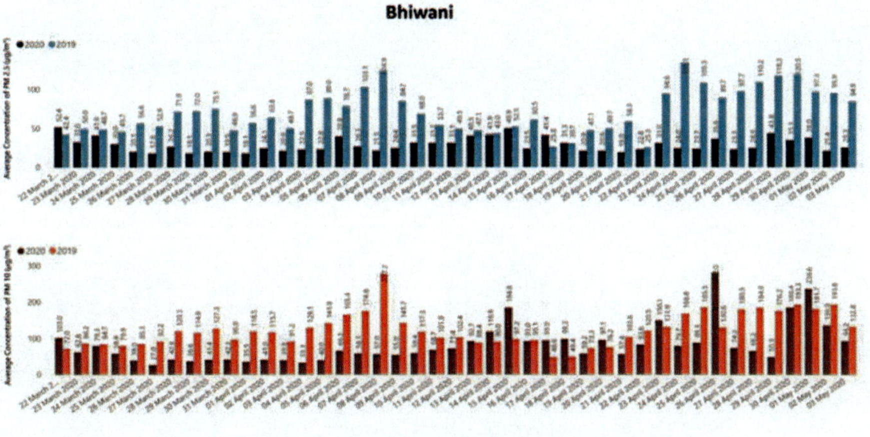

Fig. 9 $PM_{2.5}$ and PM_{10} emissions in Bhiwani, India before and after lockdown

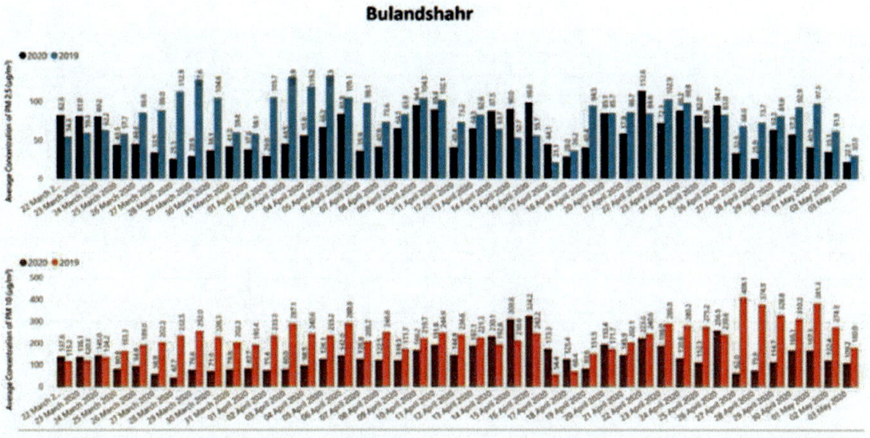

Fig. 10 $PM_{2.5}$ and PM_{10} emissions in Bulandshahr, India before and after lockdown

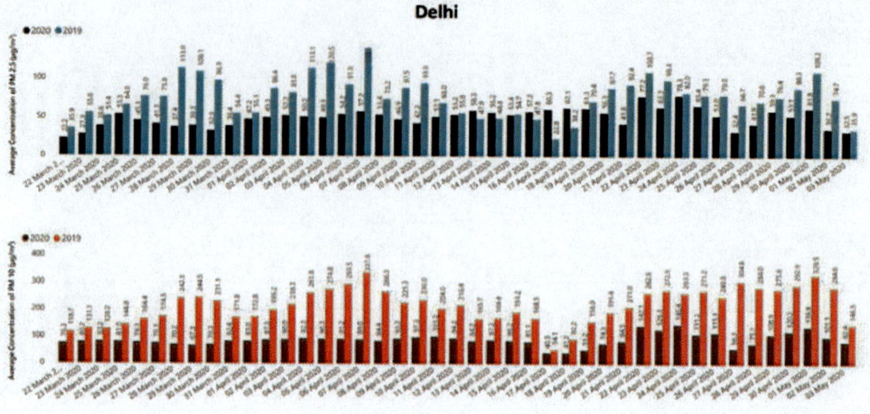

Fig. 11 $PM_{2.5}$ and PM_{10} emissions in Delhi, India before and after lockdown

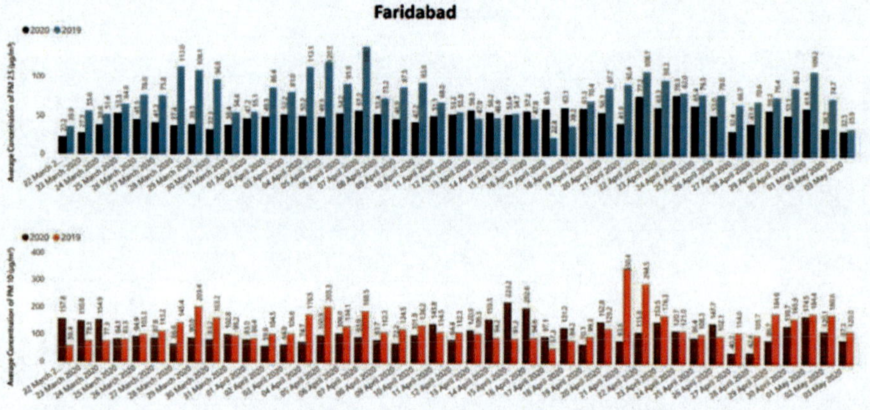

Fig. 12 $PM_{2.5}$ and PM_{10} emissions in Faridabad, India before and after lockdown

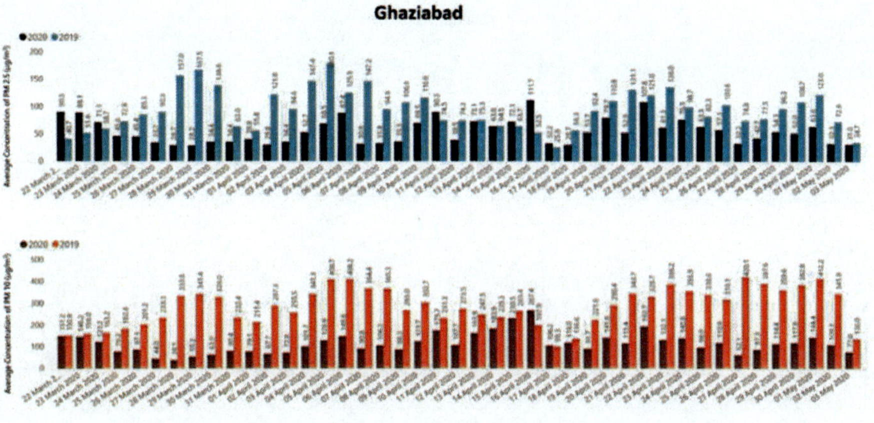

Fig. 13 $PM_{2.5}$ and PM_{10} emissions in Ghaziabad, India before and after lockdown

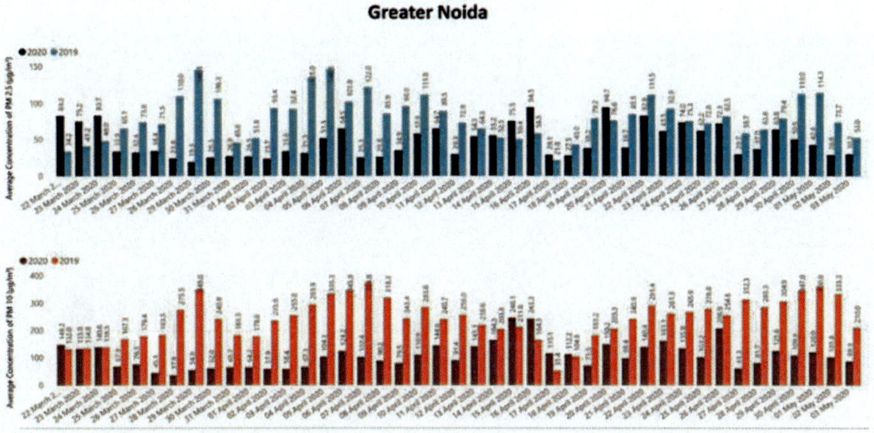

Fig. 14 $PM_{2.5}$ and PM_{10} emissions in Greater Noida, India before and after lockdown

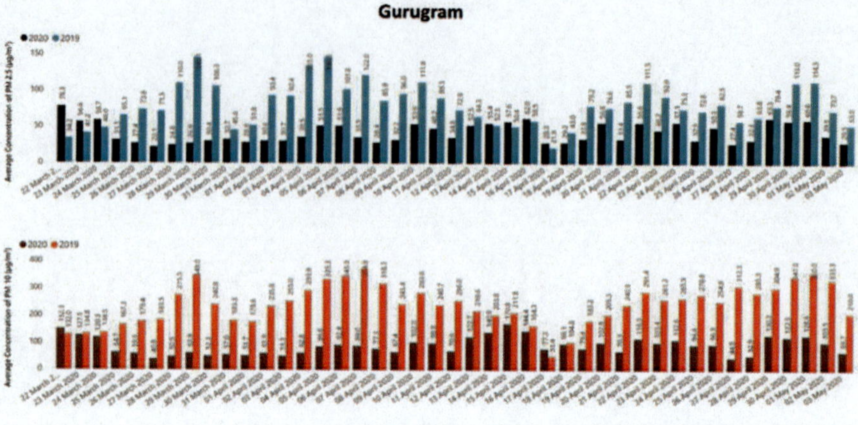

Fig. 15 $PM_{2.5}$ and PM_{10} emissions in Gurugram, India before and after lockdown

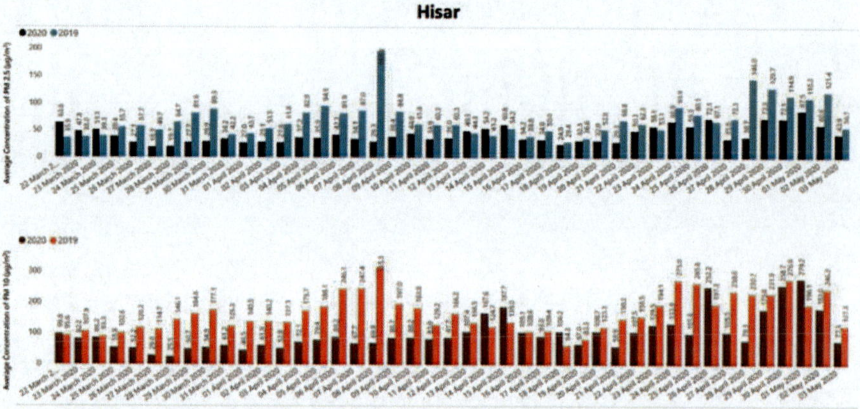

Fig. 16 $PM_{2.5}$ and PM_{10} emissions in Hisar, India before and after lockdown

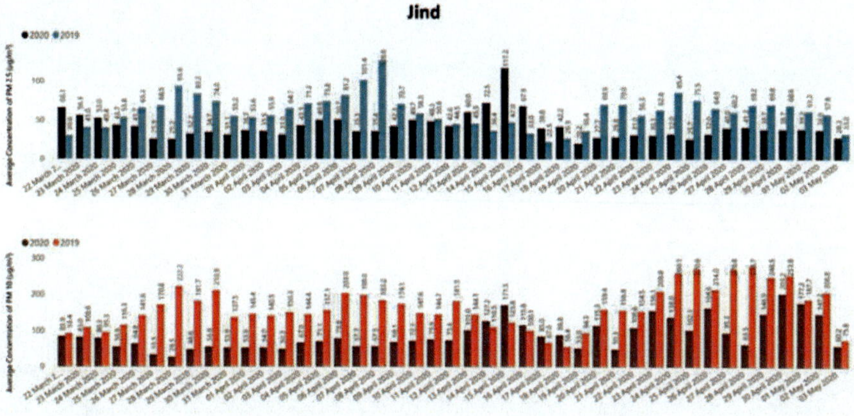

Fig. 17 $PM_{2.5}$ and PM_{10} emissions in Jind, India before and after lockdown

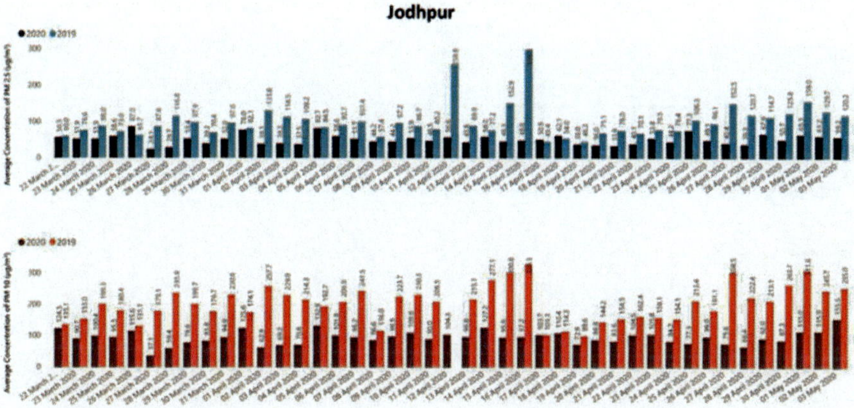

Fig. 18 $PM_{2.5}$ and PM_{10} emissions in Jodhpur, India before and after lockdown

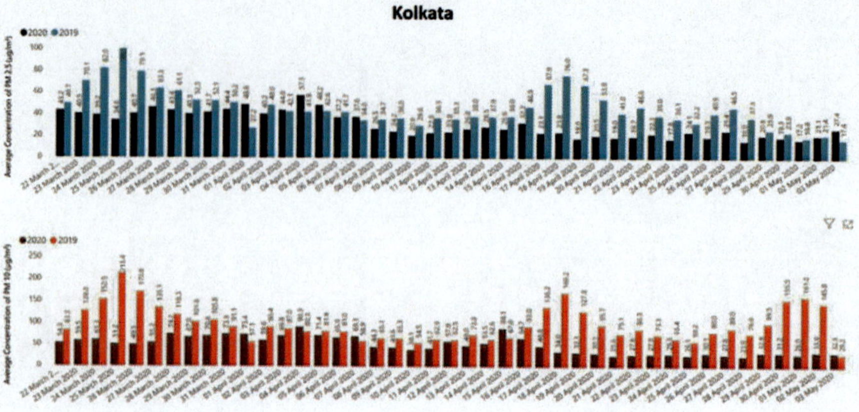

Fig. 19 $PM_{2.5}$ and PM_{10} emissions in Kolkata, India before and after lockdown

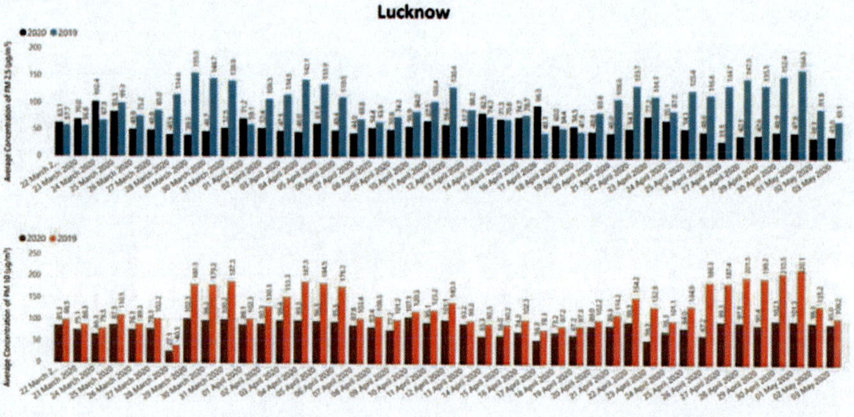

Fig. 20 $PM_{2.5}$ and PM_{10} emissions in Lucknow, India before and after lockdown

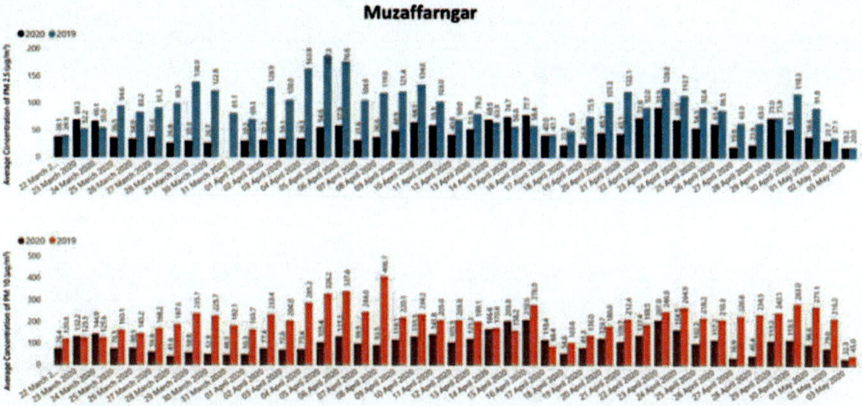

Fig. 21 $PM_{2.5}$ and PM_{10} emissions in Muzaffarnagar, India before and after lockdown

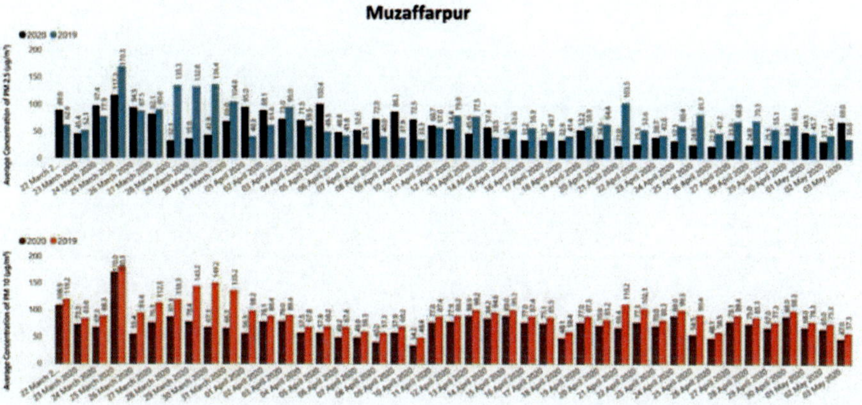

Fig. 22 $PM_{2.5}$ and PM_{10} emissions in Muzaffarpur, India before and after lockdown

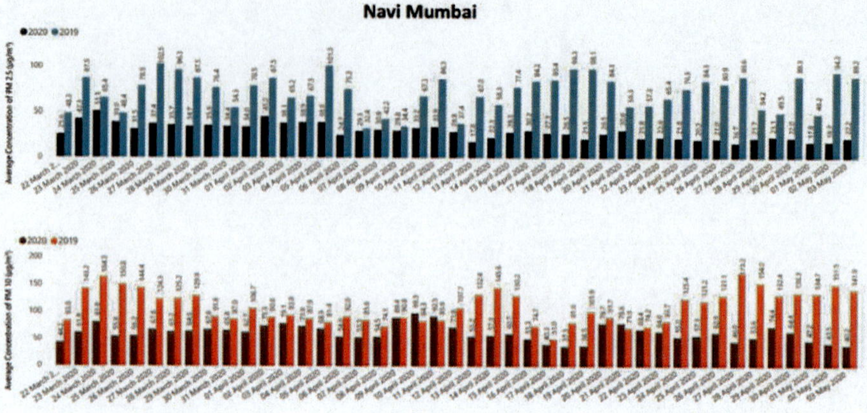

Fig. 23 $PM_{2.5}$ and PM_{10} emissions in Navi Mumbai, India before and after lockdown

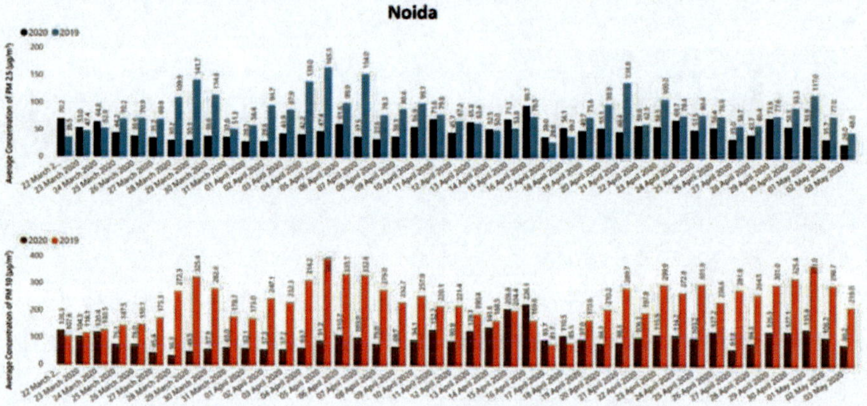

Fig. 24 $PM_{2.5}$ and PM_{10} emissions in Noida, India before and after lockdown

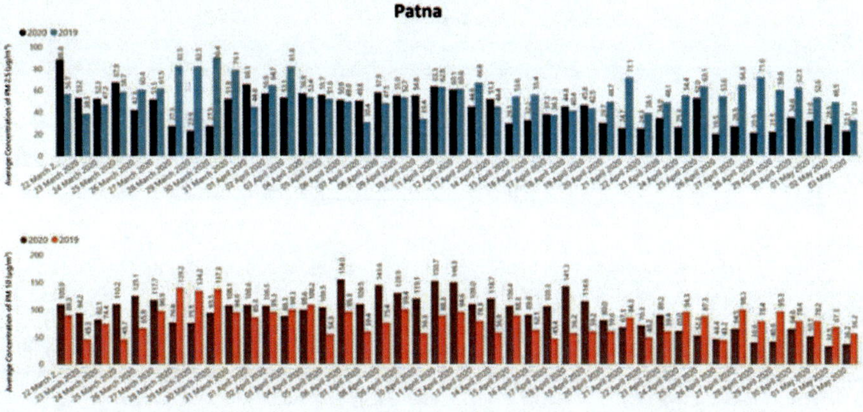

Fig. 25 $PM_{2.5}$ and PM_{10} emissions in Patna, India before and after lockdown

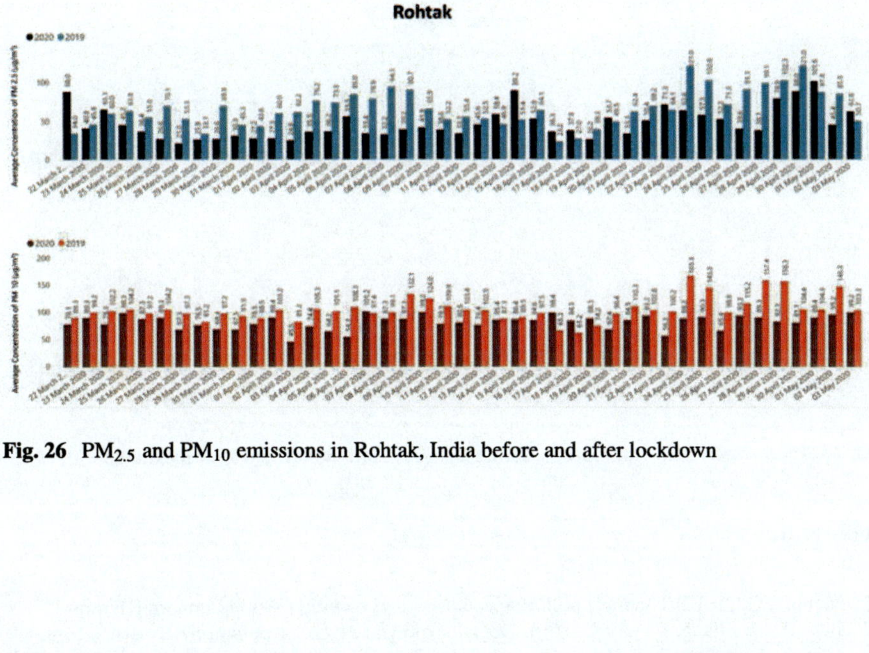

Fig. 26 $PM_{2.5}$ and PM_{10} emissions in Rohtak, India before and after lockdown

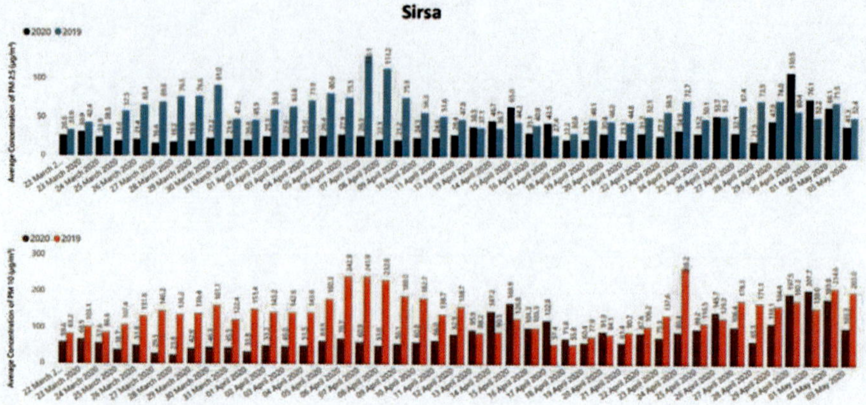

Fig. 27 $PM_{2.5}$ and PM_{10} emissions in Sirsa, India before and after lockdown

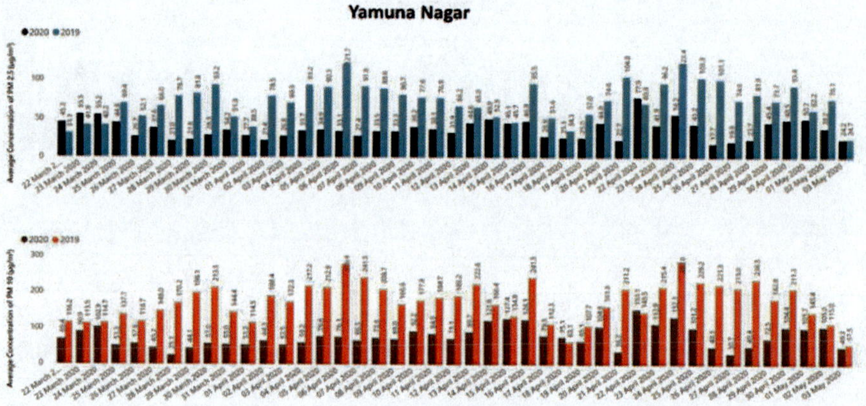

Fig. 28 $PM_{2.5}$ and PM_{10} emissions in Yamuna Nagar, India before and after lockdown

References

1. Moreno, G. (2017) The most polluting countries in the world | World Economic Forum.
2. Bao, R., & Zhang, A. (Aug. 2020). Does lockdown reduce air pollution? Evidence from 44 cities in Northern China. *Science of the Total Environment, 731*. https://doi.org/10.1016/j.scitotenv.2020.139052
3. Mahato, S., Pal, S., & Ghosh, K. G. (Aug. 2020). Effect of lockdown amid COVID-19 pandemic on air quality of the megacity Delhi, India. *Science of the Total Environment, 730*. https://doi.org/10.1016/j.scitotenv.2020.139086
4. Sharma, S., Zhang, M., Anshika, Gao, J., Zhang, H., & Kota, S. H. (2020). Effect of restricted emissions during COVID-19 on air quality in India. *Science* of the *Total Environment, 728*, 138878. https://doi.org/10.1016/j.scitotenv.2020.138878.
5. Wei, F. et al. (1999). Ambient concentrations and elemental compositions of PM10 and PM2.5 in four Chinese cities. *Environmental Science and Technology, 33*(23), 4188–4193. https://doi.

org/10.1021/es9904944
6. Salvador, S., & Salvador, E. (2012). Air quality communication workshop.
7. Li, L., et al. (2020). Air quality changes during the COVID-19 lockdown over the Yangtze River Delta Region: An insight into the impact of human activity pattern changes on air pollution variation. *Science of the Total Environment, 732*. https://doi.org/10.1016/j.scitotenv.2020.139282
8. Muhammad, S., Long, X., & Salman, M. (2020). COVID-19 pandemic and environmental pollution: A blessing in disguise? *Science* of the *Total Environment, 728*. https://doi.org/10.1016/j.scitotenv.2020.138820.
9. He, G., Fan, M., & Zhou, M. (2016). The effect of air pollution on mortality in China: Evidence from the 2008 Beijing Olympic Games. *Journal of Environmental Economics and Management, 79*, 18–39. https://doi.org/10.1016/j.jeem.2016.04.004
10. Muller, N. Z., & Mendelsohn, R. (2007). Measuring the damages of air pollution in the United States. *Journal of Environmental Economics and Management, 54*(1), 1–14. https://doi.org/10.1016/j.jeem.2006.12.002
11. Ghose, M. K., & Majee, S. R. (2001). Air pollution caused by opencast mining and its abatement measures in India. *Journal of Environmental Management, 63*(2), 193–202. https://doi.org/10.1006/jema.2001.0434
12. Paital, B., Das, K., & Parida, S. K. (2020). Inter nation social lockdown versus medical care against COVID-19, a mild environmental insight with special reference to India. *Science of the Total Environment, 728*. Elsevier B.V., p. 138914. https://doi.org/10.1016/j.scitotenv.2020.138914.
13. Chowdhury, Z. et al. (2007). Speciation of ambient fine organic carbon particles and source apportionment of PM2.5 in Indian cities. *Journal of Geophysical Research: Atmospheres, 112*(15). https://doi.org/10.1029/2007JD008386.
14. Guttikunda, S. K., Goel, R., & Pant, P. (2014). Nature of air pollution, emission sources, and management in the Indian cities. *Atmospheric Environment, 95*, 501–510. https://doi.org/10.1016/j.atmosenv.2014.07.006
15. What's Polluting Delhi's Air?, 2016.
16. IQAir, "World's Most Polluted Countries in 2019—PM2.5 Ranking," https://www.iqair.com/world-most-polluted-countries, 2019.
17. CPCB, "CPCB | Central Pollution Control Board," 2020.
18. Zhu, Y., Xie, J., Huang, F., & Cao, L. (2020). Association between short-term exposure to air pollution and COVID-19 infection: Evidence from China. *Science of the Total Environment, 727*,. https://doi.org/10.1016/j.scitotenv.2020.138704

Ergonomic Study in Domestic Sewing Machine

R. Dinesh and K. Muthukumar

1 Introduction

A sewing machine was used to join two pieces of cloth with the help of a thread. In the first industrial revolution to decrease manual sewing work, mechanical sewing machines were introduced by Englishman Thomas Saint in 1790. Mechanical sewing machines are single stitch type sewing machines. The increasing population leads to the development of sewing machines in order of garments growth. Still, the mechanical sewing machine is being used in the domestic sector because of its low cost, it stands as the backbone of the economically backward peoples (Fig. 1).

The nature of the operation is in such a way that, the operator is working with forwarding moved body posture, tilting head, bent trunk, and static sitting posture. During work, the sewer has to simultaneously move hands, arms side by operating the foot pedal continuously. The awkward posture of the lower extremities, upper extremities, and repetitive movements result in a higher prevalence of musculoskeletal complaints. NASA Task Load Index (NASA-TLX) questionnaire, which was a simple technique to identify perceived physical and mental workload, was used as a basic tool for the start of the study. Sealetsa and Thatcher [1] using a spinning cone and cushion can reduce MSD. The armrest on the chair reduces the ergonomic problem [2]. Scientifically designed sewing workstations reduce muscular-skeletal diseases [3]. Every part of the sewing operation is ergonomically designed to reduce work-related musculoskeletal disorder [4]. The designing workstation according to operator anthropometric value [5]. An ergonomic design chair [6] controls the forward bend position. Armrests on sewing machines [7] support arms. Field test on a sewing

R. Dinesh (✉) · K. Muthukumar
Industrial Safety Engineering, Department of Mechanical Engineering, Bannari Amman Institute of Technology, 638401 Sathyamangalam, Tamil Nadu, India
e-mail: Dinesh.is19@bitsathy.ac.in

K. Muthukumar
e-mail: muthukumarkm@bitsathy.ac.in

N. A. Siddiqui et al. (eds.), *Advances in Behavioral Based Safety*,
https://doi.org/10.1007/978-981-16-8270-4_11

Fig. 1 Awkward postures

machine by risk assessment and design ergonomically [8]. Providing ergonomics can increase productivity. It reduces fatigue [9]. Forcing people on different tasks and environments reduces productivity [10]. The ergonomic risk factors are contributing to musculoskeletal disorder [11].

In this many research papers were reviewed and from which we got to know variously ergonomically study techniques which had been done at the industrial level to reduce occupational health hazard but not at the domestic level.

2 Materials and Methods

2.1 Theoretical Background

Ergonomics study is done in order to study human capabilities in relationship to work demands.

Classification of Ergonomics
The ergonomics are broadly classified into three categories they are:

1. **Physical Ergonomics**:
   Physical activities related to anatomical, anthropometric, and physiological factors [12].
2. **Cognitive Ergonomics**:
   Mental workload like decision-making, skilled performance, and work stress factors [12].
3. **Organizational Ergonomics**:
   Communication, resource management, and design of working times [12].

Ergonomic Risk Factors
See Table 1.

Table 1 Ergonomic risk factors

Ergonomics factor	Example
Physical hazard	Static muscle load Awkward posture
Psychosocial	Cognitive stress Social relationship Psychological factors Lighting Noise
Organizational factors	Excessive work rates Duration of work Inadequate work break
Individual risk factor	Obesity, BMI, sex
Athletic activities/hobbies	Knitting and sewing Musical instrument

Rapid Entire Body Assessment (REBA)

REBA was the most popular and widely used observational ergonomic tool for postural assessment of jobs both in industrial and service sectors [13]. Different working postures can be measured easily and quickly. No formal ergonomic skill and no training are required to use this tool, only pen and pencil is required. The whole body evaluation in both static and dynamic work can be assessed by REBA.

The worksheet is divided into two sections as

- Section A—trunk, neck, and leg (Table A)
- Section B—upper arm, lower arm, and wrist (Table B)

Initially, find a table a (trunk, neck, and leg) and table B (upper arm, lower arm, and wrist) values. By adding load and coupling value to table A and table B we get Score A and Score B. Finally, from score A and score B, we obtain a score of C and then the activity score is added to get the final REBA score [14] (Fig. 2).

Based on the REBA score, the level of MSD risk was identified. The table shows the level of MSD risk and the required action to be taken (Table 2).

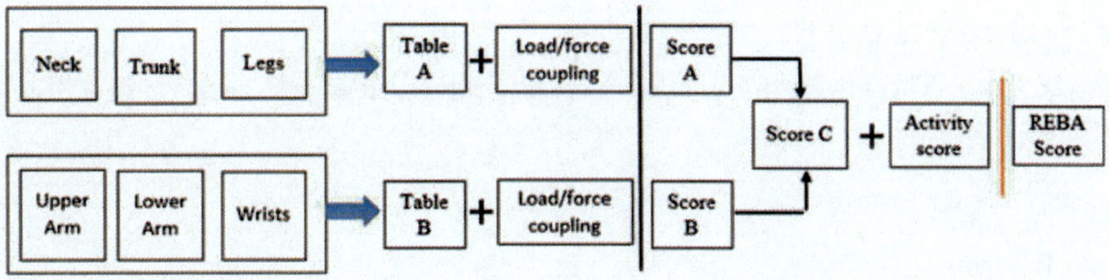

Fig. 2 Flow chart of REBA

Table 2 Risk level chart

Score	Level of MSD risk
1	Negligible risk, no action required
2–3	Low risk, change may be needed
4–7	Medium risk, further investigation, change soon
8–10	High risk, investigate and implement change
11+	Very high risk, implement change

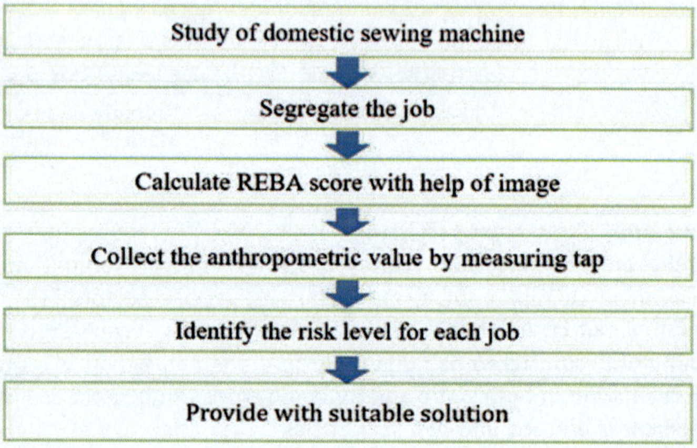

Fig. 3 Methodology

2.2 Methodology

See Fig. 3.

2.3 Evaluation

Ten different people who use the sewing machine for domestic purposes were taken for the study. The sewing job was divided into three categories based on function. they are.

1. Sitting
2. Threading
3. Tailoring

Rapid Entire Body Assessment (REBA) was ten people for the job segregated. Thus based on the job segregation, 30 different work postures at different angles were taken into consideration for a total of 10 different persons for this study.

Sitting

Totally ten postures were taken into consideration concerning 10 different people and the REBA score was evaluated.

A person 1 example for REBA calculation of sitting posture (Fig. 4).

Figure 5 shows REBA scoring for sitting posture. The value obtained from table A was 2 which indicates that the trunk value as 1, the neck value as 1 and for legs, its 2. The load/force coupling score as 0 thus the overall score A as 2. The value obtained from table B was 1 which indicates that the upper arm value as 1, lower arm value as 1, and wrist value as 1. The coupling score as 0 thus the overall score B as 2. From score A and score, B obtained the score C value was 1 hence the total REBA score, concerning activity score 0 was 1. The REBA score indicates that the job is done at a negligible risk level hence no action is required.

Fig. 4 Example for Sitting Posture

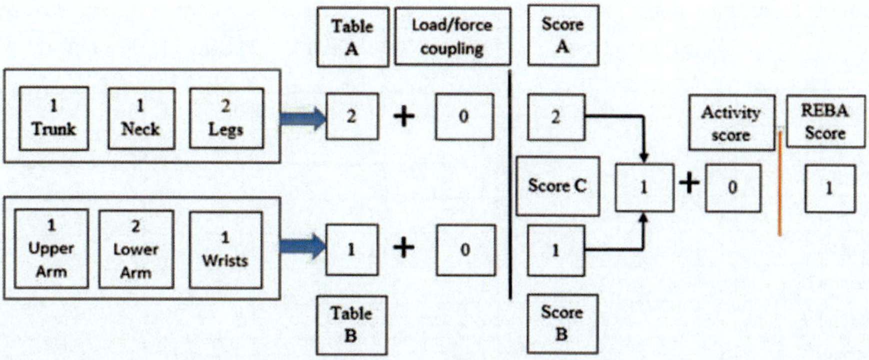

Fig. 5 REBA Score for person 1 sitting

The same procedure was repeated to identify the risk level for sitting posture of other persons also (Tables 3, 4 and 5).

From the study, it was analyzed that the average REBA score obtained for 10 people was 1 which means MSD risk level was a negligible risk and no action is required.

Threading

Totally ten postures were taken into consideration concerning 10 different people and the REBA score was evaluated.

A Sample example for REBA calculation of Threading posture (Fig. 6).

Figure 7 shows REBA scoring for threading posture. The value obtained from table A was 7 which indicates that the trunk value as 4, the neck value as 3 and for legs, its 2. The load/force coupling score as 0 thus the overall score A as 7. The value

Table 3 Score A for sitting

	Neck	Trunk	Leg	Table A	Load	Score A
Person 1	1	1	2	2	0	2
Person 2	1	1	2	2	0	2
Person 3	1	1	2	2	0	2
Person 4	1	1	2	2	0	2
Person 5	1	1	2	2	0	2
Person 6	1	1	2	2	0	2
Person 7	1	1	2	2	0	2
Person 8	1	1	2	2	0	2
Person 9	1	1	2	2	0	2
Person 10	1	1	2	2	0	2

Table 4 Score B for sitting

	Upper arm	Lower arm	Wrist	Table B	Load	Score B
Person 1	1	2	1	1	0	1
Person 2	1	1	1	1	0	1
Person 3	1	1	1	1	0	1
Person 4	1	1	1	1	0	1
Person 5	1	1	1	1	0	1
Person 6	2	2	1	2	0	2
Person 7	1	2	1	1	0	1
Person 8	1	2	1	1	0	1
Person 9	1	2	1	1	0	1
Person 10	1	2	1	1	0	1

Table 5 REBA score for sitting

	Score C	Activity	REBA
Person 1	1	0	1
Person 2	1	0	1
Person 3	1	0	1
Person 4	1	0	1
Person 5	1	0	1
Person 6	2	0	2
Person 7	1	0	1
Person 8	1	0	1
Person 9	1	0	1
Person 10	1	0	1

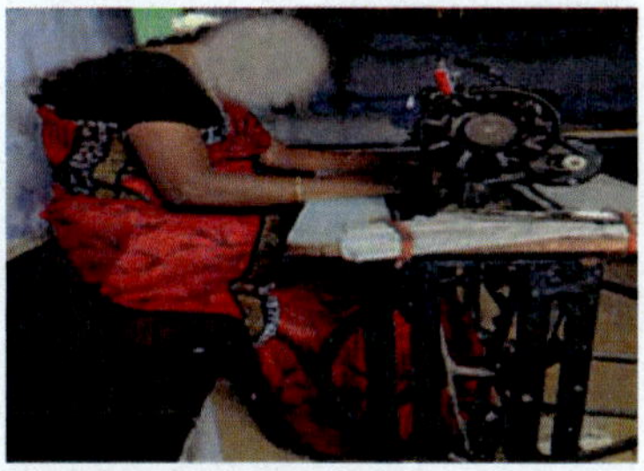

Fig. 6 Example for threading posture

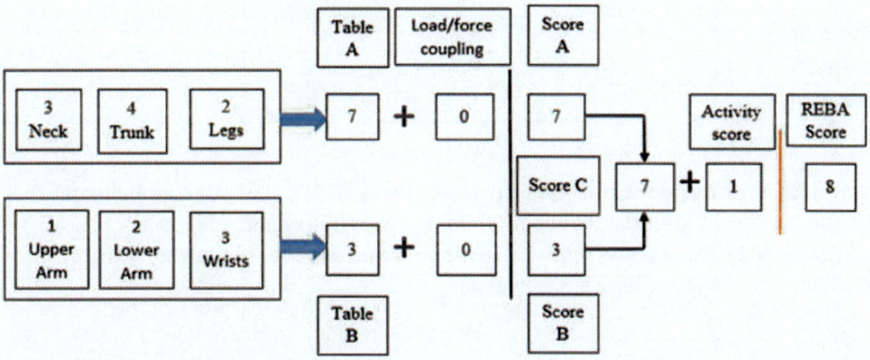

Fig. 7 REBA Score for person 1 Threading

obtained from table B was 3 which indicates that the upper arm value as 1, lower arm value as 2, and wrist value as 3. The load/force coupling score as 0 thus the overall score B as 3. From score A and score B obtained the score C value was 7 hence the

Table 6 Score A for threading

	Neck	Trunk	Leg	Table A	Load	Score A
Person 1	3	4	2	7	0	7
Person 2	3	3	2	6	0	6
Person 3	3	2	2	5	0	5
Person 4	3	3	2	6	0	6
Person 5	3	4	2	7	0	7
Person 6	3	4	2	7	0	7
Person 7	3	4	2	7	0	7
Person 8	3	4	2	7	0	7
Person 9	3	4	2	7	0	7
Person 10	3	4	2	7	0	7

Table 7 Score B for threading

	Upper Arm	Lower Arm	Wrist	Table B	Load	Score B
Person 1	1	2	3	3	0	3
Person 2	2	1	3	3	0	3
Person 3	1	1	3	2	0	2
Person 4	1	2	3	3	0	3
Person 5	2	2	3	4	0	4
Person 6	1	2	3	3	0	3
Person 7	1	2	3	3	0	3
Person 8	1	2	3	3	0	3
Person 9	1	2	3	3	0	3
Person 10	1	2	3	3	0	3

total REBA score, concerning activity score 1 was 8. The REBA score indicates that the job is done at high risk hence implement change immediately.

The same procedure was repeated to identify the risk level for the threading posture of other persons also (Tables 6, 7 and 8).

Table 8 REBA score for threading

	Score C	Activity	REBA
Person 1	7	1	8
Person 2	6	1	7
Person 3	4	1	5
Person 4	6	1	7
Person 5	8	1	9
Person 6	7	1	8
Person 7	7	1	8
Person 8	7	1	8
Person 9	7	1	8
Person 10	7	1	8

From the study, it was analyzed that the average REBA score obtained for 10 people for threading operation was 7.6–8 which means MSD risk level was a high risk, investigation, and implement change is needed immediately. In the threading process, the person was worse in their appearance. Therefore, it has a higher risk of musculoskeletal disorder. In this activity, investigate and implement changes to reduce the risk level of MSD.

Tailoring

Totally ten postures were taken into consideration concerning 10 different people and the REBA score was evaluated.

A Sample example for REBA calculation of tailoring posture (Fig. 8).

Figure 9 shows REBA scoring for tailoring posture. The value obtained from table A was 4 which indicates that the trunk value as 2, the neck value as 2 and for legs, its 2. The load/force coupling score as 0 thus the overall score A as 4. The value obtained from table B was 3 which indicates that the upper arm value as 1, lower arm value as 2, and wrist value as 3. The load/force coupling score as 0 thus the overall score B as 3. From score A and score B obtained the score C value was 4 hence the

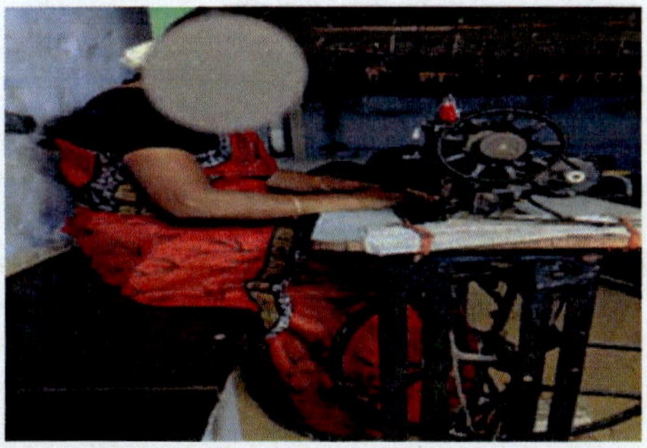

Fig. 8 Example for tailoring posture

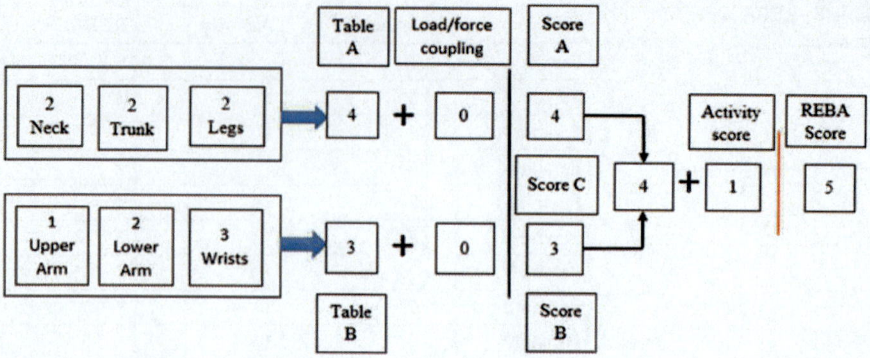

Fig. 9 REBA score for sample 1 tailoring

total REBA score, concerning activity score 1 was 5. The REBA score indicates that the job was done at medium risk and further investigation and changes to be bought soon to reduce MSD.

The same procedure was repeated to identify the risk level for tailoring the posture of other persons also (Tables 9, 10 and 11).

From the study, it was analyzed that the average REBA score obtained for 10 people for tailoring operation was 5.2–6 (approximate) which means MSD risk level is medium risk and further investigation and changes to be bought soon to reduce MSD.

Table 9 Score A for tailoring

	Neck	Trunk	Leg	Table A	Load	Score A
Person 1	2	2	2	4	0	4
Person 2	2	1	2	2	0	2
Person 3	2	2	2	4	0	4
Person 4	2	2	2	4	0	4
Person 5	2	3	2	5	0	5
Person 6	2	2	2	4	0	4
Person 7	2	3	2	5	0	5
Person 8	2	3	2	5	0	5
Person 9	2	2	2	4	0	4
Person 10	2	4	2	4	0	4

Table 10 Score B for tailoring

	Upper arm	Lower arm	Wrist	Table B	load	Score B
Person 1	1	2	3	3	0	3
Person 2	1	2	3	3	0	3
Person 3	3	2	3	5	0	5
Person 4	1	2	2	3	0	3
Person 5	1	2	3	3	0	3
Person 6	1	1	3	2	0	2
Person 7	1	2	3	2	0	2
Person 8	2	2	3	4	0	4
Person 9	1	2	3	3	0	3
Person 10	1	2	3	3	0	3

Table 11 REBA score for tailoring

	Score C	Activity	REBA
Person 1	4	1	5
Person 2	2	1	3
Person 3	5	1	6
Person 4	4	1	5
Person 5	4	1	5
Person 6	4	1	5
Person 7	4	1	5
Person 8	5	1	6
Person 9	4	1	5
Person 10	6	1	7

3 Results and Discussions

The study was about analyzing MSD risk level, for a total of 10 persons using the sewing machine for domestic purposes with job segregation as sitting, threading, and tailoring for 30 different postures using the REBA tool. The REBA score obtained for sitting posture for 10 different people was 1 hence the risk level was found negligible. The REBA score obtained for threading posture for 10 different people was 8 hence the risk level was found high. The most affected body part identified was the trunk due to bending while inserting thread in the needle. Thus to reduce the risk level magnifier was kept near the needle in the sewing machine. A sewing machine magnifier makes it easy to insert thread and examine fine details including stitch quality, without bending the trunk, thus reducing stress and strain. Thus finally the risk level would also be reduced (Fig. 10).

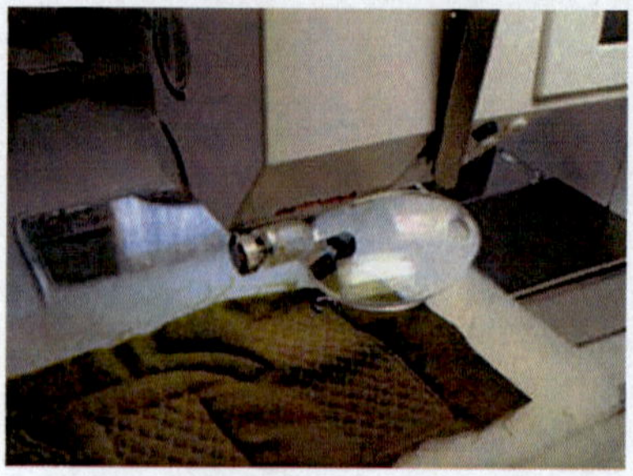

Fig. 10 Magnifier

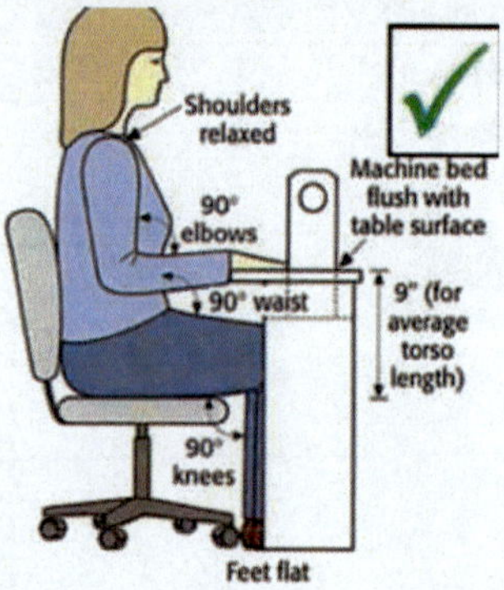

Fig. 11 Adjustable chair

The REBA score obtained for tailoring posture for 10 different people was 6 hence the risk level was found medium. A major job for sewer was tailoring thus further investigation to be done to reduce the medium risk level. The most affected body part was found to be the wrist. Therefore, either the sewing machine or chair should be of the adjustable type, in order to change concerning proper anthropometric values. In this study, the mechanical sewing machine was taken into consideration in which the table cannot be adjustable hence the chair should be of the adjustable type to reduce the risk (Fig. 11).

An adjustable chair was of high cost which cannot be afforded by all domestic sewers hence cushion can be provided to increase height such that the wrist is in proper alignment with the sewing machine.

4 Conclusion

The highlight of this study was to rise ergonomic awareness among the domestic sewer. The study majorly concentrates on economically poverty people. Following the Occupational Health and Safety Administration (OSHA) degree of freedom from risk and hazard in all environments [15], to provide a better working atmosphere for domestic sewer this study will be a better source on further use to reduce the MSD risk factor of people using the sewing machine. Thus suitable ways to reduce the risk level are,

- Develop good visibility to the needle area.
- To create an easy-to-operate foot pedal mechanism.
- Appropriate adjustable chair to work.
- Maintain space for movement while standing and sitting.
- Special arm supports to be attached.

References

1. Sealetsa, O. J., & Thatcher, A. (2011). Ergonomics issues among sewing machine operators in the textile manufacturing industry in Botswana. *Work (Reading, Mass.) 38*, 279–289. https://doi.org/10.3233/WOR-2011-1131.
2. Ahmed, F. (2018). Improving workplace safety through ergonomic intervention—A case study of a readymade garment industry. Bangladesh University Of Engineering & Technology.
3. Polaner, A., Leber, M., & Vujica Herzog, N. (2010). Muscular-skeletal diseases require scientifically designed sewing workstations. *Strojniški Vestnik—Journal of Mechanical Engineering, 56*(1), 31–40.
4. Parker, S. (2018). Ergonomic industrial sewing station. Publication in the Humber Digital Library (Open Access).
5. Kalınkara, V., Çekal, N., Akdoğan, I., & Kacar, N. (2011). Anthropometric measurements related to the workplace design for female workers employed in the textiles sector in Denizli, Turkey. *Eurasian Journal of Anthropology, 2*(2), 102–111.
6. Stevens, E. M., Jr. (2004). Design guidelines and evaluation of an ergonomic chair feature capable of providing support to forward-leaning postures. Texas A&M University.
7. MEWA Textil-Service AG & Co. Management OHG, Systematic ergonomic workplace design in sewing work, EU-OSHA—European agency for safety and health at work.
8. Hignett, S., & McAtamney, L. (2000). Rapid entire body assessment (REBA). *Applied Ergonomics, 31*(2), 201–205.
9. Mart and Telsang. (2011). Ergonomics. In *Industrial engineering and production management*, 12th ed. (pp. 157–174). Chand & Son Publisher.
10. Kolgiri, S., Hiremath, R., & Bansode, S. (2016). Literature review on ergonomics risk aspects association to the power loom industry. *IOSR Journal of Mechanical and Civil Engineering (IOSR-JMCE), 13*, 56–64. https://doi.org/10.9790/1684-13135664.
11. Bridger, R. S. (2003). Anatomy posture and body mechanics. In *Introduction to ergonomics*, 2nd ed. (pp. 31–57). Taylor & Francis.
12. Keyserling, W. M., Brouwer, M., & Silverstein, B. A. (1992). A checklist for evaluating ergonomic risk factors resulting from awkward postures of the legs, trunk, and neck. *International Journal of Industrial Ergonomics, 9*(4), 283–301. ISSN 0169-8141.
13. Middlesworth, M. (2015). A step-by-step guide rapid entire body assessment (REBA). Ergonomic plus, 2015. [Online]. Retrieved Jan 10, 2018, from www.Ergo-plus.com.
14. David, M. W., & Gloss, S. (1983). Introduction. In *Introduction to safety engineering*, 1st ed. (pp. 3–19). Wiley.
15. Bridger, R. (2003). Design of repetitive task. In *Introduction to ergonomics*, 2nd ed. (pp. 121–157). Taylor & Francis

Ergonomic Risk Assessment of Students in Digital Learning

Zachariah C. Eldho and K. Muthukumar

1 Introduction

College and school education delivery systems have a problem due to the COVID-19 lockdown, and online education has taken off in a big way. To manage during the lockdown, schools have been holding real-time online classes and those students that do not have the required infrastructure are suffering. Some progressive schools were already using the flipped classroom where the lessons are made available to students online and the classroom time was used for discussion only. This is being done online quite effectively. With classes online, children have to spend as much time as possible on computers, tablets, and mobile phones. Students working from home find they are tired of it, and they are facing many health challenges. Health problems that are seen in those who sit for a long time are therefore common even in children. Old problems have got aggravated and new ones have appeared. Dry eyes, neck pain, backache, pain in the joints of the hands, inability to relax and sleep through the night are the most common health problems that can occur when sitting for long periods of time.

Risk identification and risk rating become a risk assessment component of the risk management process to determine the major risks faced by an organization, project, or strategy. The reason for risk assessment is to address the identified risk at the basic or current level [29]. This study intends to explore the ergonomic risk assessment of students in digital learning. Ergonomics concerned with understanding the relationship between humans, technology, and organization to create an environment tailored to a user's physical needs [12]. The advantage of using ergonomics is important to the industry also, so safety and health evaluation should be the initial stride in this

Z. C. Eldho (✉) · K. Muthukumar
Industrial Safety Engineering, Department of Mechanical Engineering, Bannari Amman Institute of Technology, 638401 Sathyamangalam, Tamil Nadu, India

K. Muthukumar
e-mail: muthukumarkm@bitsathy.ac.in

N. A. Siddiqui et al. (eds.), *Advances in Behavioral Based Safety*,
https://doi.org/10.1007/978-981-16-8270-4_12

task [8, 23]. The motive of ergonomic risk assessment is to recognize the current or potential risks that could lead to MSD's and also to obliterate work-related health risks. When pointing out dangerous parts, some effort should be made to lessen them. Risk assessments can be gainful in assessing interaction in the place of work [2]. Musculoskeletal disorders (MSD) are associated with inflammation and damage to ligaments, tendons, muscles, and peripheral nerves [21]. Work environment exposures can cause or exacerbate these musculoskeletal disorders, it is essential to identify ergonomic exposures in risk assessment. Intensity, frequency, and duration are the most important indicators of exposure, so it is essential to consider the risk for the development of MSDS [2]. To get over these problems, the sitting position of the students was investigated. The Rapid Upper Limb Assessment (RULA) method is one of several suitable work posture analyses.

The RULA method, introduced in 1993, evaluates the function of static muscles and the force applied to the upper extremities [22, 35]. This assessment method is mostly used to determine the load on the musculoskeletal system due to the posture of the neck, waist, upper extremities, muscle function, and excess weight on the body [20, 35]. By avail of RULA method, we can Perform a preparatory analysis to determine worker risk affected by factors of injury also include posture, static muscle contraction, and repetitive motion Style, determine employment preferences hinged on injury risk factor also compare the impact of work progress on evaluation before and after application [1, 36]. The risk assessment study was conducted on 15 students by observing the videos and photos taken. Different postures are selected and RULA scores are given to assess the risk level. Some economics tips are provided for those posters, which will help reduce the risk.

2 Materials and Methods

2.1 Literature Review

At present, most research activities in the field of education focus on e-learning and the risk associated with it. Let's go to various research journals full of research papers focusing on different aspects of e-learning, as well as ergonomic risk assessment of desktop, laptop, or mobile users. As the digital platform plays an important role in the education system, a study was conducted by reviewing forty-seven published studies and research on digital education and learning then provides some useful tips for those who want to develop online courses [32]. The effectiveness, efficiency, and satisfaction of the introduction of electronic technologies in education have been evaluated among students for two consecutive years by using the ergonomic card, which can identify positive trends in the field of e-learning according to the evaluation [19]. To know the effectiveness of online learning, a study says that 92% of all distance learning and online education, distance online education is better and more effectual than traditional education. Approximately 3% of studies have proven that the face to face, traditional format is more effectual and showing about 4% mixed findings [28]. From the review of 38 documents, it is stated that e study has several benefits like learning from anywhere at any time, study without the same preconditions, speed and process studying for personal needs, collaborative learning saves time and expense along with personal study significantly, availability of fast results also learning with multimedia helps to conserve resources and reduce the environment and audio pollution [25]. Online learning can be interrupted when any five economic factors are unsatisfactory and the five ergonomic factors are physical, environmental, organizational, cognitive, and social ergonomic all of these are dominant factors in smoothening the student's online learning [34].

Different ergonomic studies take place in the school environment. A study was conducted by reducing school bag weight, correct poor seating in children, to reduce the musculoskeletal problems. The result shows some improvements in children and concluded that some ergonomic education program is effective in children to reduce musculoskeletal pain [6, 33]. Postural correction can also be done through exercise. There has been a significant reduction in neck pain and neck deformity in rural areas school-going children, proper correction was given through postural corrections and exercise [31]. Based on the Moyer projection phenomenon, it has been shown that prolonged seating leads to extensive lopsided trunk and scoliosis, as well as lumbar lordosis and kyphosis of the entire spine in school children [11]. To overcome the musculoskeletal problems faced by the students in school's researchers have suggested innovative and economical classroom chairs that meet all the basic needs of students in the classroom environment [2, 3].

Working on devices such as phones, tablets, laptops, and desktops can cause long-term physical problems for users. Using the RULA (Rapid Upper Limb Assessment) tool studies on the long-term use of smartphone users by collecting samples through videos, photographs, and also by providing musculoskeletal questionnaire, researchers mainly conduct the studies in school-aged children since they are mainly involved in it. The appearance of smartphone use among students is particularly vulnerable. In some studies, it is moderately dangerous while in others it is a high risk, the researchers point out that lying down posture is more dangerous than sitting posture [9, 24]. Risk assessment is done not only on smartphones but also on laptop users. Headache and back injuries are the most common symptoms correlated with the long-term usance of laptops and desktops [7].

2.2 Problem Description

Due to long-standing COVID-19 restrictions, students are reluctant to attend classes from home and face many health challenges. The old problems got worse and the new ones appeared. Common problems include dry eyes, neck pain, back pain, and inability to rest and sleep at night. Experts from many fields have a lot of advice on this subject. Tips from physiotherapists, gym instructors, yoga teachers, nutritionists, and psychotherapists are available online. Much of this advice is related to exercise, yoga, meditation, following a proper routine, eating and drinking the right things, managing your responsibilities well, and sharing them. While all these recommendations make sense, one basic problem remains unaddressed—the kind of equipment one is using during learning. Most students use phones, tablets, laptops, and desktops while studying. Used improperly, this device can cause long-term physical problems for users. Frequent mobile phone use also creates new problems. Home furniture like sofas, dining chairs, dining tables, and coffee tables are not intended for continuous use. They also complicate issues related to phones, tablets, laptops, and desktop usage.

Students in everyday studies spend a lot of time on desktops and laptops without thinking about the impact on their bodies. Stretching the wrists, drooping, sitting without leg support, and poorly placed monitors can make it difficult for the body to recognize on a daily basis. These methods can lead to musculoskeletal disorders or stress injuries, which can greatly affect health, leading to pain, muscle fatigue, hypoesthesia, paraesthesia, and loss of performance. In one study, 20% of college students experienced musculoskeletal problems while working on a laptop and desktops. Students occasionally experience various musculoskeletal health problems like headache (20%), wrist and arm injury (35%), shoulder and arm injury (10%), neck discomfort (15%), backache (15%), eye strain (30%) after using the laptop and desktops, which is stated in a study [14]. There is a link between musculoskeletal disorders (MSD) and poor posture alignment. Studies show that a student with poor posture can develop a variety of musculoskeletal problems [14, 15]. Improper laptop use can lead to forwarding flexion of the neck and head, which can lead to certain biomechanical reactions and then to abnormal conditions like kyphosis [4, 14].

The mobile phone causes health problems even during normal times, but during the lockdown, phone calls are long and it is always in use. While the laptop is a problem only for those who use it, the phone creates postural problems for everyone. The user bends his / her neck to look at the phone. The ergonomic risk level of smartphone users is very high. Most smartphone usage levels are at 91.5% aeronautical risk [13]. One study found that musculoskeletal disorders were 90% more common in the neck, 73.30% in the shoulder, 63.30% in the upper back, 36.70% in the wrist and arm, and 30% in the lower back. Musculoskeletal disorders are rare, almost 13.30% in the hips and thighs, 13.30% in the knees, 10% in the ankles and feet, and 6.70% in the elbows [27]. Neck injuries from smartphone slump, a study found that the force on the neck increases as the human head jumps forward [5, 9]. There are several disadvantages to the long-term use of a smartphone. Tendonitis of the thumb and forefinger is a condition where injury, warmth, redness, swelling pain, and deterioration will occur to one or more tendons. In addition to these problems, it can also affect the grip of users' fingers, wrists, and thumbs [9, 10]. Carpal tunnel syndrome is a disease, examples of activities that involve repetitive and difficult hand movements, such as typing for messages or internet browsing using mobiles, can cause numbness, tingling, and weak grip [9, 30].

The chair and table need to work well together for optimal results. A poorly matched table and chair is the main cause of poor posture. Coffee tables cannot be used for online learning with sofas or even while sitting on the floor. The anthropometric features of the users are essential for achieving various jobs that are safe and economical. If there are inconsistencies between human anthropometric data, tools, equipment, and furniture may cause this to decrease productivity, restlessness, accidents, biomechanical stresses, exhaustion, injuries, and cumulative impact. This is why a high percentage of students complain of neck and lower back pain as a result of health [26]. According to Mayo Clinic, Excessive seating deficiencies include obesity, high blood pressure, high blood sugar, body fat around the waist, and abnormal cholesterol—all of which can lead to metabolic syndrome. Overall sitting too much and sitting too long seems to increase the risk of death from heart disease and cancer. The seating includes not only the work area but also the dining table, sofa, and transport.

2.3 Methodology

An economist at the University of Nottingham in England, Lynn McTamney, and Professor E Nigel Colette was developed the Rapid Upper Limb Assessment (RULA) method. This is a posture sample device used to examine the risk associated with the neck, trunk, and upper eternities in individual workers. This tool uses a single page worksheet (Fig. 1) to assess the required body position, force, and repetition.

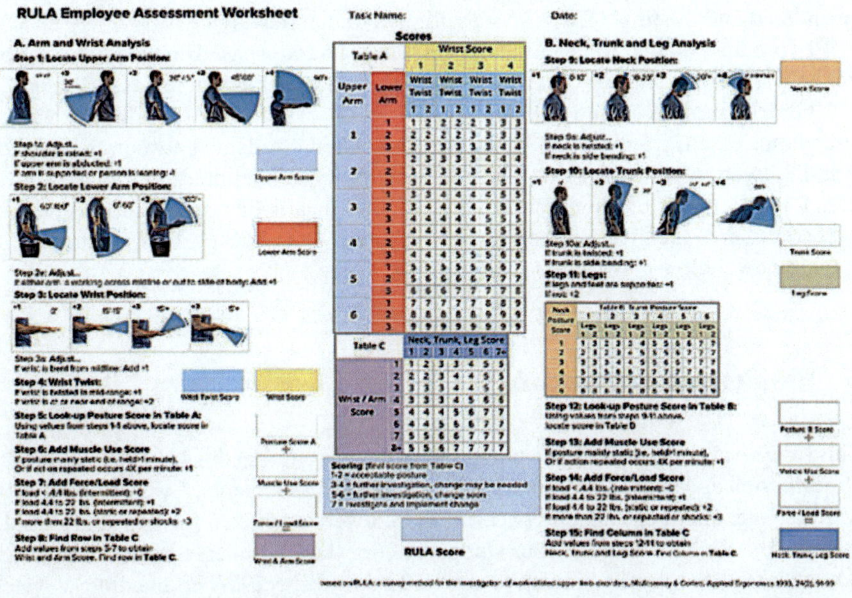

Fig. 1 Single page RULA worksheet (*Source* McAtamney & Corlett [22])

Score	Level of MSD Risk
1-2	neglibible risk, no action required
3-4	low risk, change may be needed
5-6	medium risk, further investigation, change soon
6+	very high risk, implement change now

Fig. 2 Score level of risk (*Source* McAtamney & Corlett [22])

Assessment scores are given for arm and wrist in A section and neck and trunk in B section [17]. Using this method there are three main steps for evaluating [18, 22].

Posture observation: Students sitting posture can be observed in different ways like direct observation, through videos, or by taking photos. The posture should be selected in such a way that the student posture is most repeated, which performs higher work tasks and the highest force load and those which are most harmful.

Giving scores: To evaluate postures, the angle between each part of the body and the vertical should be measured. While using the RULA method, we should consider only the upper limb not the lower limb but it will be considered to an extent. While working on it, load factors and muscle activity of that particular work also should be considered and the final RULA score will give to that posture.

Performance status: Appropriate action should be taken based on the RULA score. For understanding the level of risk RULA differentiates it into four levels [16] (Fig. 2).

The risk assessment study was conducted on 15 students by observing the videos and photos taken, from that 25 posters are evaluated. Students between the ages of 5 and 25 were selected for this study, especially those attending their classes on a digital platform in this pandemic situation. Other details like time interval, number of sections, etc., are given in the following Table 1. Out of 25 posture, a sample of 4 is mentioned below (Figs. 3, 4, 5 and 6).

3 Results and Discussions

This study evaluated students sitting in digital learning during this long-term COVID-19 restriction and to assess their risk level. Of the 25 participants, 5 were male and 10 were female, and those aged between 5 and 25 were evaluated for this study. Table 2 shows the RULA scores of four students out of the 25 students mentioned above, it contains all the details similar to the RULA worksheet. When handling devices such as mobiles and laptops, the force/load score is always zero because their weight

Table 1 Data's of participants

Number of subjects	Gender	Time/Section	Section/Day
1 (Fig. 3)	Male	1 h	2
2 (Fig. 4)	Female	1 h	3
3 (Fig. 5)	Male	1.5 h	3
4 (Fig. 6)	Female	2 h	3
5	Female	1 h	6
6	Male	1 h	3
7	Male	1 h	4
8	Female	1 h	3
9	Female	1 h	1
10	Female	2 h	1
11	Female	1 h	2
12	Female	3 h	1
13	Female	2 h	1
14	Female	2 h	1
15	Male	1 h	4

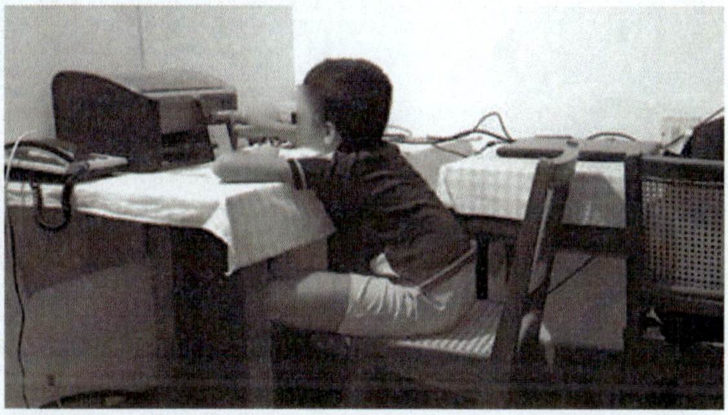

Fig. 3 Subject number 1

is less than 4.4 lbs., but this does not apply to 4 students with a laptop on the body which is mentioned in Table 3.

The consolidated rule score of 25 posers of 15 students is mentioned in Table 3. Out of 25 posture observed 11 posture have RULA score as 4, for 8 posture score as 5, for 4 posture score as 6, and its shown RULA score as 7 for 2 posture (Fig. 7). The results show that no student is in the right position, so the risk-free posture is 0%. 44% of postures show a low-risk, changes in those postures may be necessary and should be investigated in future. 48% of postures show a moderate-risk, which should be investigated and replaced as soon as possible. Those with a RULA score of 7 must implement the change, and 8% of the posture falls into this category (Fig. 8).

Out of fifteen participants observed, almost more than 50% of them fall under the high-risk category. Both male and female participants worked one to six hours

Fig. 4 Subject number 2

Fig. 5 Subject number 3

each day on their computer, laptop, and mobile on an average. The neck, trunk, and leg postures had a uniting effect on neck musculoskeletal disorder. In this case, it is essential that students and parents understand how to deal with the health problems caused by prolonged sitting also they should learn to sit properly and follow healthy rules. People who work from home and children who attend their online classes and who relax on their mobiles and tabs should definitely make the seat healthier, as improper sitting can cause neck discomfort, backache, shoulder, and knee injury. At home, students convert the space available to them into a classroom with limited

Fig. 6 Subject number 4

resources, while providing them with suitable seats and other facilities at the school. There is no other reason for health problems as the table and chairs are not made for long sitting activities.

The seat should conform to the natural curvature of the spine. Sitting upright is not good. The buttocks should rest on the back of the shore. The seat of the chair should be soft. Seating should be such that the body weight is shared equally on both sides. Leaning and tilting can cause back and lower back pain. Once a good office chair and table have become an important part of the home, provide children with a table and chair according to their height to sit in an online class. Proper cushioning should be provided for it. It is better to sit with a thin pillow on the curved part of the waist. Do not bend the shoulders forward or backward. The position of the shoulders when sitting should be the same as when standing.

When sitting in front of a computer, the height of the table and chair should be adjusted so that the neck is straight and the gaze is in the center of the computer. The table can be raised through some blocks and the computer can be raised with some books or boxes. Prolonged computer uses while sitting incorrectly can cause neck pain. Since computer and mobile use is common to all, it is important to exercise the neck for at least some time before the pain starts. Do neck exercises that strengthen the neck muscles. Students can also do a few light exercises to strengthen their muscles. The chin should not be close to the chest; the neck should bend forward

Table 2 RULA Scores of the students

Parameters	Subject 1 (Fig. 3)		Subject 2 (Fig. 4)		Subject 3 (Fig. 5)		Subject 4 (Fig. 6)	
	Right	Left	Right	Left	Right	Left	Right	Left
Upper arm position	4 (90° +)	4 − 1 = 3 (90° +)	2 (20–45°)	2 − 1 = 1 (20–45°)	3 − 1 = 2 (45–90°)	2 − 1 = 1 (20–45°)	1 − 1 = 0 (20°)	2 − 1 = 1 (20–45°)
Lower arm position	2 + 1 = 3 (100°+)	2 + 1 = 3 (100°+)	2 + 1 = 3 (100°)	2 + 1 = 3 (100°)	2 (100° +)	2 + 1 = 3 (100°+)	1 (60–100°)	1 + 1 = 2 (60–100°)
Wrist position	3 + 1 = 4 (15°+)	2 (15°)	3 + 1 = 4 (15°+)	3 (15°+)	2 + 1 = 3 (15°)	2 (15°)	2 + 1 = 3 (15°)	2 (15°)
Wrist twist	1		1		1		1	
Posture score (Table A)	6		5		4		3	
Muscle use score	0		0		0		0	
Force/Load score	0		0		0		0	
Wrist and arm score	6		5		4		3	
Neck position	4 (In extension)		3 (20°+)		2 + 1 = 3 (10–20°) (Side bending)		3 (20°+)	
Trunk position	3 (20–60°)		3 (20–60°)		2 (0–20°)		2 (0–20°)	
Legs	2 (Not supported)	2 (Not supported)	2 (Not supported)	2 (Not supported)	2 (Not supported)	2 (Not supported)	2 (Not supported)	2 (Not supported)
Posture score (Table B)	7		5		4		4	
Muscle use score	1		1		1		1	

(continued)

Table 2 (continued)

Parameters	Subject 1 (Fig. 3)		Subject 2 (Fig. 4)		Subject 3 (Fig. 5)		Subject 4 (Fig. 6)	
	Right	Left	Right	Left	Right	Left	Right	Left
Force/Load score	0		0		0		0	
Neck, trunk and leg score	8		6		5		5	
RULA score	7		7		5		4	
Level of MSD risk	Very high risk, implement change		Very high risk, implement change		Medium risk, future investigation, change soon		Low risk, future investigation, change may be needed	

Table 3 Consolidated RULA scores of the students

Number of subject		Posture score A	Force/Load score	Wrist and arm score	Posture score B	Force/Load score	Neck, trunk and leg score	RULA
Subject 1	Posture 1 (Fig. 3)	6	0	6	7	0	8	7
	Posture 2	4	0	4	5	0	6	6
Subject 2	Posture 1 (Fig. 4)	5	0	5	5	0	6	7
	Posture 2	4	0	4	4	0	5	5
Subject 3	Posture 1 (Fig. 5)	4	0	4	4	0	5	5
	Posture 2	3	0	3	5	0	6	5
Subject 4	Posture 1 (Fig. 6)	3	0	3	4	0	5	4
	Posture 2	3	0	3	3	0	4	4
Subject 5	Posture 1	3	0	3	3	1	5	4
	Posture 2	3	0	3	5	0	6	5
Subject 6	Posture 1	4	0	4	3	0	4	4
	Posture 2	4	0	4	5	0	6	6
Subject 7	Posture 1	4	0	4	4	0	5	5
	Posture 2	3	0	3	4	0	5	4
Subject 8	Posture 1	3	0	3	4	1	6	5
	Posture 2	4	0	4	6	0	7	6
Subject 9	Posture 1	3	0	3	4	0	5	4

(continued)

Table 3 (continued)

Number of subject		Posture score A	Force/Load score	Wrist and arm score	Posture score B	Force/Load score	Neck, trunk and leg score	RULA
	Posture 2	3	0	3	3	0	4	4
Subject 10	Posture 1	3	0	3	3	0	4	4
	Posture 2	3	0	3	5	0	6	5
Subject 11	Posture 1	3	0	3	3	0	4	4
Subject 12	Posture 1	3	0	3	5	1	7	6
Subject 13	Posture 1	3	0	3	3	1	5	4
Subject 14	Posture 1	2	0	2	5	0	6	5
Subject 15	Posture 1	2	0	2	3	0	4	4

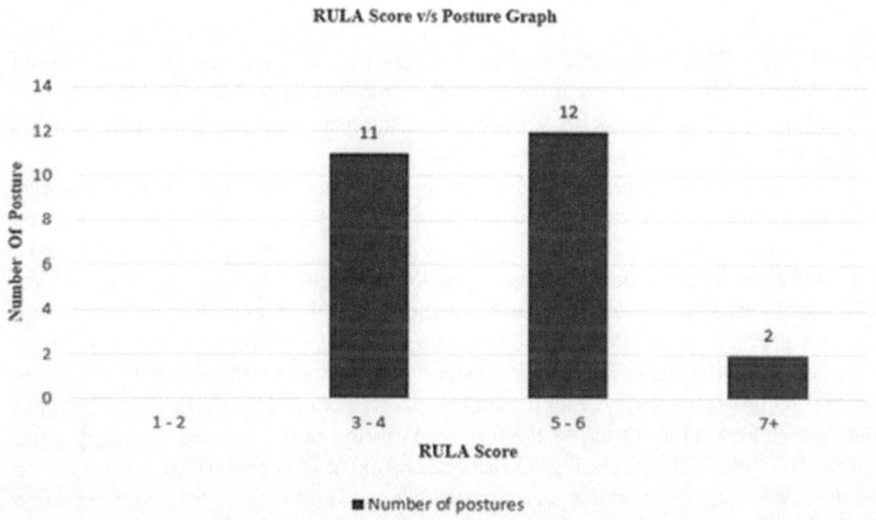

Fig. 7 RULA score v/s posture graph

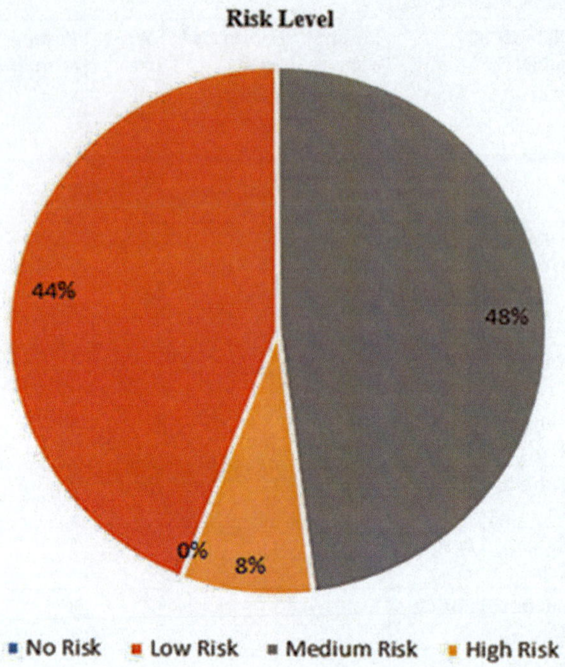

Fig. 8 The risk level of postures

in such a way. Bend alternately to either side by moving forward slightly. Now hold both hands together and clench the fists. The chin can be pressed into the fist for a while. No need to find a specific time for such exercises. This can only be done when sitting at regular intervals during the study.

In the best of circumstances, the laptop, whatever the brand, has an intrinsic ergonomic issue—it makes the user slouch. The keyboard and the display are too close to each other and one can optimize the height of the device either for the eyes or for the arms, but not for both. The display ought to be placed so that its top is level with your eyes. The keyboard should be an inch lower than your elbows. This should not be a problem if students are working on desktops or laptops in schools. Some experts say that the laptop can be used properly only when you lie down with a pillow for back support and place the laptop on your thighs, but it is hard to see the keyboard that way and even harder to use the mouse. Working in the bed creates many other problems as well. To get the right posture at work, it is necessary to lift the laptop display by 6–8 inches and use an external keyboard and mouse. Students can use a stack of books or a shoebox to raise the laptop and check the result instantly. Improper keyboard use can cause pain in the hands, knees, wrists, and fingers. Working on a desk with a keyboard and mouse can cause hand and knee pain. Those who type a lot should do short arm exercises every hour. Typing-mouse units on a laptop are on the same surface, which is not good for health at all. It is best to work with the keyboard and mouse attached.

Most students use their smartphone to attend their daily classes during this time, the mobile phone creates health issues even in normal times. The user bends his / her neck to look at the phone. The only way to use this device is to raise it to eye level so that the neck can be held straight while using the phone. One way to reduce phone usage is to get social media on computers. When sitting at a table, one can use easily available phone supports to keep the phone at the table. To help hold the phone up near eye level, students can use a selfie stick, with the bottom supported on the lap. Holds the phone up at a convenient level if they are sitting at a table.

Keep both legs firmly on the floor while sitting. Height-adjustable chairs are mandatory for those who sit for more than an hour in class. Even after adjusting the height the legs cannot be fixed to the ground, set a stand to place the legs so that the legs fall vertically from the knees. The knees should be raised straight or slightly. It is not good to lower the knees or stretch the legs forward or backward. Perform knee strengthening exercises before going to bed or getting up in the morning.

4 Conclusions

Most of the work of the new age can be done by sitting anywhere in the world. Since classes are online, children need to spend as much time as possible on computers, tablets, and mobile phones. Health problems that are seen in those who sit for a long time are therefore common even in children. Common health problems that can occur when sitting for long periods are neck pain, back pain, and pain in the joints of the hands. This study was conducted to dictate the severity of the risk in students studying on a digital platform in this contagious situation (COVID-19). A study was conducted on 15 students and 25 different sitting posters were evaluated during the study. For the study, Rapid Upper Limb Assessment (RULA), the most widely used monitoring method, is used to determine the risk in students. The results obtained after the study were unexpected. Of the 25 posture observed, not a single one appeared to be safe. About 44% of postures show a low-risk, 48% of postures show a moderate-risk, and 8% of them are showing a high risk. After assessing the risk, the students are given some economic tips which will help them to overcome the health problems they face during their studies, thus reducing the risk. This research may provide useful information for students who experience neck pain, back pain, or pain in the joints of the hands by the prolonged use of phone, tablet, laptop, and desktop during the study.

Researchers will be able to study these issues in detail as digital learning will last longer due to COVID-19 regulation, so the risk associated with digital learning is greatly increased if students use existing resources. A study can be conducted in designing new devices for holding laptops or mobiles. Designing efficient ergonomic chairs and tables for students while some of them are available in the market, the new design should overcome all the disadvantages of existing ergonomic chairs and tables. Since the high usage of mobile has much higher radiation problems than computers and laptops, a new device can be introduced to replace the mobile. All you need to do when designing or developing a new thing is to make it affordable not only for the rich but also for the poor students.

References

1. Adiputra, N., Sutjana, D. P., & Manuaba, A. (2000). Ergonomicss intervention in small scale industry in Bali, Dalam, In K. Y. Lim (Ed.), *Proceeding of the Joint Conference of APCHI and ASEAN Ergonomics*, Singapore.
2. Al-Hinai, N., Al Kindi, M., & Shamsuzzoha, A. (2018). An ergonomic student chair design and engineering for classroom environment. *International Journal of Mechanical Engineering and Robotics Research, 7*, 534–543. https://doi.org/10.18178/ijmerr.7.5.534-543.
3. Ansari, S., Nikpay, A., & Varmazyar, S. (2018). Design and development of an ergonomic chair for students in educational settings. *Health Scope, 7*(4). https://doi.org/10.5812/jhealthscope.60531
4. Arora, S., Gupta, M., & Gupta, V. K. (2012). Computer-related illnesses and Facebook syndrome: What are they and how do we tackle them?.
5. Berry, M. (2017). Cell phone ergonomics: How to avoid the "Smart Phone Slump".
6. Bettany-Saltikov, J., & McSherry, R. et al. (2019). School-based education programmes for improving knowledge of back health, ergonomics and postural behaviour of school children aged 4–18: A systematic review. *Campbell Systematic Reviews, 15*(1–2). https://doi.org/10.1002/cl2.1014.
7. Borhany, T., Shahid, E., Siddique, W. A., & Ali, H. (2018). Musculoskeletal problems in frequent computer and internet users. *Journal of Family Medicine and Primary Care, 7*(2), 337–339. https://doi.org/10.4103/jfmpc.jfmpc_326_17
8. Burdorf, A. (2010). The role of assessment of biomechanical exposure at the workplace in the prevention of musculoskeletal disorders. *Scandinavian Journal of Work, Environment & Health, 36*(1), 1–2.
9. Canaria, J., Croox, J., Dayao, A., Macatangay, J., Vinluan, R., & Mariano L. (2019). Ergonomic risk assessment for the prolonged usage of smartphones on students. https://doi.org/10.1007/978-3-319-94484-5_16.
10. De Quervains Tendonitis, www.physioadvisor.com.au/injuries/wrist-hand.
11. Drzał-Grabiec, J. et al. (2015). Effects of the sitting position on the body posture of children aged 11 to 13 Years. 1 Jan. 2015 (pp. 855 – 862).
12. Dul, J., Bruder, R., Buckle, P., Carayon, P., Falzon, P., Marras, W. S., et al. (2012). A strategy for human factors/ergonomics: Developing the discipline and profession. *Ergonomics, 55*(4), 377–395.
13. Eardley, R., Roudaut, A., Gill, J., & Stephen J. T. (2018). Investigating how smartphone movement is affected by body posture. Publication: CHI'18: Proceedings of the 2018 CHI conference on human factors in computing systems, Paper No.: 202, pp. 1–8. https://doi.org/10.1145/3173574.3173776.
14. Chavda, E., Parmar, S., & Parmar, M. (2013). Current practice of laptop computer and related health problems: A survey based on ergonomics. *International Journal of Medical Science and Public Health, 2*(4), 1024–1026. https://doi.org/10.5455/ijmsph.2013.120820131
15. Gerbaudo, L., & Violante, B. (2008). Relationship between musculoskeletal disorders and working posture. *Medicina del Lavoro, 99*(1), 29–39.
16. Gomez-Galan, M. et al. (2020). Musculoskeletal risks: RULA bibliometric review. *International Journal of Environmental Research and Public Health, 17*(12), 4354. https://doi.org/10.3390/ijerph17124354.
17. Chaudhary, H., & Singh, J. (2013). A literature review on MSDs using ergonomic body assessment tools: RULA and REBA, 2(8). ISSN 2277-8179.

18. Instituto de Ergonomía (INERMAP) Ergomet 3.0. Retrieved Jan 22, 2020, from http://www.inermap.com/software/ergomet.html.
19. Karamanska D. Y (2015). Ergonomics. In E- Learning, PhD student Todorova, M.V .2 University of Chemical Technology and Metallurgy, Bulgaria1 year XXIII, vol. 3, pp. 62–65.
20. Kee, D., & Karwowski, W. (2007). A comparison of three observational techniques for assessing postural loads in industry. *International Journal of Occupational Safety and Ergonomics, 13*, 3–14.
21. Middlesworth, M. (2014). Definition of Musculoskeletal Disorder (MSD) | ErgoPlus. Retrieved Jan 29, 2019, from https://www.ergo-plus.com/definition-ofmusculoskeletal-disorder-msd/.
22. McAtamney, L., & Corlett, E. N. (1993). RULA: A survey method for the investigation of work-related upper limb disorders. *Applied Ergonomics, 24*, 91–99.
23. Mohd. Nasrull Bin, & Rahman, A. (2014). Development of an ergonomic risk assessment tool for work postures; A thesis submitted in fulfilment of the requirements for the award of the degree of Doctor of Philosophy.
24. Mongkonkansai, J., Madardam, U., & Veerasakul, S. (2020). Smartphone usage posture (sitting and lying down) and musculoskeletal symptoms among school-aged children (6–12 years old) in Nakhon Si Thammarat, Thailand. https://doi.org/10.21203/rs.3.rs-51245/v1.
25. Somayeh, M., & Dehghani, M. et al. (2016). The effectiveness of E- learning in learning: A review of the literature. *International Journal of Medical Research & Health Sciences, 5*(2), 86–91.
26. Musa, A., Ismaila, O., Adejuyigbe, S. B., & Akinyemi, O. (2011). Ergonomic evaluation and design of tables and chairs for use in tertiary institutions in Nigeria.
27. Namwongsa, S., Puntumetakul, R., Neubert, M. S., Chaiklieng, S., & Boucaut, R. (2018) Ergonomic risk assessment of smartphone users using the Rapid Upper Limb Assessment (RULA) tool. *PLoS ONE, 13*(8), e0203394.
28. Nguyen, T. (2015). The effectiveness of online learning: Beyond no significant difference and future horizons. *MERLOT The Journal of Online Teaching and Learning, 11*, 309–319.
29. Peddada, K. (2013). Risk assessment and control. *Journal of Governance and Regulation, 2*. https://doi.org/10.22495/jgr_v2_i2_p4.
30. Postural Syndrome—Bad Posture, www.physioadvisor.com.au/injuries/neck-head/postural-syndrome/.
31. Perooru, R., Malarvilzhi, D., Prasanth, D., & Sivakumar, V. P. R. (2016). Effects of sitting posture modification and exercises in school going children with neck pain in rural area in Tamil Nadu. *International Journal of Clinical Skills., 10*. https://doi.org/10.4172/Clinical-Skills.1000105
32. Sun, A., & Chen, X. (2016). Online education and its effective practice: A research review. *Journal of Information Technology Education: Research, 15*. 157–190. https://doi.org/10.28945/3502.
33. Syazwan, A., Azhar, M. M., Anita, A., et al. (2011). Poor sitting posture and a heavy schoolbag as contributors to musculoskeletal pain in children: An ergonomic school education intervention program. *Journal of Pain Research, 4*, 287–296. https://doi.org/10.2147/JPR.S22281
34. Walker, R., Jansz, J., Bay, J., Paudel, N., Swapan, A., & Smith, R. (2018). How do ergonomic factors affect students' online learning in tertiary education? *World safety Journal, 17*(2), 26–33.
35. Yazdanirad, S. et al. (2018). Comparing the effectiveness of three ergonomic risk assessment methods-RULA, LUBA, and NERPA-to predict the upper extremity musculoskeletal disorders. *Indian Journal of Occupational and Environmental Medicine, 22*(1), 17–21. https://doi.org/10.4103/ijoem.IJOEM_23_18.
36. Yusuf, M. et al. (2016). The improvement of work posture using RULA (Rapid Upper Limb Assessment) analysis to decrease subjective disorders of strawberry farmers in Bali. *International Research Journal of Engineering, IT & Scientific Research, 2*(1). https://doi.org/10.21744/irjeis.v2i9.163.

A Short Review on the Development of Novel Face Masks During COVID-19 Pandemic

Gorakh Sawant, Kumar Gaurav, Gaurav Mittal, and Ashish Karn

1 Introduction

Global outbreak of disease, such as Viral hemorrhagic fevers, Zika virus epidemic, small pox, influenza including swine flu, Asian flu, Hong Kong Flu, Spanish flu etc., has been encountered by the humankind in the past. Currently world is facing COVID-19 pandemic, which is due to the spread of a coronavirus called SARS CoV-2. Coronaviruses are a family of RNA viruses which is responsible for causing diseases in mammals and birds. These viruses have been transferred from animals to human and have caused respiratory diseases ranging from mild diseases like common cold to severe diseases like SARS, MERS, and COVID-19. Coronavirus is of spherical structure with diameter around 125 microns. It has an envelope of diameter around 85 micron which is surrounded by 20 microns long spikes [1]. Coronavirus has around 74 spikes on the surface of its envelope [2].

Presently, eight categories of coronaviruses have been found which can cause illness to human i.e., HCoV-NL63, HCoV-229E, HCoV-OC43, HCoV-HKU1, SARS-CoV, MERS-CoV, SADS-CoV, and SARS CoV-2. Bat is the natural host for all the coronaviruses except HCoV-OC43 and HCoV-HKU1 which was first found in rats [3]. Generally, the coronaviruses are responsible for causing mild diseases such as cold, diarrhea etc., but SARS CoV, MERS CoV, and SARS CoV-2 were found responsible for severe fatal respiratory related diseases. Severe acute respiratory syndrome (SARS) was first detected in February 2003, during global outbreak of severe pneumonia associated with human deaths. Middle East Respiratory Syndrome (MERS) was started from Saudi Arabia in 2012. At present, the world is facing the outbreak of SARS CoV-2. The virus attacks the upper respiratory system of human,

G. Sawant · K. Gaurav · G. Mittal · A. Karn (✉)
Department of Mechanical Engineering, School of Engineering, University of Petroleum and Energy Studies, 284007 Bidholi, Dehradun, India
e-mail: akarn@ddn.upes.ac.in

N. A. Siddiqui et al. (eds.), *Advances in Behavioral Based Safety*,
https://doi.org/10.1007/978-981-16-8270-4_13

which can also be fatal in some cases. The disease is named by WHO as COronaVIrus Disease 2019 (COVID-19). Initial cases of COVID-19 were reported in Wuhan, China in Dec 2019.

During various pandemics, governments around the world have taken serious protective measures to stop the transmission of virus among humans. Maintaining social distancing, proper hand hygiene, use of face mask, and proper cough etiquettes are proven to be effective and advised by global health organizations such as WHO and CDC. Face masks play key role in minimize the transmission of virus from one person to another where social distancing cannot be maintained. Government of various countries announces the compulsorily use of masks in public places. Masks have proven to be very effective in restricting the virus from reaching respiratory track via mouth or nose. Mask only allow air particles to go through it, and the virus gets stuck at the outer surface of mask.

Mask uses various air filtration mechanism such as straining, inertial impaction, interception, diffusion, and electrostatic attraction. Straining, inertial impaction, and interception are the dominant collection mechanisms for particles greater than 0.2 microns whereas diffusion is dominant for particles less than 0.2 microns. Straining occurs when the gap between the media members is kept smaller than the diameter of the particulate, therefore, these particles can be captured outside the mask as shown in Fig. 1a. Inertial Impaction uses a rapid change in air direction and the principles of inertia to separate the particulate from the air stream. Particles at a certain velocity tend to remain at velocity and travel in a continuous direction as shown in Fig. 1b. This principle is normally applied when there is a high concentration of coarse particulate. Figure 1c shows the phenomenon of interception in which the particulate makes physical contact and remain attached to the media fiber. The particulate which is intercepted is smaller and its inertia is not strong enough to cause the particle to continue in a straight line. It, therefore, follows the air stream until it comes into contact with a fiber. The phenomenon of diffusion is shown in Fig. 1d in which the gas molecules collide with the smallest particles, especially those below 100 microns in diameter, which are thereby impeded and delayed in their path through the filter, increasing the probability of capturing of that particle. Electrostatic attraction, also called electrostatic precipitation involves using certain resins, waxes, and plastics as coatings on the filter material to attract particles with an electrostatic charge that holds them on the filter surface as shown in Fig. 1e. In most of the masks, a layer of electret is kept to filter the substrate by electrostatic attraction. This technique is used for the removal of very fine particulates such as dust and smoke. [4].

The face masks can be broadly divided into three different categories: (i) cloth face masks, (ii) surgical mask, and (iii) N95 mask.

i. Cloth mask: Cloth mask can be made at home with the use of available cloth material. These masks are helpful restricting an infected person to transmit their viruses during talking, coughing, or sneezing to another person or nearby objects. However, it possesses very low effectiveness in restricting the virus to reach wearer nose or mouth. Still wearing a cloth face mask is a far better choice than not using any mask. Due to sudden increase in the demand of N95

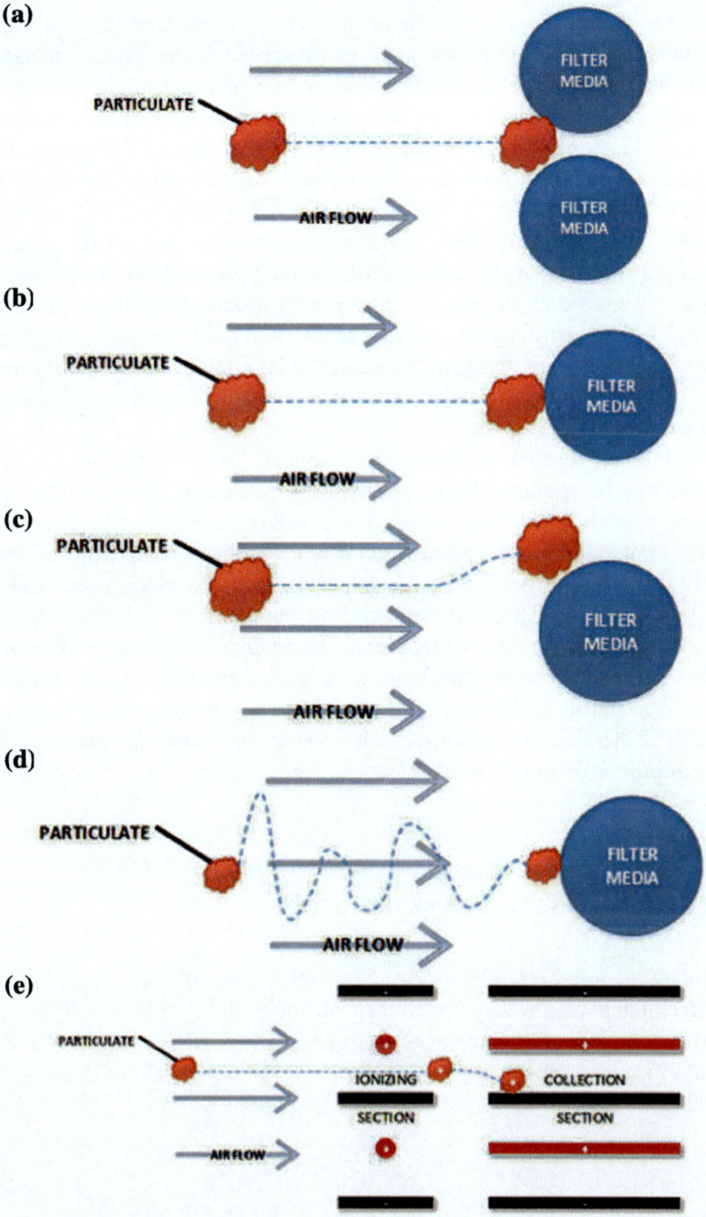

Fig. 1 A schematic demonstrating various air filtration phenomenon: **(a)** Straining, **(b)** Inertial impaction, **(c)** Interception, **(d)** Diffusion, and **(e)** Electrostatic attraction [4]

respirator, CDC has recommended the use of cloth face mask by people in public areas, so that N95 masks can be preserved for healthcare professionals.

ii. Surgical Mask: These masks are also known as medical mask or procedure mask. It is supposed to be wore by healthcare professionals during surgery in order to stop bacteria in liquid droplets and aerosols emerged from the surgeon's mouth or nose to enter the patients' open wound. These masks cover nose, mouth, and chin of the wearer. Although these masks are better than a cloth mask still very less effective in preventing the wearer from getting infected because of its inability to filter small particles and lose fitting causing leakage from sides. Its main purpose is to restrict the spread of large virus loaded droplets coming out from the mouth of an infected wearer during cough or sneeze. A surgical mask with an electret layer also has the ability to capture small particles by electrostatic attraction and thus is more effective than the one without electret layer.

iii. N95 Masks: These are circular or oval shaped masks with a tight fit to restrict air leakage from sides. These masks are made up of tough yet flexible non-woven polypropylene fiber. These masks can remove 95% of the contaminated particles of size more than 3 micron if wore properly. These masks have been proved very infective in protecting the wearer from getting infected. Due to limited supply and high demand during pandemic, these masks are recommended only for healthcare workers. The major disadvantage of these masks is that, it makes breathing difficult due to less supply of oxygen and hence can cause discomfort to the wearer, if used for a longer duration of time. Another variant of N95 masks with exhalation valve for quick dissipation of exhaled air minimizes breathing related issues (Fig. 2).

2 Effect of Coronaviruses (SARS-CoV, MERS-CoV, and SARS-CoV-2) Across the Globe

Coronaviruses are a family of RNA viruses which is responsible for causing diseases in humans related to respiratory system [3]. Although there were various coronavirus have been found in the past three coronavirus namely SARS-CoV, MERS-CoV, and SARS-CoV-2 caused serious impact on society. The details about these viruses are shown in Table 1.

3 Influence of Current Pandemic COVID-19 Across the Globe

Present pandemic COVID-19 is due to the spread of SARS CoV-2. This virus can be transferred from one person to other through respiratory droplet and contact routes [12]. Transmission of infected droplets can occur when a person is in close contact

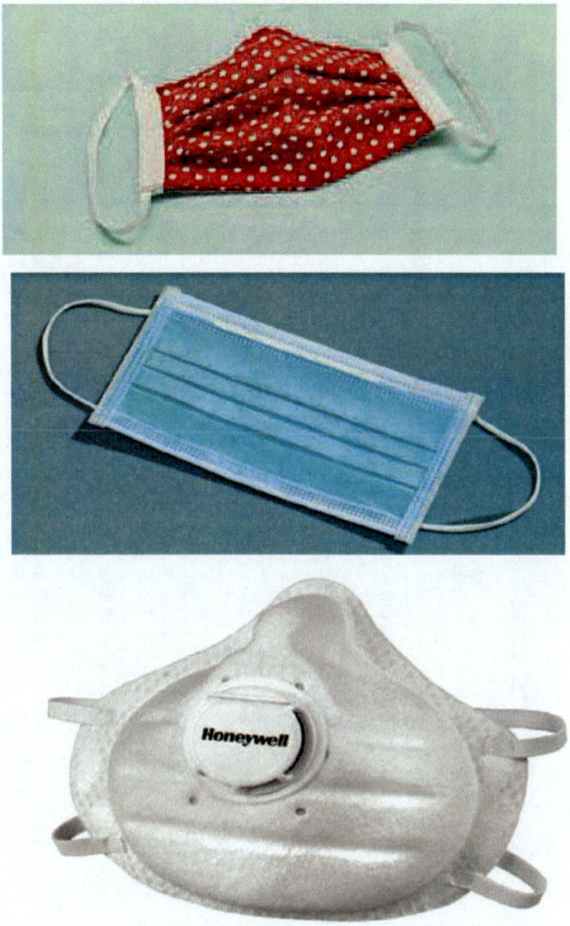

Fig. 2 (**a**) Cloth mask, (**b**) Surgical mask, and (**c**) N95 mask [5–7]

(less than 1 m) with an infected person during coughing or sneezing. These droplets can enter into the body of a healthy person though mouth, nose, and eyes. Transmission may also occur through indirect contact with surfaces in the immediate environment or with objects used on the infected person [13]. The life span of SARS CoV-2 at different surface is given in Table 2. Although the virus Present pandemic COVID-19 is due to the spread of SARS CoV-2. This virus can be transferred from one person to other through respiratory droplet and contact routes [12]. Transmission of infected droplets can occur when a person is in close contact (less than 1 m) with an infected person during coughing or sneezing. These droplets can enter into the body of a healthy person though mouth, nose, and eyes. Transmission may also occur through indirect contact with surfaces in the immediate environment or with objects

Table 1 Details of SARS-CoV, MERS-CoV, and SARS-CoV-2 [8–11]

Virus	SARS CoV		MERS CoV		SARS CoV-2	
First case	2002		2012		2019	
Last case	2003		Few (17) cases were last reported in Jan 2020		Ongoing	
Origin country	China		Saudi Arabia		China	
Hosts	Natural host	Bat	Natural host	Bat	Natural host	Bat
	Intermediate host	Civet cat	Intermediate host	Dromedary camels	Intermediate host	–
Mode of transmission	Person to person through respiratory droplets and close contact		Person to person through respiratory droplets and close contact		Person to person through respiratory droplets and close contact	
Symptoms	Fever, malaise, myalgia, headache, diarrhea, and shivering		Fever, chills, myalgia, cough, shortness of breath, nausea, vomiting, and diarrhea		Fever, cough, shortness of breath, and headache	
Incubation Period	Range	1–13 days	Range	2–15 days	Range	2–14 days
	Average	4 days	Average	5 days	Average	5 days
No. of countries affected	26		27		More than 180[a]	
Majorly affected regions	China, Toronto in Canada, Hong Kong, Singapore, Hanoi in Viet Nam		Saudi Arabia, United Arab Emirates, and the Republic of Korea		USA, Russia, Spain, Italy, UK, Brazil, France, Germany, Turkey, Iran, and India[a]	
No. of cases	8000+		2519		Approx 5.6 Million[a]	
No. of deaths	800+		866		Approx 348 Thousand[a]	
Infection fatality rate	Approx. 10%		Approx. 35%		3.4% as on March 3, 2020	

[a] Data as on May 23, 2020

Table 2 Life span of SARS CoV-2 on different surface [15]

S. no	Surface	Life span of SARS CoV-2[a]
1	Air	3 h
2	Copper	4 h
3	Cardboard	24 h
4	Stainless Steel	2–3 Days
5	Polypropylene Plastic	3 Days

[a] At 21–23 °C and 40% relative humidity

used on the infected person [13, 14]. The life span of SARS CoV-2 at different surface is given in Table 2.

The SARS CoV-2 was first found in Wuhan City, Hubei Province of China in December 2019. As of January 3, 2020, a total of 44 patients with pneumonia of unknown etiology have been reported to WHO by the national authorities in China. In the next few week, the coronavirus cases were reported from different counties all over the globe. On February 11, 2020, WHO named this novel coronavirus disease as COVID-19. On March 7, 2020, the total number of cases due to COVID-19 outbreak reaches 100,000 mark. On March 11, 2020, WHO characterizes COVID-19 as a pandemic. Total number of COVID-19 cases and associated death throughout the world with time is shown in Fig. 3a and 3b, respectively. The percentage distribution of cases in various countries is shown in Fig. 3c.

4 Importance of Adopting Preventive Measures During Pandemic

WHO, CDC, and government of different countries have issued advisory to the public for taking preventive measures during current pandemic. The advisory was majorly focused on wearing mask, maintaining social distance, and washing hand frequently. In order to ensure social distancing, government of various countries, including USA, China, Italy, Spain, France, UK, India etc., has imposed complete lockdown during which civilians (excluding healthcare professional and those who are associated with essential services) cannot go outside their house except for essential activities like buying food items, medicine, or in case of medical emergency etc. The importance of taking precautions can be well understood from Fig. 4.

The capacity of health care system of a country is limited and cannot handle sudden increase in the number of patients. This situation can cause many deaths due to lack of medicine and attention of medical personnel. By taking proper protective measures the peak of this curve can be flattened, such that the magnitude of requirement of the medical facilities remain under the health care system capacity. Protective measures also delay the of peak of the curve, which ensure that the government have enough time to increase the health care capacity by developing new testing facility, increasing number of beds in hospitals, opening new isolation center etc. The total time to end a pandemic does increase by taking precaution measures but the reduction in total number of cases are far important than decreasing the time duration of the pandemic. A systematic review of physical intervention to reduce the spread of respiratory viruses (Table 3) clearly shows that the hand washing, wearing mask, and gloves and maintaining social distance effectively reduce the spread of viruses. [17].

(a)

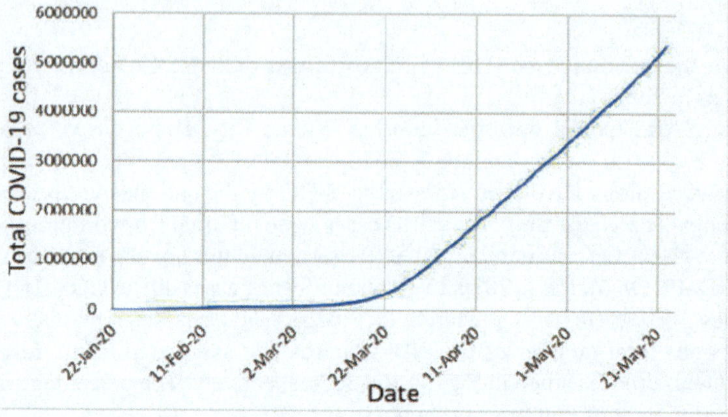

(b)

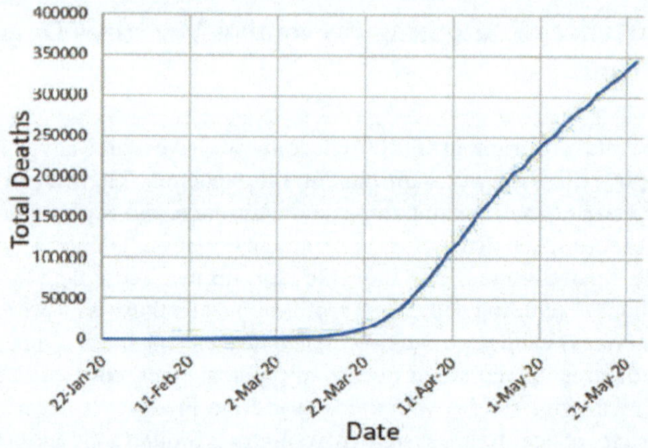

Fig. 3 Statistical graph on COVID-19 outbreak (**a**) Increase in total number of COVID-19 cases across the globe with time*, (**b**) Increase in number of death due to COVID-19 across the globe with time*, (**c**) Country-wise distribution of COVID-19 cases across the globe* [16].* Data as on May 23, 2020

(c)

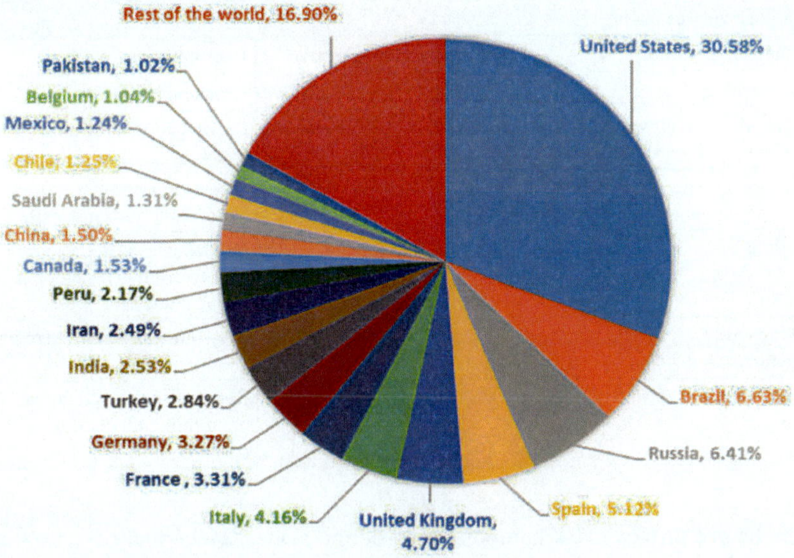

Fig. 3 (continued)

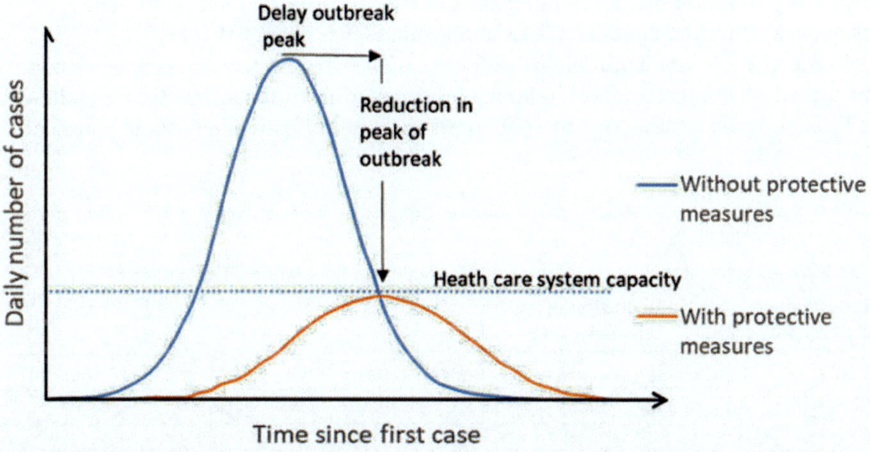

Fig. 4 Variation in daily number of cases with and without protective measures. *Source* CDC

5 Effectiveness of Face Mask During a Pandemic

The use of face masks along with proper hand hygiene and social distancing can be very effective in controlling a pandemic. WHO, CDC, and different government has

Table 3 Main finding of systematic review of physical intervention to reduce the spread of respiratory viruses [17]

Physical intervention	Intervention effective	Interpretation
Handwashing	Yes	Physically removes virus
barriers (masks, gloves, gowns, goggles)	Yes	Prevent contact or inhalation of virus
Social distancing	Probably	Alters environmental conditions for transmission
Gargling	Probably	Dilutes or neutralizes virus (observation is based on a single study)
Adding antiseptics to barriers and hygiene measures	Unknown	May dilutes or neutralize virus
Combined intervention	Yes	Removes virus and alters environmental conditions for transmission

made the use of face mask compulsory in public places. Meta-analysis of observational studies provided evidence of a protective effect of masks and respirators against severe acute respiratory syndrome (SARS). Meta-analysis of randomized controlled trials (RCTs) indicated a protective effect of masks and respirators against clinical respiratory illness (CRI) and influenza- like illness (ILI). Compared to masks, N95 respirators conferred superior protection against CRI, but not ILI. [18].

Researchers have quantified the influence of wearing N95 facemasks in reducing the spread of influenza H1N1. The result obtained from the simulation are shown in Table 4. If we assume that the N95 mask are 50% effective in reducing suscepti-

Table 4 Reduction in percentage of population infected in case of wearing N95 mask during pandemic influenza [19]

N95 Mask effectiveness		Percentage of population wear N95 mask (%)	Percentage of population infected (%)
Reduction in susceptibility (%)	Reduction in infectivity (%)		
0	0	None	73
20	20	10	56
		25	45
		50	38
20	50	10	37
		25	14
		50	7.5
50	50	10	20
		25	5
		50	2.7

Table 5 Reduction in percentage of population infected in case of wearing surgical mask during pandemic influenza [20]

Surgical Mask effectiveness		Percentage of population wear surgical mask (%)	Percentage of population infected (%)
Reduction in susceptibility (%)	Reduction in infectivity (%)		
0	0	None	74.61
2	2	10	73.13
		25	72.68
		50	72.34
2	5	10	71.85
		25	71.12
		50	70.49
5	5	10	70.75
		25	69.40
		50	68.55

bility and infectivity and 50% of the population wear them on the first day, the final percentage of infected people reduced from 73% to 2.7%. [19].

The effect of wearing surgical mask to reduce spread of influenza H1N1 was also quantified and compared.

with N95 mask. The result shows that if we consider the effectiveness of surgical mask in reducing susceptibility and infectivity is 5% and 50% of the total population wear it, then the percentage of infected population decreases from 74 to 68% as shown in Table 5. It can also be concluded that N95 masks are far more effective than surgical masks in controlling pandemic influenza. [20].

A mathematical model was constructed for a population comprised of three different age groups and assume that some individuals wear facemasks. The impact face masks could have had on the spread of pandemic (H1N1) 2009 was quantified and their cost effectiveness was examined. Based on present value of future earnings, hospital costs, and lost income estimates due to illness, it was found that an unmitigated pandemic could result in losses of nearly $832 billion in the United States. It was concluded that the use of face masks by 10%, 25%, and 50% of the population could reduce economic losses by $478 billion, $570billion, and $573billion, respectively. [21].

The guidelines of WHO and CDC on wearing face masks during various outbreak of virus is shown in Fig. 5a and 5b, respectively. Due to the shortage of N95 masks during the COVID-19 pandemic, CDC have recommended the use of cloth face mask in community setting to control the spread of virus. On the other hand, WHO has not recommended the use of cloth mask because of the reason that it is not medically tested.

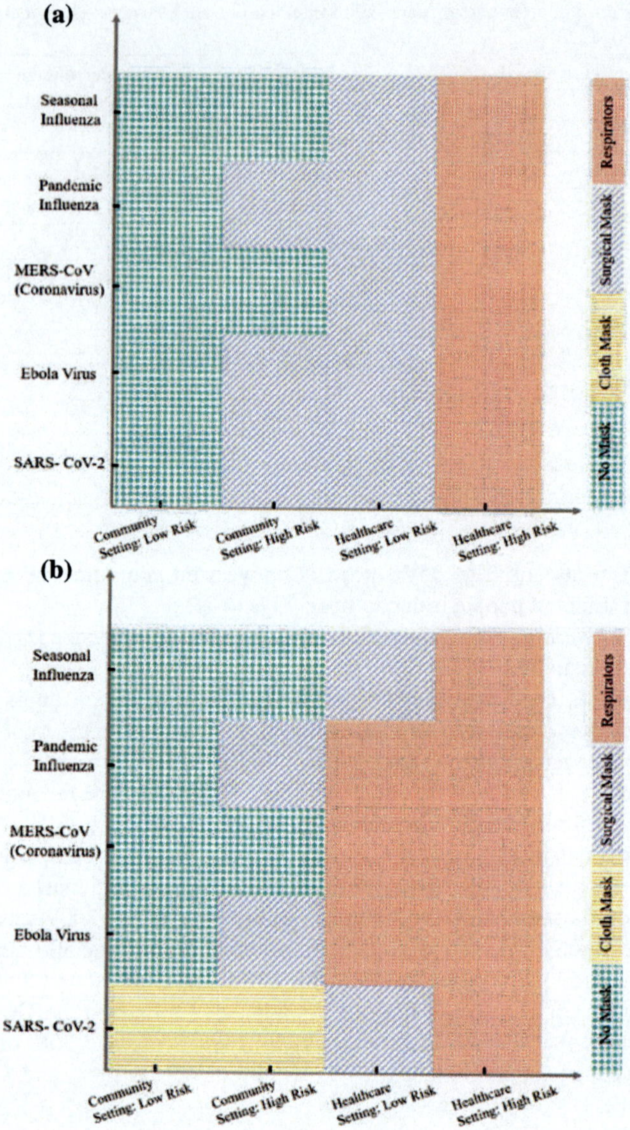

Fig. 5 Infographic of guidelines on usage of face masks in various community settings during virus outbreak by (**a**) WHO and (**b**) CDC [22–24]

6 Novel Face Mask Design Emerged During COVID-19 Pandemic

During COVID-19 pandemic colleges, universities, and research labs from all over the world have focused their attention on advancement of convention face mask. During this period, many novel face masks were designed by various organizations. Due to the lack of availability of face masks, reusable silicon face masks, and washable cloth face mask were manufactured by various industries [25]. Japan based Mitsufuji Corporation has manufactured reusable mask (Fig. 6) that can be washed more than 50 times and used repeatedly to help alleviate the nationwide shortage of sanitary masks [26].

LIGC Applications have developed the Guardian G-Volt face mask (Fig. 7) with a graphene filtration system with integrated electrical charge technology to repel viruses. These masks can be sterilized and safely re-used. [27].

Microbiologist from University of Massachusetts Amherst invented a reusable facemask made out of copper, a material known to kill bacteria and inactivate respiratory viruses including the coronavirus [28]. An ear guard was developed to take the pressure off health care workers' ears from wearing masks all day [29]. Stanford University bioengineering associate professor Manu Prakash has modified full-face snorkel masks into reusable PPE for health care workers shown in Fig. 8 [30].

Hyo-Jick Choi, a biomedical engineer and assistant professor at the University of Alberta, has developed special salt-coated filters that can deactivate a pathogen like the coronavirus in few minutes [31]. Dr. Ashish Karn, faculty from University

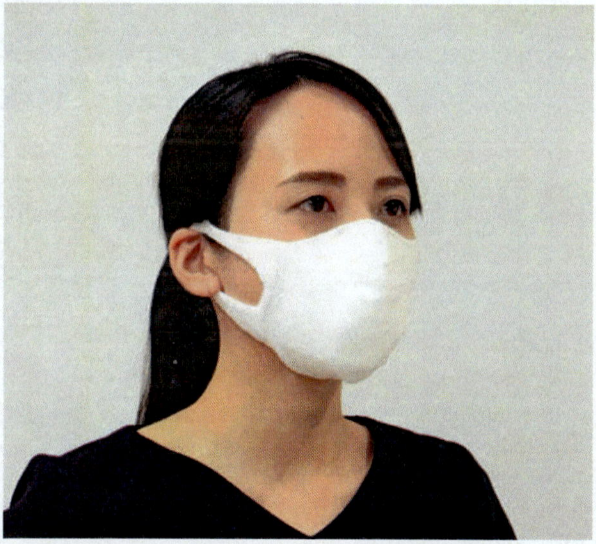

Fig. 6 Reusable face mask designed by Mitsufuji Corporation [26]

Fig. 7 Guardian G-Volt face mask designed by LIGC Applications [27]

Fig. 8 The reusable PPE design of the Pneumask project [30]

of Petroleum and Energy Studies has designed a comfortably Vented, Indigenously Designed (COVID) Fabric Helmet (Fig. 9) to curb infection spread in Education, Healthcare, and other community settings [32]. The fabric helmet is integrated with many innovative design features to make it better fitted, comfortable, cheap, more protective, and well ventilated.

Efforts have been made toward designing customized face mask for all age groups because of the reason that an adult face mask is not effective for children due to poor

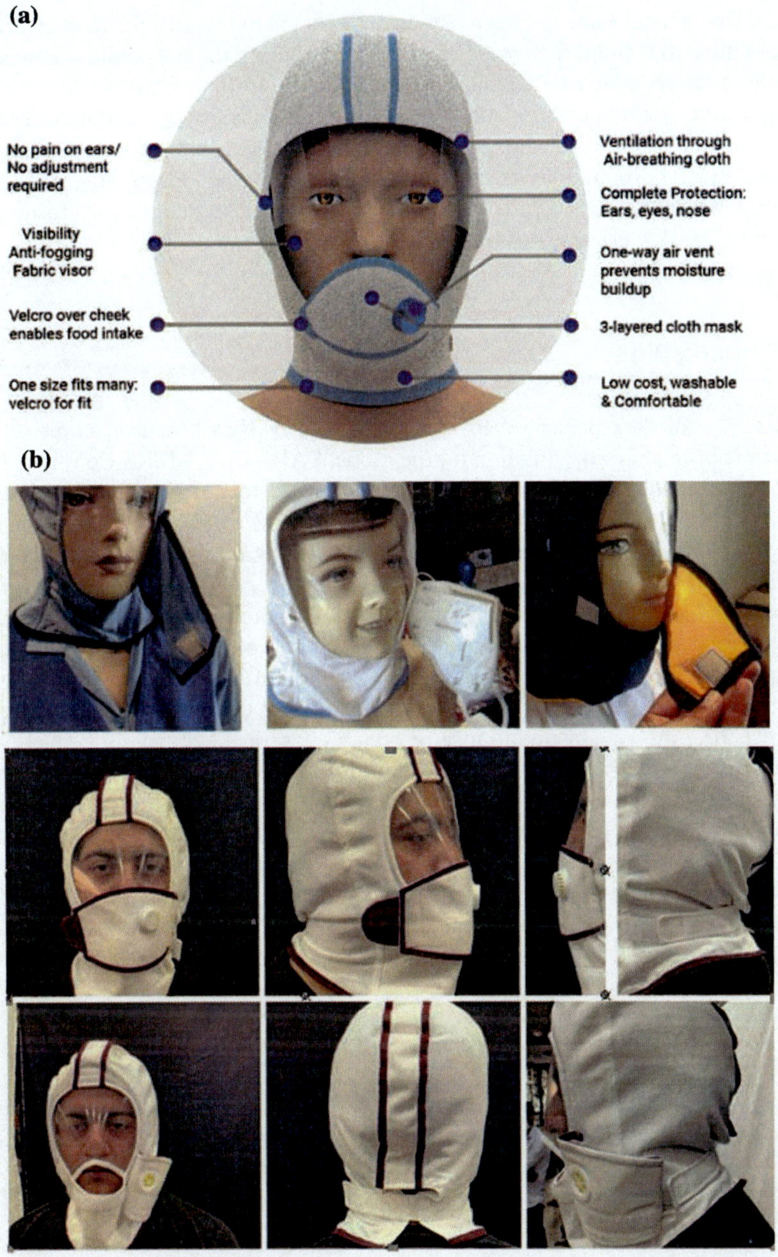

Fig. 9 (**a**) A computer model showing a mannequin wearing a COVID Fabric Helmet and (**b**) an actual prototype of COVID Fabric Helmet [32]

fit. Mobile applications have been launched to get the 3D map of the face, in order to manufacture well-fitted face mask for each face. During the pandemic, various DIY tutorials were made to easily make face masks using stuff which is readily available at home like paper towel, plastic bottle, cotton clothes etc. Researchers from the University of Cambridge and the University of Queensland have collaborated to create a special origami design for a face shield called the "HappyShield" [33]. As the mask became a part of daily wardrobe during the pandemic, various fashionable design for mask were also observed.

7 Conclusions

This statistical data presented in this paper clearly shows that an outbreak of a pandemic is an alarming situation for the globe. SARS CoV, MERS CoV, and SARS CoV-2 can cause severe respiratory related disease to human. SARS CoV-2 has infected around 5.6 million persons across the world and is responsible for more than 348 thousand death. Preventive measures can significantly decrease the transmission of virus from one human to another and is very effective in controlling the outbreak of a pandemic. Maintaining social distance, proper usage of mask, and hand hygiene reduces the risk of spreading of infection in community. N95 face masks can remove 95% the contamination of size more than 3 micron and are most effective in preventive transmission of virus. Surgical masks on the other hand are ineffective in preventing the wearer due to lose fit which cause leakage from the sided, these masks are also incapable of filtering small droplets. Cloth masks although cannot provide higher degree of safety to the wearer but is effective in restricting transmission of virus from wearer to others. Due to lack of N95 masks, during the outbreak of disease, cloth masks can be used in community settings in order to preserve N95 masks for healthcare workers. During the pandemic, various novel design of face masks were developed. Some of the design emerged during COVID-19 pandemic are Guardian G-Volt Mask, Pneumask, COVID Fabric helmet and Happyshield.

References

1. Neuman, B. W., et al. (2006). Supramolecular architecture of severe acute respiratory syndrome coronavirus revealed by electron cryomicroscopy. *Journal of Virology, 80*(16), 7918–7928.
2. Neuman, B. W. et al. (2011). A structural analysis of M protein in coronavirus assembly and morphology. *Journal of Structural Biology*, no. November 2018.
3. Cui, J., Li, F., & Shi, Z.-L. (2019). Origin and evolution of pathogenic coronaviruses. *Nature Reviews Microbiology, 34*.
4. "Mechanisms of Filtration," *Air quality engineering*, 2018. [Online]. https://www.air-quality-eng.com/air-cleaners/filtration-mechanisms/.
5. MU Health Care. Cloth face masks: What you need to know. www.muheath.org. [Online]. https://www.muhealth.org/our-stories/cloth-face-masks-what-you-need-know.

6. Healthline. Can face masks protect you from the 2019 coronavirus? What types, when and how to use. www.healthline.com. [Online]. https://www.healthline.com/health/coronavirus-mask.
7. Honeywell. N95 masks explained. www.honeywell.com. [Online]. https://www.honeywell.com/en-us/newsroom/news/2020/03/n95-masks-explained.
8. WHO. SARS (Severe Acute Respiratory Syndrome). [Online]. https://www.who.int/ith/diseases/sars/en/.
9. Johns Hopkins Center for Health Security, "Coronaviruses: SARS, MERS, and 2019-nCoV," 2020.
10. WHO. Middle East respiratory syndrome coronavirus (MERS-CoV). [Online]. https://www.who.int/emergencies/mers-cov/en/.
11. Worldometer. COVID 19 updates. [Online]. https://www.worldometers.info/coronavirus/.
12. Liu, J., Liao, X., Qian, S., Yuan, Z. Z., Wang, F., Liu, Y., Wang, Z., Wang, F. S., & Liu, L. (2020). Community transmission of severe acute respiratory syndrome coronavirus 2, *26*(6).
13. Ong, S., Tan, Y., Chia, P., Lee, T., Ng, O., & Wong, M. S. (2020). Air, surface environmental, and personal protective equipment contamination by severe acute respiratory syndrome coronavirus 2 (SARS-CoV-2) From a symptomatic patient. *JAMA*, 3–5.
14. WHO. (2020). *Modes of transmission of virus causing COVID-19: implications for IPC precaution recommendations* [Online]. https://www.who.int/news-room/commentaries/detail/modes-of-transmission-of-virus-causing-covid-19-implications-for-ipc-precaution-recommendations.
15. Aerosol and surface stability of SARS-CoV-2 as compared with SARS-CoV-1, 2020.
16. Worldometer. Coronavirus worldwide graphs. [Online]. https://www.worldometers.info/coronavirus/worldwide-graphs/.
17. Jefferson, T. et al. (20009). Physical interventions to interrupt or reduce the spread of respiratory viruses: Systematic review. *British Medical Journal, 339*.
18. Offeddu, V., Yung, C. F., Sheau, M., Low, F., & Tam, C. C. (2017). Effectiveness of masks and respirators against respiratory infections in healthcare workers: A systematic review and meta-analysis. *Clinical Infectious Diseases, 65*(11), 1934–1942.
19. Cui, J., Zhang, Y., Feng, Z., Guo, S., & Zhang, Y. (2019). Influence of asymptomatic infections for the effectiveness of facemasks during pandemic influenza. *Mathematical Biosciences and Engineering, 16*(5), 3936–3946.
20. S. M. Tracht, S. Y. Del Valle, and J. M. Hyman, "Mathematical Modeling of the Effectiveness of Facemasks in Reducing the Spread of Novel Influenza A (H1N1)," *PLoS One*, vol. 5, no. 2, 2010.
21. Tracht, S. M., Del, S. Y., & Edwards, B. K. (2012). Economic analysis of the use of facemasks during pandemic (H1N1) 2009. *Journal of Theoretical Biology, 300*, 161–172.
22. WHO. Advice on the use of masks the community , during home care and in health care settings in the context of the novel coronavirus (2019-nCoV) outbreak—Interim Guidance, 2020.
23. WHO. Advice on the use of masks in the context of COVID-19—Interim Guidance, 2020.
24. CDC. Recommendation regarding the use of cloth face coverings, especially in areas of significant community-based transmission, 2020. [Online]. https://www.cdc.gov/coronavirus/2019-ncov/prevent-getting-sick/cloth-face-cover.html.
25. Dorn, L. (2020). Kitchen utensil company makes reusable silicone face masks to help personal protective gear shortage. Laughingsquid.com. [Online]. https://laughingsquid.com/gir-reusable-silicone-face-mask/.
26. Mitsufuji Corporation. Wearable Mitsufuji Launches 'hamon AG Mask,' Mitsufuji.co.jp, 2020. [Online]. https://www.mitsufuji.co.jp/en/information/hamonagmask_20200317/.
27. Guardian G-Volt masks would use graphene and electrical charge to repel viruses and bacteria, *de zeen*, 2020.
28. University of Massachusetts Amherst. Copper masks—A hopeful weapon against COVID-19, University of Massachusetts Amherst, 2020. [Online]. https://www.cns.umass.edu/news-events/news/copper-masks-hopeful-weapon-against-covid-19.
29. Ozdemir, D. (2020). Boy Scout 3D prints numerous 'Ear Guards' to help medical staff suffering from face masks, Interestingengineering.com, 2020. [Online]. https://interestingengi

neering.com/boy-scout-3d-prints-numerous-ear-guards-to-help-medical-staff-suffering-from-face-masks.
30. The Pneumask Project. [Online]. https://www.pneumask.org/.
31. Grant, L. (2020). Inventor of virus-killing mask needs funding, 660citynews.com. [Online]. https://www.660citynews.com/2020/04/07/inventor-of-virus-killing-mask-needs-funding/.
32. Karn, A., Deo, S. S., & Kumar, A. (2020). A COmfortably Vented, Indigenously Designed (COVID) fabric helmet to curb infection spread in Education, healthcare and other community settings. SSRN.
33. HappyShield, "HappyShield," happyshield.github.io, 2020. [Online]. https://happyshield.github.io/en/.

Design and Development of an Automated Monitored Hand Hygiene System to Curb Infection Spread in Institutional Settings During COVID-19 Pandemic

Ashish Karn, Rithvik Kanchi, and Shashank Singh Deo

1 Introduction

In the event of a pandemic, such as the COVID-19 outbreak, it requires a multi-pronged approach to address the issues including both pharmacological and non-pharmacological interventions, apart from the massive education of the general populace and communities. While a pharmacological intervention may refer to a drug or a vaccine, non-pharmacological interventions could refer to a number of strategies such as hand hygiene or facemasks case isolation in home, social distancing, voluntary home quarantines, and closure of educational institutions [1]. Of the different kinds of such non-pharmacological interventions, hygienic hand antisepsis is one of the most important intervention to prevent the transmission of pathogenic microorganisms and can reduce infection spread, even among high-risk populations [2]. Evidence from the literature showed that frequent hand-washing would reduce the risk of viral transmission by 55% [3, 4]. Further, an appropriate hand washing intervention can even reduce the risk of infection by 6–44%, by impairing the transmission cycle and thus hand hygiene education has an indispensable role in the prevention of infectious diseases. [5] Particularly, in an institutional setting, such as an educational institution, industry, offices, healthcare facilities etc., ensuring the correct hand hygiene promotion program is extremely important. Such a hand hygiene scheme should preferably be "touch-free," so that the chances of viral transmission are minimized. A plausible solution to ensure the hand hygiene of individuals is using an automated hand sanitizer. However, this is not enough in an institutional setting where individuals throng in large numbers, particularly during peak hours, exacerbated by the apathy, and unwillingness of individuals to comply to a standard hand hygiene protocol. There is

A. Karn (✉) · R. Kanchi · S. S. Deo
Department of Mechanical Engineering, School of Engineering, University of Petroleum and Energy Studies, Bidholi, India
e-mail: akarn@ddn.upes.ac.in

N. A. Siddiqui et al. (eds.), *Advances in Behavioral Based Safety*,
https://doi.org/10.1007/978-981-16-8270-4_14

clearly a need for a technological intervention which not only facilitates hand sanitization, but also maintains sufficient physical distance between people, and necessitates compliance by different members of institution. Such a multimodal hand hygiene antisepsis program can also identify institution workers with poor compliance to hand hygiene regulations, and thus assist in the development of educational interventions at a targeted sub-population within an institution. [6] During the outbreak of a pandemic, social distancing, lockdowns, and work-from-home regulations have been released by a majority of the countries, in a bid to reduce the infection spread of the Coronavirus pandemic. However, as the lockdowns are lifted, there is an increased apprehension and concern among organizations to safeguard their workers against the pandemic by instituting systems which can protect individuals against the viral transmission. Under such a situation, the "touch-based" hand sanitizers are susceptible to cause human-to-human transmission, and even small-scale automated sanitizers are not enough to bridge the gap, without a large-scale hand antisepsis system. Further, these large-scale systems, should ensure enough physical distance between different users, should monitor the usage, and should be cheap enough to be installed at multiple locations within an organization. Although there are some manufacturers of automatic hand hygiene monitoring systems in USA, such as Biovigil, DebMed, Gojo Industries etc., there are no such existing industry in India. These systems typically employ some sort of wireless communications technology such as a WiFi, RFID badges, both of these or even none. Many of these systems require the subject to wear a badge/tag or to wear it on the wrist. However, there are no currently existing systems which are low-cost and that institutions can deploy without having the workers to wear a badge/tag/ wrist bands etc., and simultaneously provide monitoring. Thus, there is clearly a need for automated monitored hand hygiene antisepsis systems which are easily customizable and so cheap that many of these units could be deployed within a single institution. Here, we propose the design and development of such a system which is extremely simple to build.

2 Considerations in the Design of Hand Hygiene Systems for Community Settings During COVID-19 Pandemic

There are various important considerations for the design of hand hygiene systems for community and health care systems, which can be outlined as under.

2.1 Automation

Coronaviruses are enveloped viruses and in particular SARS-CoV-2, the virus that causes COVID-19 is believed to spread mainly through human–human transmission via respiratory droplets or through other routes such as contact with contaminated fomites and inhalation of aerosols, produced during aerosol generating procedures. [2] A recent review during Coronavirus pandemic has shown that corona viruses survive on surfaces for a period ranging from 2 h to 9 days. [7] The conventional hand sanitizers are touch-based and its use in institutional settings may pose a significant health hazard by promoting infection spread.

2.2 Compliance and Monitoring

Although the hand hygiene has been acknowledged as the most important component of an adequate infection control scheme, the major bottleneck to its success comes from the paucity of adherence to the proper protocols. The poor compliance has often been attributed to such factors such as lack of time, hand dryness, hand irritation, skeptical attitude of people regarding its usefulness, lack of monitoring, perceptions regarding low risk of cross-infection, inconvenience, and the misconceived belief that protection is offered by gloves alone [8]. To increase the level of adherence to hand hygiene regimen, the hand hygiene compliance should be monitored, and the prior literature suggests that a monitored protocol yields much higher compliance. [6].

2.3 The Need to Ensure "Social Distancing"

The implementation of a hand hygiene program with a single sanitizer unit, whether "touch-based" or automated is fraught with risk, since it does not promote physical distancing between individuals. Any possible design solution must ensure that different "automated" hand hygiene units are separated and institution workers can only approach through a queue. Further, such systems must be compact and so can be placed at a suitable entry location in the institutions such as reception, main entry points etc.

2.4 Format

There are various formats of hand hygiene strategies such as a medicated soap solution, antiseptic agent, detergents, waterless antiseptic agents, alcohol-based hand rub (ABHRs), or it could be a generic preparation containing one or more types of alcohols, humectants, excipients etc., in the form of a gel, foam, or liquid [9]. Gels and foams provide a less clean-feeling and slower to dry at higher doses, whereas liquids yield a smoother, cleaner, a greater moisturized feel, although there is a greater difficulty in handling these liquids which outweighs these benefits. Thus, the key desirable properties are fast absorption, moisturized, and clean hand feel, lack of stickiness, and low odor, and hence gels and foams are more preferred over liquids. [10] Using ABHR over handwashing with soap has several key advantages including better effectiveness, faster action, convenience in use, and also lesser and milder reactions by the skin compared to soap and water. Contrary to the antimicrobials used in soaps, bacterial, or viral resistance to alcohol at the typical concentrations used in ABHRs is much smaller. Usage of ABHRs have another distinct advantage: it reduces the water requirements and is advantageous in water-deficient locations. [11].

2.5 Formulation

There are various formulations proposed by WHO which can be used as an antiseptic or disinfectants against viruses including soap and water, ABHRs, chlorine solutions, Iodine and Iodophors, Hydrogen peroxide etc. It has been reported that the virucidal efficiency of alcohol against certain viruses, such as a Norovirus varied depending upon the contact time, formulation, and concentration. [12] Typically, washing hands with soap and water works best when hands are visibly dirty, and it should be done for 40–60 s. However, if hands are not visibly dirty, using ABHRs for 20–30 s is the preferred method of cleaning. WHO recommended formulations contain either 75% v/v isopropanol or 80% v/v ethanol. [9] ABHRs containing 60–90% ethyl alcohol or isopropyl solutions in water can be used for cleaning since it contains strong virucidal agents inactivating all the lipophilic viruses. A recent review by Kampf (2020) reported that 62–71% ethanol solution can efficiently inactivate human coronaviruses within 1 min. WHO typically recommends 70% ethyl alcohol to disinfect small areas. Chlorine solutions such as 0.05% chlorinated water can be used as a suitable disinfectant for hand washing. Kampf et al. (2020) reported that 0.1% sodium hypochlorite solutions can efficiently inactivate human coronaviruses within 1 min. Similarly, Chlorhexidine (0.5–4% concentration) can also be used to decrease the infectivity of the enveloped viruses and thus are suitable for hand washing, although some studies have reported its inability to disinfect enveloped human coronavirus [13]. Commercially available formulations of Povidone Iodine are effective virucidal activities, for instance, Iodophors (0.5–10% concentration) can be used for hand washing to reduce the infection potential of enveloped viruses. Even Hydrogen Peroxide at 0.5% concentration has been reported to decrease the infection capability of human coronaviruses, when used for a minute [14].

2.6 Volume

Some studies have reported that in a hospital setting, the volume of ABHR dispensed by the hospital's automated dispenser is close to 1.1 mL per dose and the ABHR dose is inversely related with the number of applications of ABHR per shift, but has no relationship with the hand size, although there is a concern that it might create a risk for people with larger hand size. [15].

2.7 *Time*

A study reported that hand rubbing for 15 s yielded results at par with hand rubbing for 30 s and no significant difference with respect to bacterial counts on hands were found. [16] Similarly, a comparison of ABHR and the traditional hand-washing with regard to the time required for complete disinfection revealed a value of 72 s and 40 s, respectively indicating that ABHRs are more time-effective as compared to the traditional hand-washing with soap and water. [8].

2.8 *Economics of Hand Hygiene*

Interestingly, a recent study conducted in the dental setting compared the cost associated with traditional hand washing against an ABHR protocol and the corresponding time required to implement acceptable hand hygiene levels. As far as the cost of hand hygiene is concerned, it was found out that the hand sanitation regimen with soap and water costs $0.052, as compared to $0.023 with ABHR. These results indicate that ABHR protocol is less expensive as compared to a traditional hand washing. [8].

2.9 *Skin Reactions*

According to WHO, there are two major types of skin reactions pertaining to hand antisepsis program. The first kind could be mild to debilitating and commonly referred to as irritant contact dermatitis. It can include irritation, dryness, itching, or even cracking and bleeding. The other kind of skin reaction, alluded to as the allergic contact dermatitis, is scarce and usually stems from an allergy to some ingredient in the disinfectant liquid. [17].

3 Proposed Design and Its Innovative Features

The proposed product "Automated Monitored Hand-hygiene system for Institutional Settings" is intended to be used in educational institutions, corporate houses, healthcare facilities, and other community settings, such as shopping complexes etc., which report a considerable amount of influx of people, particularly at some peak hours during the day. The proposed design is a compact, low-cost alternative which not only dispenses disinfecting liquids at multiple ports near the entrance gates, but also monitors the compliance by different users. In addition, the volume of the dispensed liquid, its format and formulation are strictly in accordance with the findings reported in the literature.

Figure 1 shows a schematic of the proposed design ant different gateways in a typical University campus—at the main entrance, near the University central office reception, outside the University gymnasium and out of a management office. Typically, these are the prime zones where the congregation of students, faculty/staff is expected, particularly during the peak hours. It is evident that several of these proposed design units could be installed in an organization such as a University. With its compact design, the product does not take a considerable space, and the presence of the tall separators between the adjacent dispensing units ensure that individuals don't throng together, breaking the social distancing norms. Rather, individuals have to walk in a queue from the four directions, to approach the dispensing units. Further, the four separators have been designed extending outwards, so that the users can walk in and out of line from the either side, reducing chaos and ensuring safe distance between individuals.

Fig. 1 A schematic of the developed hand hygiene system product deployed at four representative locations in a University campus where there is a large influx of students, such as (**a**) The main gate, (**b**) A management office, (**c**) In front of University reception and (**d**) in front of University gymnasium and auditorium. Note that the adjacent dispensing units have tall separators between them, extending outwards

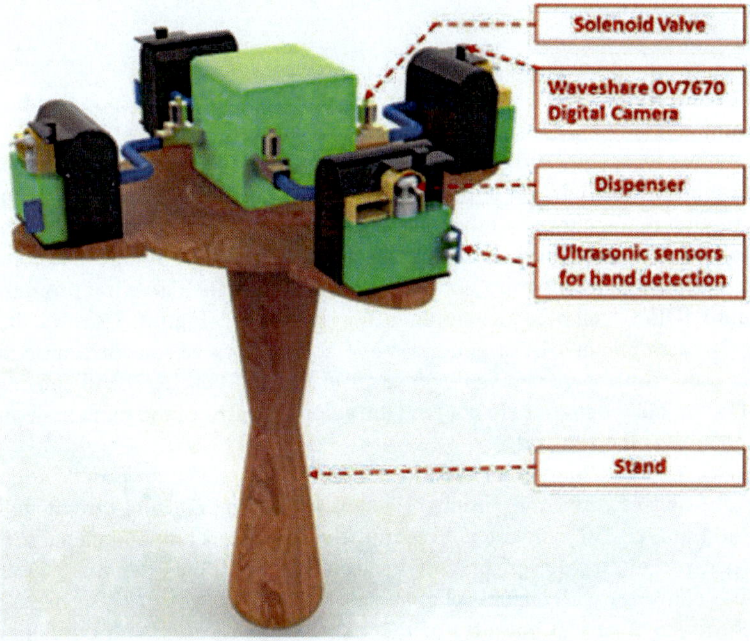

Fig. 2 An isometric view of the proposed automated and monitored hand hygiene system. The main tank is shown smaller in height, for the sake of clarity

Figure 2 shows a closer view of the designed product. For getting the disinfectant liquid, the user has to place their hand under the nozzle of the dispenser, whereupon the ultrasonic sensors detect the presence of a hand and triggers the dispensing process out of a sub-tank. Figure 3 shows a top view of the product through which it is clearly evident that the four dispensing units with their sub-tanks lie at the periphery of a central sanitizer tank (also referred to as "main tank"). The main tank is a large tank and is expected to stock up the sanitizer solution for almost a week. The volume of the main tank is estimated to be 60 L, based upon a simple calculation. Assuming a single person uses the same product at the same location thrice in a day, and based on the reported research literature, it has been determined that 1.2 ml of solution is considered to suffice for most efficient disinfection. Thus, for an organizational strength of 3000 people per product, this consumption amounts to 10.8 L for every single day, or 54 L per work week. An excess of 6 L solution is ensured to account for the sanitization by the guests and visitors. The main tank can be refilled easily, simply by replacing the top cover.

The proposed product is recommended to work with ABHRs, although it can work with any other solution. The amount of the ABHR dispensed per usage has been fixed at 1.2 ml, which has been found to be more than the required dosage for complete hand disinfection.

4 Working

The proposed design uses a basic control system with a feedback loop for the actuation/ automation of the device. An ultrasonic sensor senses the hand and a signal is relayed to the camera module and the servomotor simultaneously, which performs an angular rotation, clockwise, or anticlockwise as shown in Fig. 4. The power acquired by this motion is transferred to the single displacement pump (or the pump on the dispenser) which ejects sanitizer or hand wash via a rack and a pinion. The control is accomplished through an Atmega328p microprocessor which is initial programmed in Arduino IDE (Integrated development environment). Figure 5 shows that the refilling is accomplished by a generic 12 V solenoid valve, an ultrasonic sensor which is used to measure the sanitizer level in the sub-tank for hand sensing and pipes of favorable length to help connect the solenoid valve of the main tank and the solenoid valve of the sub-tank.

Figure 6 shows a comprehensive electrical circuitry of the proposed product. It broadly consists of three sub-circuits: dispensing and monitoring circuit, refilling circuit, and an AC-DC converter. Since all the required components such as the microprocessor, ultrasonic sensor, and the servo motor all operate on a 5 V supply, an AC-DC converter with capability to step down the voltage to 12 V is used. Next, Fig. 7 shows an electrical circuitry of the dispensing process. With the help of an

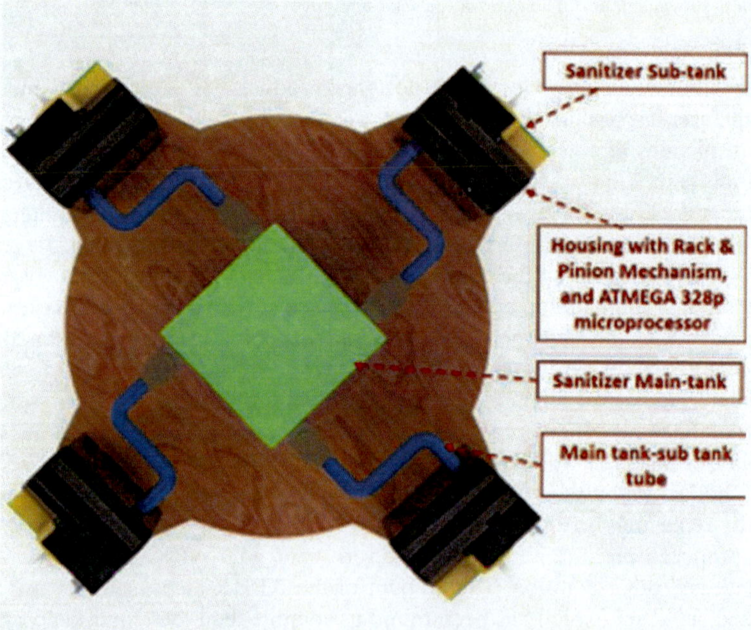

Fig. 3 A top view of the proposed automated and monitored hand hygiene system

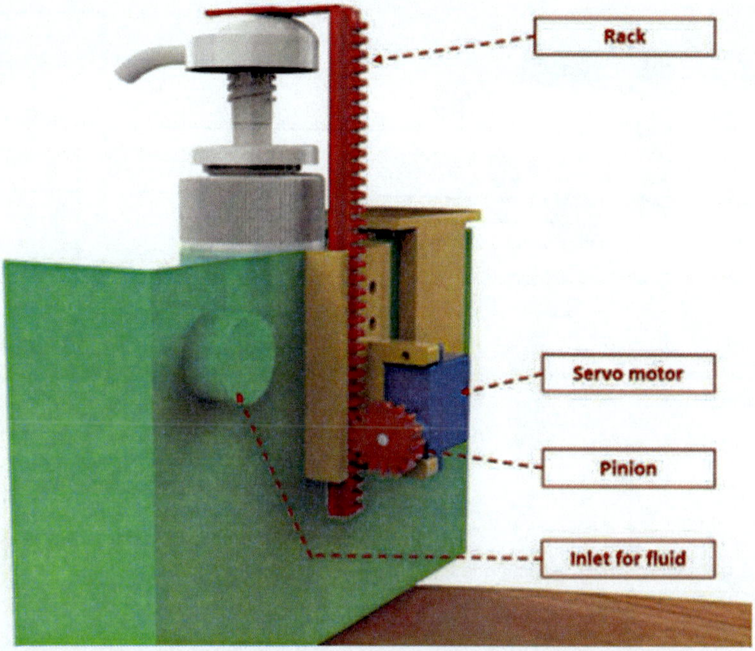

Fig. 4 A backside isometric view of the dispenser showing rack and pinion mechanism

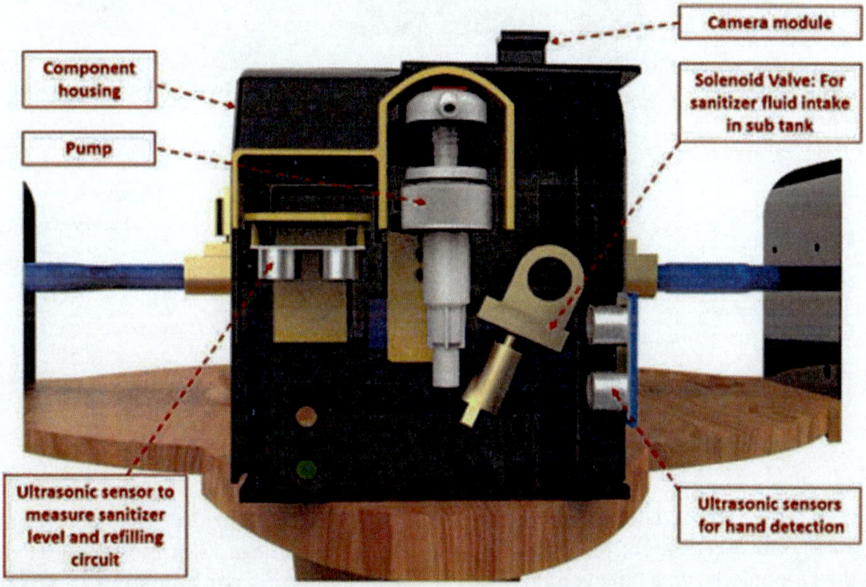

Fig. 5 A cut-section view inside the housing, near the dispenser

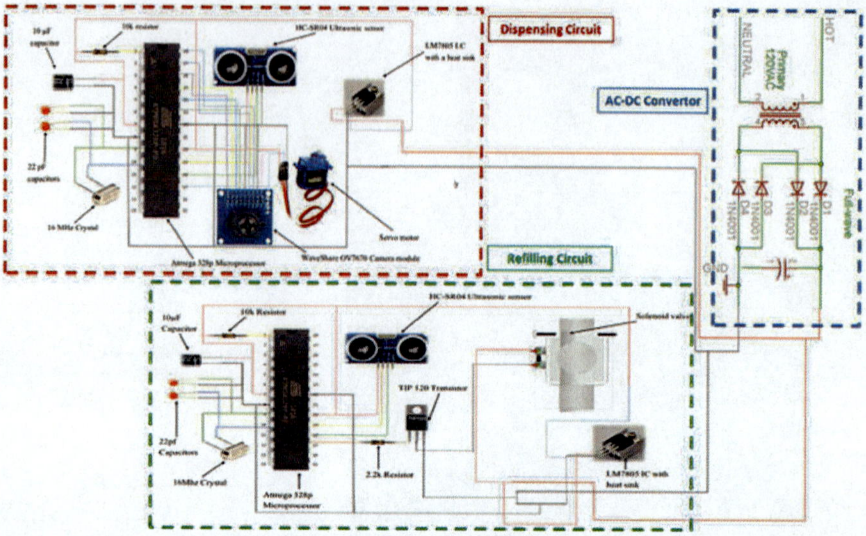

Fig. 6 A comprehensive electrical circuitry of the proposed product

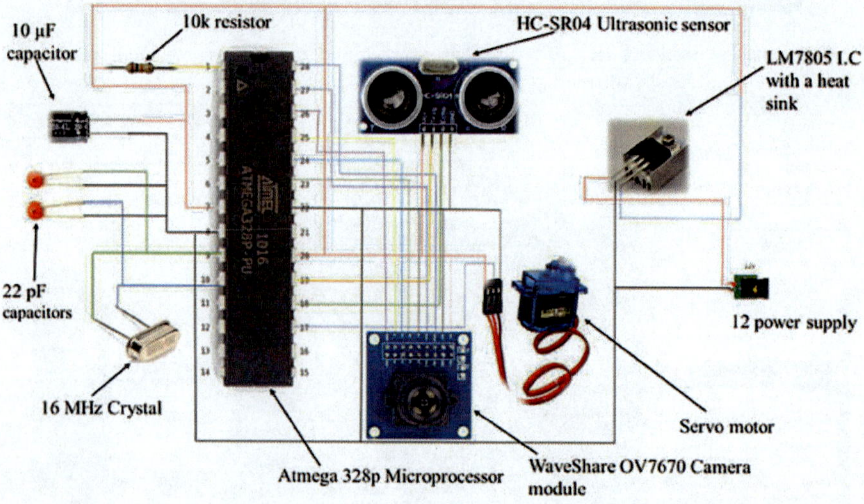

Fig. 7 Electrical circuitry of the dispensing process

LM7805 IC (which is nothing but a voltage divider circuit) with a heat sink, this voltage is further dropped to 5 V operating voltage. The Atmega 328p microprocessor is then mounted over a honeycomb board and with supporting units such as a 16 MHz crystal to keep track of time, while some capacitors and resistors are employed to maintain the required current input for the processor to work. The ultrasonic sensor is used for the detection of the proximity of any kind and the servomotor responds to the output generated by the ultrasonic sensor. Further, with the appropriate code the values for desired motor rotation and at a desired proximity can be set.

The monitoring is accomplished by a camera module (Waveshare OV7670 digital camera), which gets triggered parallel to the servomotor and captures a photograph of the individual using the dispenser. It is expected that the images are further passed through a face recognition algorithm and is used to identify and record the hand hygiene events against each individual of an institution. This data may be crucial to check compliance of the hand hygiene protocols followed by each individual in an institution. The solenoid valve is used for regulating the flow of mixture at times of refilling the sub-tank/tanks. The valve, being an electromechanical device, consumes 12 V from the battery. A TIP120 transistor is connected in series with the solenoid valve for energizing and de-energizing the valve as per the input signal given by the ultrasonic sensor inside the sub-tank. An ultrasonic sensor perfectly measures the level of a liquid and its presence in dark surrounding won't cause any difference in its readings as the sensor working is based on ultrasonic waves (which have nothing to do with light). As soon as the liquid level falls below a threshold value, the Atmega 328p signals the solenoid valve to normally close and after refilling, it would signal the valve to close. Figure 8 presents a schematic of the electrical connections relevant for the refilling circuit.

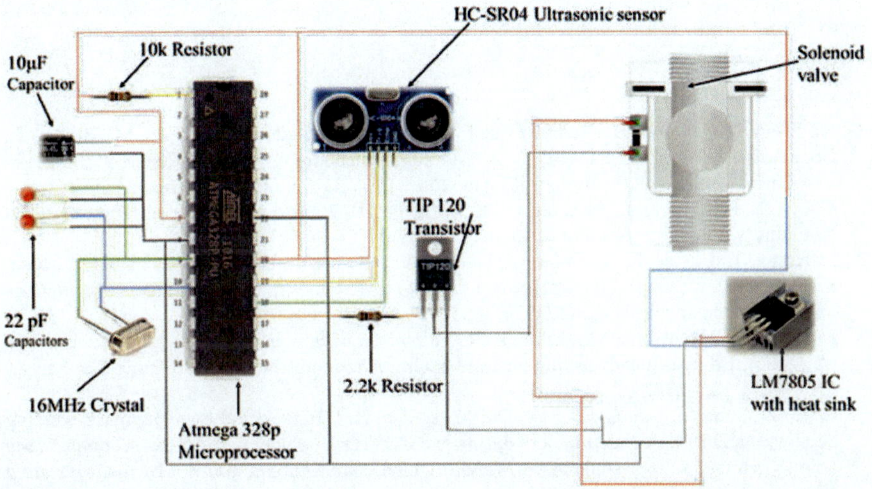

Fig. 8 Electrical circuitry of the refilling process

5 Conclusions

We propose the design of a fully automated and monitored hand hygiene system, which can be befittingly used in an institutional setting. The proposed design is low-cost, compact and can be installed at the entry points of various important buildings in an educational/healthcare setting. By the integration of a camera module, which triggers in sync with each dispense event, the current design monitors and tracks the hand hygiene regimen followed by each individual in an institutional setting. Further, the proposed product provides for an automated sanitization scheme, in such a way that social distancing norms are not highlighted and thus can effectively help curb the infection spread in community settings.

6 Funding Sources

No funding source till the submission of this manuscript.

Conflict of Interest The authors declare no conflict of interest in the publication of this manuscript.

References

1. Ferguson, N., Laydon, D., Nedjati Gilani, G., Imai, N., Ainslie, K., Baguelin, M., Bhatia, S., Boonyasiri, A., Cucunuba Perez, Z., & Cuomo-Dannenburg, G. (2020). Report 9: Impact of nonpharmaceutical interventions (NPIs) to reduce COVID19 mortality and healthcare demand.
2. Guner, R., Hasanoglu, I., & Aktas, F. (2020). COVID-19: Prevention and control measures in community. *Turkish Journal of Medical Sciences, 50*(SI-1), 571–577.
3. Jefferson, T., Foxlee, R., Del Mar, C., Dooley, L., Ferroni, E., Hewak, B., Prabhala, A., Nair, S., & Rivetti, A. (2008). Physical interventions to interrupt or reduce the spread of respiratory viruses: Systematic review. *BMJ, 336*(7635), 77–80.
4. Luby, S. P., Agboatwalla, M., Feikin, D. R., Painter, J., Billhimer, W., Altaf, A., & Hoekstra, R. M. (2005). Effect of handwashing on child health: A randomised controlled trial. *The Lancet, 366*(9481), 225–233.
5. Chen, X., Ran, L., Liu, Q., Hu, Q., Du, X., & Tan, X. (2020). Hand hygiene, mask-wearing behaviors and its associated factors during the COVID-19 epidemic: A cross-sectional study among primary school students in Wuhan, China. *International Journal of Environmental Research* and *Public Health, 17*(8).
6. Boyce, J. M. (2011). Measuring healthcare worker hand hygiene activity: Current practices and emerging technologies. *Infection Control and Hospital Epidemiology, 32*(10), 1016–1028.
7. Kampf, G., Todt, D., Pfaender, S., & Steinmann, E. (2020). Persistence of coronaviruses on inanimate surfaces and its inactivation with biocidal agents. *Journal of Hospital Infection.*
8. Huber, M. A., Holton, R. H., & Terezhalmy, G. T. (2005). Cost analysis of hand hygiene using antimicrobial soap and water versus an alcohol-based hand rub. *Oral Surgery, Oral Medicine, Oral Pathology, Oral Radiology, and Endodontology, 99*(4), 424.
9. Organization, W.H. (2006). WHO guidelines on hand hygiene in health care (advanced draft): global safety challenge 2005–2006: clean care is safer care., World Health Organization.

10. Greenaway, R. E., Ormandy, K., Fellows, C., & Hollowood, T. (2018). Impact of hand sanitizer format (gel/foam/liquid) and dose amount on its sensory properties and acceptability for improving hand hygiene compliance. *Journal of Hospital Infection, 100*(2), 195–201.
11. Vermeil, T., Peters, A., Kilpatrick, C., Pires, D., Allegranzi, B., & Pittet, D. (2019). Hand hygiene in hospitals: Anatomy of a revolution. *Journal of Hospital Infection, 101*(4), 383–392.
12. Cheng, V., Tai, J., Ho, Y., & Chan, J. (2009). Successful control of norovirus outbreak in an infirmary with the use of alcohol-based hand rub. *Journal of Hospital Infection, 72*(4), 370–371.
13. Wood, A., & Payne, D. (1998). The action of three antiseptics/disinfectants against enveloped and nonenveloped viruses. *Journal of Hospital Infection, 38*(4), 283–295.
14. Kampf, G., Efficacy of ethanol against viruses in hand disinfection. J Hosp Infect, 2018. 98(4): p. 331–338.
15. Martinello, R. A., Arbogast, J. W., Guercia, K., Parker, A. E., & Boyce, J. M. (2019). Nursing preference for alcohol-based hand rub volume. *Infection Control & Hospital Epidemiology,40*(11), 1248–1252.
16. Pires, D., Soule, H., Bellissimo-Rodrigues, F., Gayet-Ageron, A., & Pittet, D. (2017). Hand hygiene with alcohol-based hand rub: How long is long enough? *Infection Control and Hospital Epidemiology, 38*(5), 547–552.
17. WHO. (2009). Skin reactions related to hand hygiene. In WHO guidelines on hand hygiene in health care: first global patient safety challenge clean care is safer care, Geneva (p. 14).

Evaluation of a Pilot-Scale SHEFROL Unit Set Up for Rapid, Inexpensive and Clean-Green Treatment of Greywater

S. Suresh, Tabassum-Abbasi, Tasneem Abbasi, N. Ramesh, and S. A. Abbasi

1 Introduction

1.1 The SHEFROL® Technology

S. A. Abbasi and co-workers have patented and trademarked SHEFROL® bioreactor [1–3] wherein the acronym SHEFROL® denotes the unique sheet-flow-root-level hydrology of the reactor. As detailed elsewhere [3–11], SHEFROL® utilizes freely and abundantly available short-statured plants—which can be aquatic, amphibious, or terrestrial—in specially designed channels stocked with one or other species of these plants. Wastewater is made to flow through these channels in the form of a sheet thick enough to cover only the roots of the plants. The dimensions are optimized to ensure maximum contact between wastewater and plant roots. This facilitates natural agitation and oxygenation as the wastewater gets turbulent while passes though the plant roots in the SHEFROL® channels. Due to a variety of mechanisms, detailed in the references mentioned elsewhere [4–11], very strong secondary treatment, together with significant primary and tertiary treatment is achieved within a single process step [12, 13].

S. Suresh
Department of Chemical Engineering, Maulana Azad National Institute of Technology, Bhopal 46 003, India

Tabassum-Abbasi (✉) · S. A. Abbasi
School of Engineering, University of Petroleum and Energy Studies, Dehradun 248007, India

T. Abbasi
Centre for Pollution Control and Environmental Engineering, Pondicherry University, Kalapet, Puducherry 605014, India

N. Ramesh
Department of Science, Technology, and Environment, Government of Puducherry, Puducherry 605005, India

N. A. Siddiqui et al. (eds.), *Advances in Behavioral Based Safety*,
https://doi.org/10.1007/978-981-16-8270-4_15

1.2 The Present Work

The present work describes exploration of the efficacy of SHEFROL® at field scale in treating greywater discharged by three hostels of Pondicherry University. Of special focus was the assessment of the ability of the SHEFROL® to handle widely varying characteristics as well as quantities of inflows because water use in the hostels went by spurts of highs followed by the lows.

2 Installation of the SHEFROL® Unit

The pattern of wastewater generation during the morning, afternoon, and evening, at 10:30, 14:00, and 17:30 h, was studied for a continuous period of 10 days (Table 1). It is seen that the flows ranged from 8 L per minute (LPM) to 13 LPM, varying by 61%. The flows were expectedly the maximum in the morning and minimum in the afternoon. Based on these flow-rates, as also making allowance for monthly variations, the SHEFROL® unit was sized to handle upto 14,000 litres per day of greywater.

The system lay-out is as shown in Fig. 1. After it was commissioned and in operation, the unit had the appearance as in Fig. 2. It consisted of one primary collection tank (A) and four channels (B, C, D, E)—each of width 0.21 m, depth 0.3 m, and length 5.5 m. The channels B and C led to a collection tank (F) while the channels D and E fed the tank G. A polishing channel each (H, I) was attached to these tanks.

In order to demonstrate that the system can be efficient and robust even if installed with minimum use of materials, and even by persons not trained in civil construction, the reactor channels and tanks were created by measured digging the soil (Fig. 3).

Table 1 Pattern of wastewater inflows into SHEFROL-II®

Day	10:30 h	14:004	17:30 h	Average
Friday	10.0	9.0	10.0	9.6
Saturday	11.0	10.0	10.0	10.3
Sunday	12.0	8.0	10.0	10.0
Monday	13.0	9.0	10.0	10.6
Tuesday	9.0	8.0	8.0	8.3
Wednesday	11.0	9.0	10.0	10.0
Thursday	12.0	8.0	8.0	9.3
Friday	12.0	9.0	9.0	10.0
Saturday	11.0	9.0	9.0	9.6
Sunday	10.0	8.0	9.0	9.0

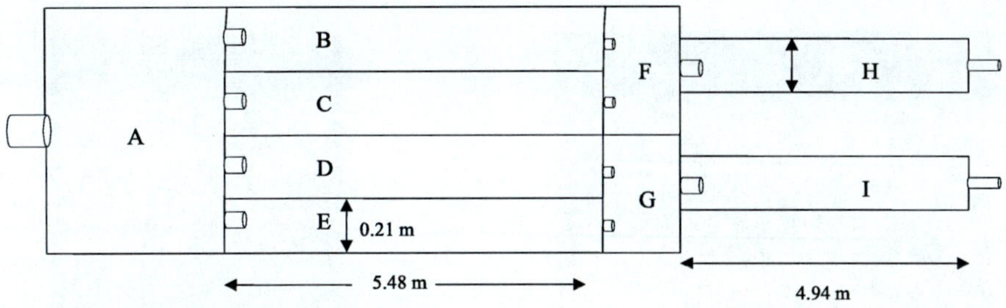

Fig. 1 Schematic of the SHEFROL-II® unit

The resultant shape was covered with high density polyethylene (HDPE) sheets to ensure bank stability as also to prevent percolation of water/wastewater.

The unit was started by allowing greywater to move through it while its channels B, C, D, and E were stocked with *M. quadrifolia, E. prostrata, A. sessilis,* and *M. quadrifolia,* respectively. The channels of the publishing component H and I were stocked with *E. crassipes* and *A. sessilis,* respectively (Fig. 2). The plants multiplied quickly and soon covered their respective tank/channel.

The system performance was monitored by drawing influent and effluent COD samples at 14.00 h every day and analysing them. This was done for a continuous period of 80 days. As already established earlier [4–11], COD serves as an all-round indicator of SHEFROL® performance and all other pollutants get attenuated in proportion as found in numerous cases earlier.

3 Results and Discussion

The results of continuous monitoring of the unit's performance over are summarized in Fig. 4.

As may be seen > 30% treatment occurred on the very first day which rose to > 55% by the tenth day. All channels came close to peak performance—as reflected in the extent of COD removal—in about 20 days and attained steady state in about 40 days. Thus the unit took very little time to start giving significant treatment and attained its peak by the 40th day. Both the *M. quadrifolia* channels achieved mean COD reduction of about 84% at the steady state. The steady state performance of *E. prostrata* and *A. sessilis* hovered around 75% (Fig. 4). The effluent of these channels was further treated in the polishing channels to CODs of 5–9 mg/L.

Even as the rate of inflow into the SHEFROL® varied over the days as well as across each day Fig. 5, showing zig-zag pattern, this variation did not effect the unit's performance or the reactor's stability. The system output remained steady. In other words, the system was able to handle the flow variations that occur daily as well as diurnally in greywater generation and could tolerate shock loads as well, without

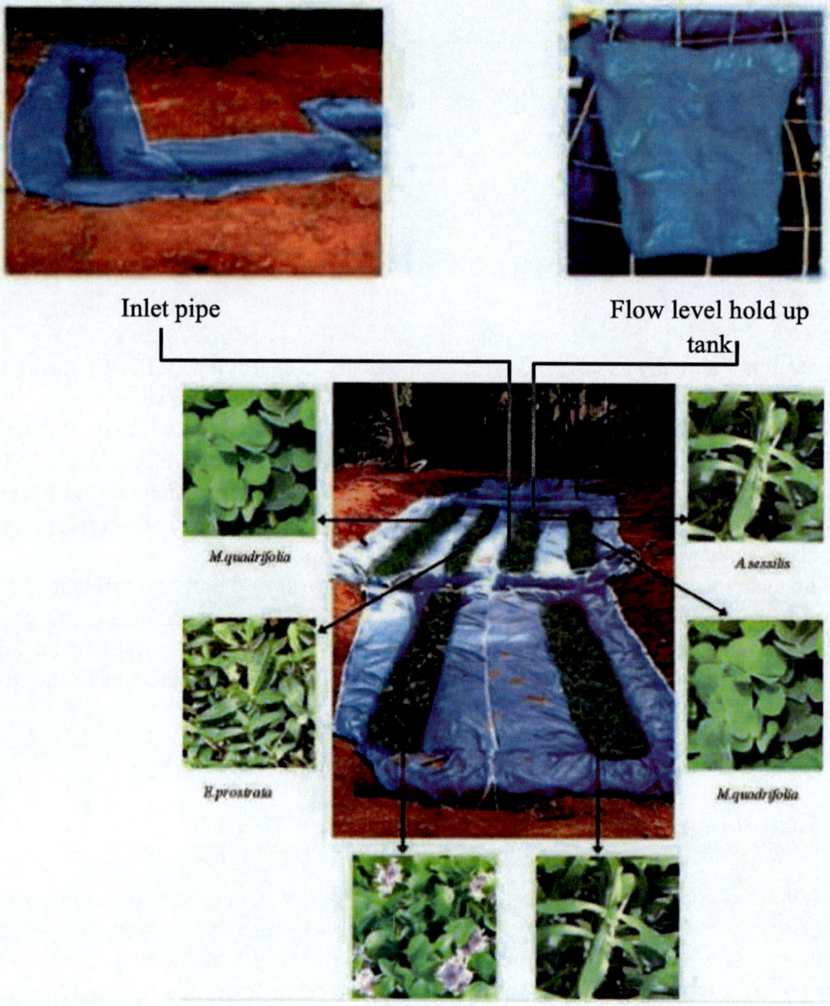

Fig. 2 The SHEFROL® unit in operation

jeopardising the quality of the system output. These attributes reflect the sturdity and sustainability of the SHEFROL technology [14–17].

4 Pest Attack in SHEFROL® and Its Management

No pest attack was encountered during the period of about a year when SHEFROL®-II was in operation. In some SHEFROL® units established elsewhere, occasional pest

Fig. 3 Making of SHEFROL® channels and tanks excavating soil

attacks were encountered but they could always be easily controlled by (a) replacing damaged plants and (b) spraying of neem oil.

Since water is constantly moving in SHEFROL units, there is no mosquito breeding possible.

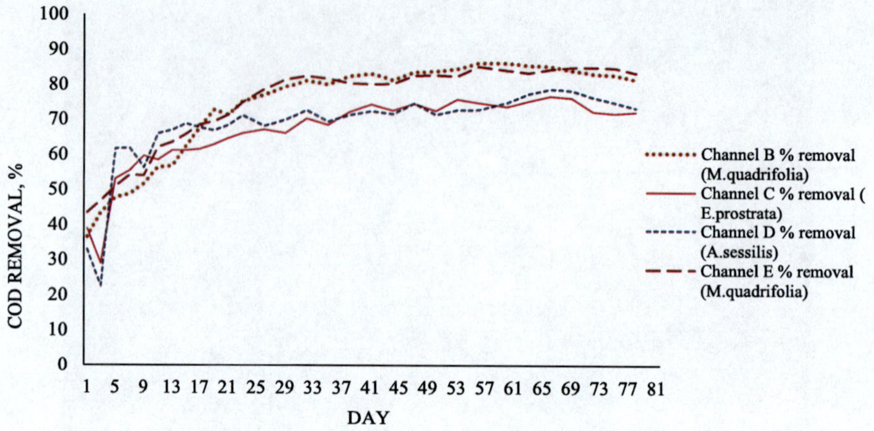

Fig. 4 Removal of COD (%) in the different channels of SHEFROL

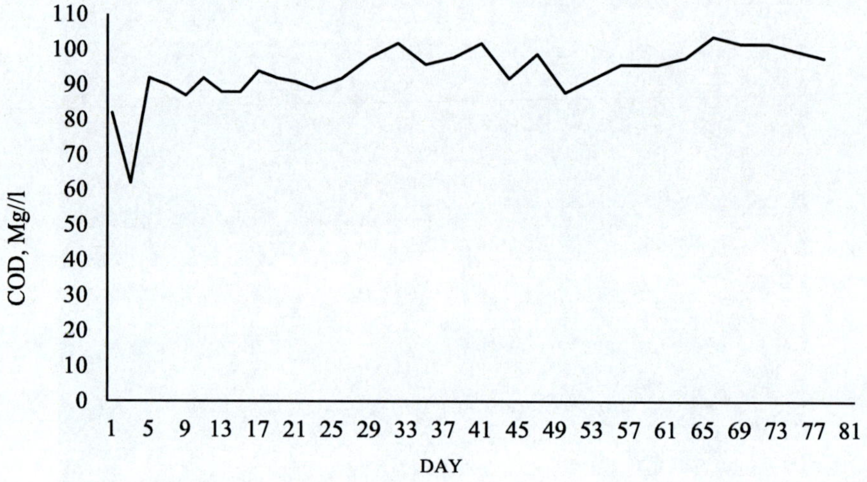

Fig. 5 COD of the greywater entering SHEFROL® unit

5 Summary and Conclusion

A 14,000 L per day pilot plant was set up on the basis of SHEFROL® technology recently patented and trademarked by S. A. Abbasi and co-workers. This paper demonstrates the ease and inexpensiveness associated with this technology as well as its efficiency and robustness. The system was quick to commission and start and soon began to provide COD removals of the order of 80 ± 5 mg/L. It was able to withstand fluctuations in the inflows on daily as well as diurnal basis.

Acknowledgements Authors thank the Ministry of Water Resources, RD & GR, Government of India, for funding an R&D project which enabled the work and Department of Biotechnology, Government of India, for supporting the patenting of SHEFROL®. SAA thanks the Council of Scientific and Industrial Research (CSIR), New Delhi, for the Emeritus Scientist grant (21(1034)/16/EMR-II) which has enabled him to document the findings presented in the paper. Thanks are also due to Dr M. Premalatha for assisting in the experiments.

References

1. Abbasi, S. A., Gajalakshmi, S., and Abbasi, T. (2018). Dynamic self sustaining and self propagating system for waste water treatment Indian Patent No. 302282 granted in 2018 with effect from 2012 as per specification. *Journal of the Patent Offices, 20*, 7611.
2. Abbasi, S. A., (2021). SHEFROL trademark certificate of registration, number 4765454, granted on 23 May 2021 with effect from 2 December 2020.
3. Bhat, M. A., Abbasi, T., & Abbasi, S. A. (2018). Soil-less use of aquatic macrophytes in wastewater treatment and the novel SHEFROL® bioreactor. In N. A. Siddiqui, S. M. Tauseef, & B. Kansal (Eds.), *Advances in Health and Environment Safety, Springer Transactions in Civil and Environmental Engineering* (pp. 297–316). Berlin: Springer.
4. Abbasi, S. A., Ponni, G., & Tauseef, S. M. (2018). Treatment of Sewage by the Weed Ipomoea Aquatica: A Feasibility Study on Bench-Scale SHEFROL® Bioreactor. In N. A. Siddiqui, S. M. Tauseef, & B. Kansal (Eds.), *Advances in Health and Environment Safety, Springer Transactions in Civil and Environmental Engineering* (pp. 353–359). Berlin: Springer.
5. Abbasi, S. A., Ponni, G., & Tauseef, S. M. (2018). *Marsilea quadrifolia*: A new bioagent for treating wastewater. *Water, Air Soil Pollution (Springer), 229*(133), 1–8.
6. Abbasi, S. A., & Tauseef, S. M. (2018). A system for rapid and inexpensive treatment of sewage using the weed false daisy *(Eclipta prostrata). Water and Environment Journal (Wiley), 32*(4), 1. https://doi.org/10.1111/wej.12355
7. Abbasi, S. A., & Tauseef, S. M. (2018). Use of the terrestrial weed *Alternanthera ficoidea* in treating greywater in soil-less SHEFROL® bioreactors. *Water Science and Technology (IWA), 77*(8), 1–11. https://doi.org/10.2166/wst.2018.093
8. Abbasi, S. A., Ponni, G., & Taussef, S. M. (2018). Proficiency of Brahmi (Indian Pennywort) Hydrocotyle asiatica in One-pot Secondary and Tertiary Treatment of Sewage in SHEFROL® System. *Nature Environment and Pollution Technology, 17*(2), 603–609.
9. Abbasi, S. A., Tabassum-Abbasi, P. G., & Tauseef, S. M. (2019). Potential of joyweed *Alternanthera sessilis* for rapid treatment of domestic sewage in SHEFROL® bioreactor. *International Journal of Phytoremediation, 21*, 1–10. https://doi.org/10.1080/15226514.2018.1488814
10. Abbasi, T., Patnaik, P., Tauseef, S. M., & Abbasi, S. A. (2020). SHEFROL® bioreactor enhances the ability of pistia *(Pistia stratiotes)* in the phytoremediation of greywater. *International Journal of Environmental Analytical Chemistry, 21*, 1. https://doi.org/10.1080/03067319.2020.1738420
11. Abbasi, S. A., & Tauseef, S. M. (2019). Rapid treatment of greywater (household sewage) by terrestrial weed *Achyranthes aspera* in SHEFROL® reactors. *Environmental Progress & Sustainable Energy, 38*(2), 467–476.
12. Bhat, M. A., Abbasi, T., & Abbasi, S. A. (2020). A household-scale on-site embodiment of the novel bioreactor SHEFROL® for treating greywater. *Taiwan Water Conservancy, 68*(2), 26–35. https://doi.org/10.6937/TWC.202006/PP_68(2).0003
13. Tabassum-Abbas, P. P., & Abbasi, S. A. (2021). Screening of twenty-three common terrestrial plants for their possible use as phytoremediators of greywater in SHEFROL®, *Taiwan Water* Conservancy

14. Anuradha, J., Abbasi, T., & Abbasi, S. A. (2015). An eco—friendly method of synthesizing gold nanoparticles using on otherwise worthless weed pistia. *Journal of Advanced Research.* https://doi.org/10.1016/j.jare.2014.03.006
15. Abbasi, T., & Abbasi, S. A. (2012). Is the use of renewable energy sources an answer to the problems of global warming and pollution? *Critical Reviews in Environmental Science and Technology (Taylor and Francis), 42*, 99–154.
16. Abbasi, S. A. (1989). Atomic absorption spectrometric and spectrophotometric trace analysis of uranium in environmental samples with Np-methoxyphenyl-2-furylacrylohydroxamic acid and 4-(2-pyridylazo) resorcinol. *International journal of environmental analytical chemistry, 36*(3), 163–172.
17. Tauseef, S. M., Rashtchian, D., & Abbasi, S. A. (2011). A method for simulation of vapour cloud explosions based on computational fluid dynamics (CFD). *Journal of Loss Prevention in the Process Industries (Elsevier), 24*, 638–647.

Need of Resilient Indian Power Sector with Ongoing COVID-19 Pandemic

Monika Yadav, Jayantika Baweja, Devender Kumar Saini, and Geetanjli Rani

1 Introduction

An infectious disease outbreak was reported in Wuhan City of China during December 2019 [1]. The WHO confirmed the widespread in China on January 12, 2020 and officially named the widespread as Corona Virus Disease (COVID-19) [2]. This disease is caused by the severe acute respiratory syndrome. The transmissions of this virus grew rapidly resulting in worldwide pandemic. India confirmed the first case of COVID-19 pandemic on January 30, 2020.

On March 22, 2020, India observed a 14-h voluntary pubic curfew followed by an unprecedented 9-Week nationwide lockdown announced by our honorable Prime Minister to fight the spread of Corona Virus Disease (COVID-19) [3]. The nationwide lockdown was set forth to ensure social distancing which is a must to curb the spread of this disease. So, in these pandemic times of social distancing the electrical sector in powering our current and future need is more evident than ever before. However, the power sector is not immune to the effects of this pandemic leading to huge power loss [1].

The major contribution is the supply of goods and raw materials to different areas. Almost all the activities related to goods and services came to a halt [4]. Even the movement of individuals had been restricted. All these factors are adversely affecting the demand and supply of electricity and hence the power that is being consumed [5]. On the other hand, consistent power supply is must for the operations especially in the COVID Test Center's and the hospitals where the patients are residing. So as the situation is becoming more critical, we need to ensure proper electricity in those centers and other essential services to cope up with the current situation. The survival of critical loads during these disaster leads a strong necessity of a resilient

M. Yadav (✉) · J. Baweja · D. K. Saini · G. Rani
Department of Electrical and Electronics Engineering, University of Petroleum & Energy Studies, Dehradun 248007, India
e-mail: m.yadav@ddn.upes.ac.in

N. A. Siddiqui et al. (eds.), *Advances in Behavioral Based Safety*,
https://doi.org/10.1007/978-981-16-8270-4_16

infrastructure. A resiliency enhancement strategy for the survival of critical loads such as hospitals, plants, and many more are significantly reported and proposed in paper [6, 7].

As a matter of fact, power supply saw a decrease of 25% to date as shown below and expected to rise more in the coming times [8] (Fig. 1).

As the power cannot be stored in the appreciable amount, the power generation and the supply for a day are planned based on the trends in demand [8].

From Fig. 2, it can be concluded that during the months of March and April 2020 there was a considerable decline in the All India Maximum Demand as compared to 2019 [9]. In mid-March, the demand was drastically reduced to 110,000 MW which itself is a major cause of concern.

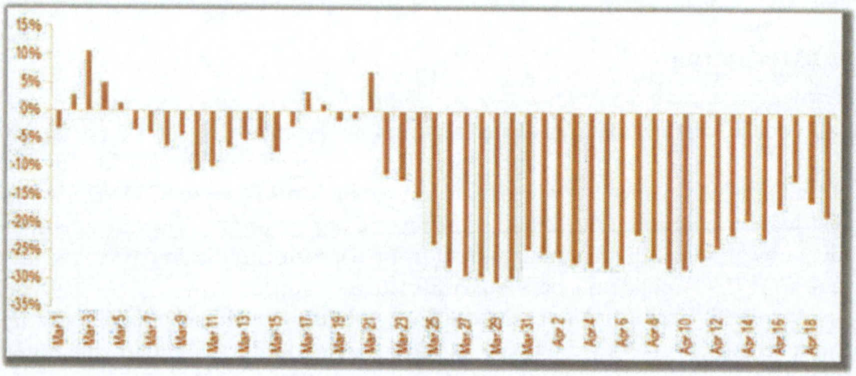

Fig. 1 Percent change in power supply during the severely affected months [8]

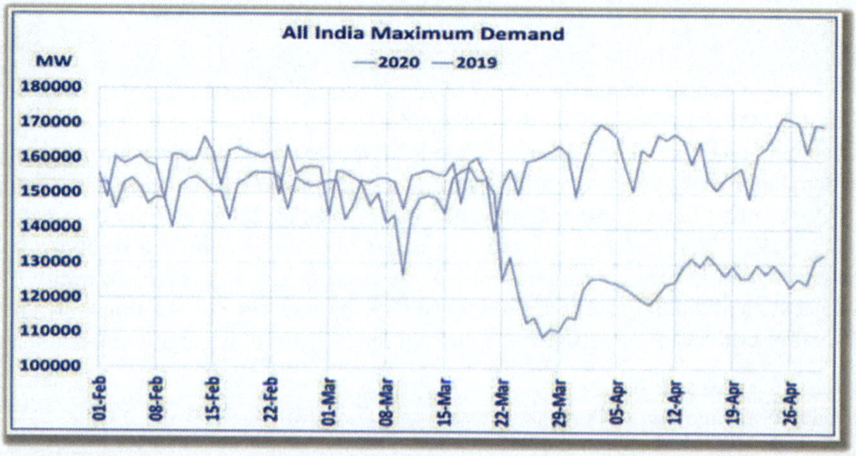

Fig. 2 Comparison of trends during 2019 and 2020 for All India Maximum Demand [8]

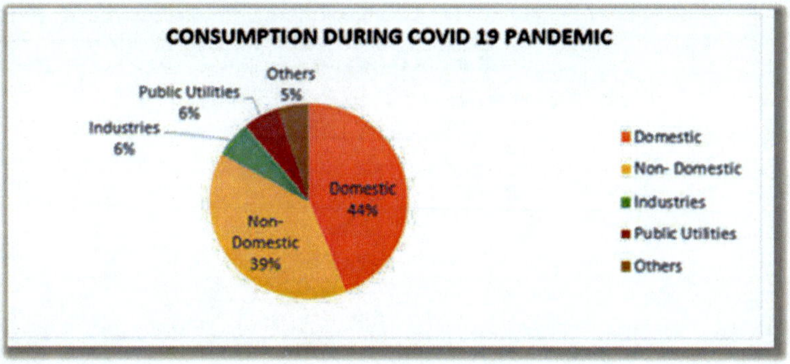

Fig. 3 Consumption in various sectors during lockdown

So according to the current scenario the domestic demand is seeing an accretion since almost all the people are staying indoors. On the other hand, the industrial and agricultural sector might have a considerable decline in demand [10] (Fig. 3).

The power sector related operations have been subcategorized as one of the essential services, so, the plant operations might not be the constraint, but the manpower or the migrant labor is a major concern [10]. The slowdown in the economy is also impacting the demand for electricity.

Also, electricity demand is expected to fall sharply in the coming days since only the "Janata Curfew" contributed between 9 and 10% reduction in the power demand as compared to March 15, 2020 [11]. The data were made available based on the trends below reported by POSOCO [1] (Fig. 4, Table 1).

Also, on April 5, 2020 another major event took place named "Pan - India lights switch off" for 9 min at 9 PM. The graph shows 9-min switch off event leads to overall reduction in 31,089 MW demand in India as reported by POSOCO [1]. This 9-min event is not exactly the same as planned, it started from 8.45 P.M. and it was until 9.10 P.M. and the minimum demand was 85,799 MW, which started to pick up and reached 114,400 MW at 10.10 P.M [2] (Fig. 5).

However, the grid frequency was within the range during this event. At 9.08 P.M., the recorded frequency is 50.259 Hz and 8.49 PM is 49.707 Hz as shown in the graph below [2] (Fig. 6).

In the current scenario, we are aiming sustainable, universal, and ambitious development by aiming at Sustainable Development Goals (SDGs) to transform our world by 2030 [12, 13]. The main idea behind this is to balance the CO2 emissions. A review on sustainable grid with the deployment of solar in remote or rural location is presented in this paper [14, 15]. However, one thing is for sure that the adversely affected global supply chain due to this pandemic threatening the power sector will not stop the industry from transitioning to net-zero CO2 emissions. Moreover, from the recent studies, we came to a conclusion that the lockdown period of approximately 9-weeks helped us in improving the emissions and the results will be much noticeable after many years [10].

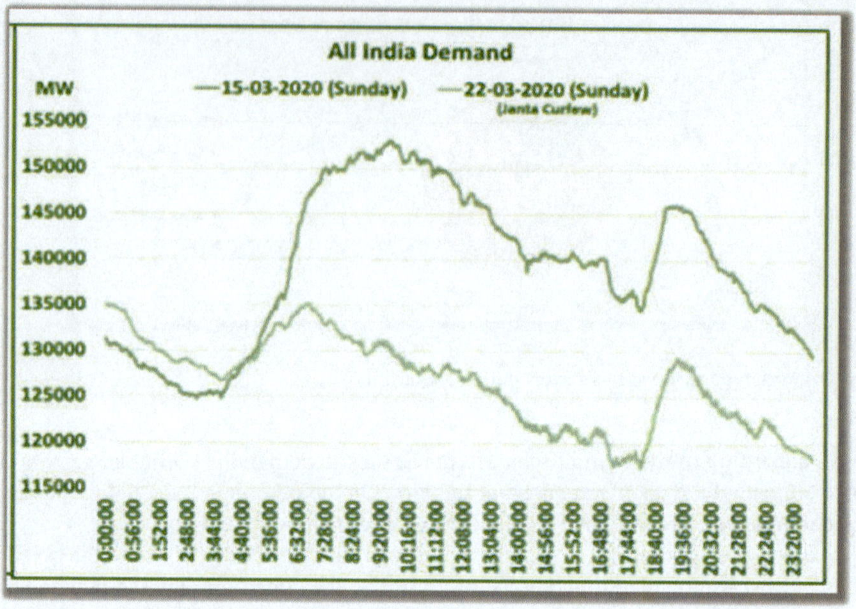

Fig. 4 All India Demand on Janata Curfew [1]

Table 1 All India region wise energy consumption (GWh) on Janata Curfew [1]

Date	Energy Consumption (GWh)					
	Northern region	Western region	Southern region	Eastern region	North Eastern region	All India
March 15, 2020	774	1119	1061	345	39.82	3339
March 22, 2020	739	977	965	319	35.10	3035

The generation, transmission, and distribution sectors all are being affected but may vary up to certain extent. The major factor is the financial status and the delay in the ongoing or the upcoming projects [16]. Certain factors that are affecting the power sector during the global widespread are discussed by analyzing the data made available to us. To get a clear understanding of the prolonged effects of the power sector, we have explained about the major impacts in the generation, transmission, and the distribution that sums up to the complete power sector.

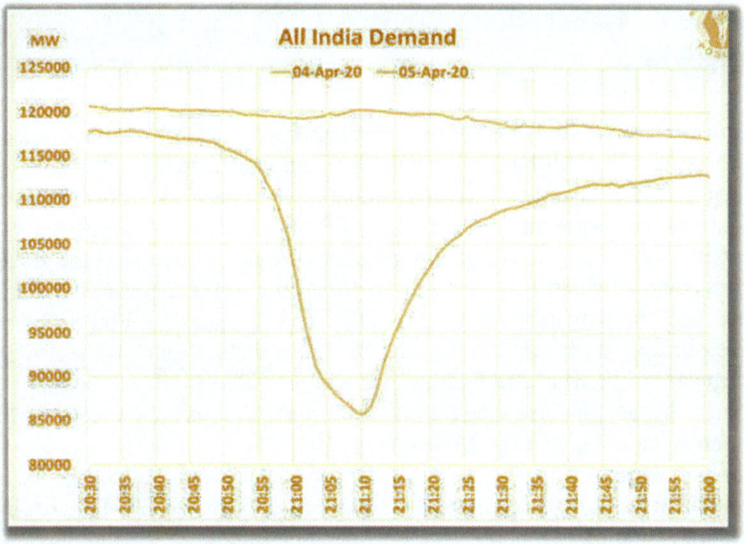

Fig. 5 All India Demand on Pan India light switch off event [1]

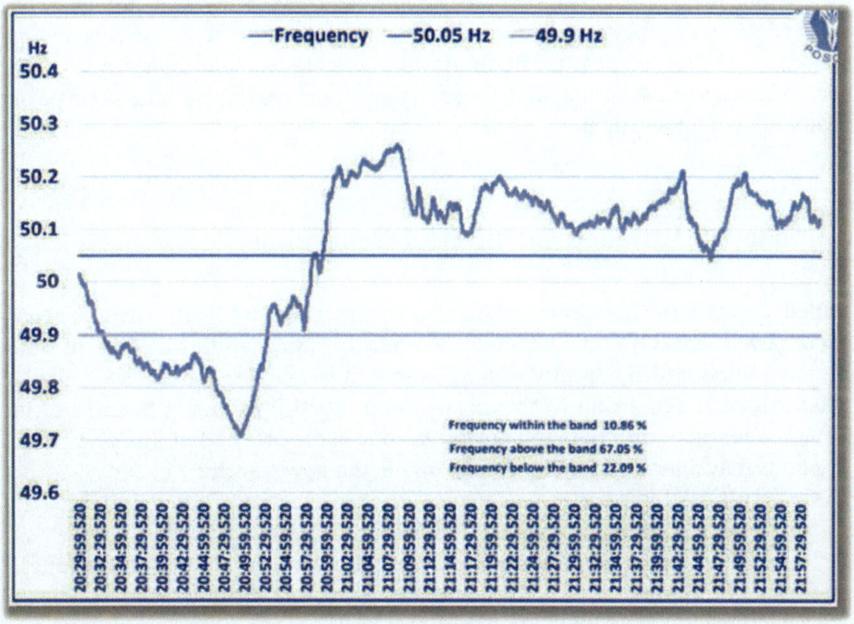

Fig. 6 Frequency trends during 9-min light switch off event [1]

2 Impact on Generation Companies and Their Fuel Suppliers

As the power sector related operations were classified as essential service since April 30, so the plant operations have not been significantly affected. But the revenues for power generation companies were affected since they need to follow the norms of the Power Purchase Agreements (PPA) [17]. PPA basically refer to electricity power contract between two parties one which generates the electricity that is the generation companies and one which is buying the electricity that is the distribution companies to further meet the demand of the nation [18]. The revenues for Power Generation companies having short-term PPAs with the state distribution utilities might get affected in case of thermal and hydropower projects. There will not be much effect on the renewable sector since they are on 'must-run' status as announced of Ministry of New and Renewable Energy (MNRE) [12, 19].

In case of thermal power plants, the fuel availability might not be the major constraint since the fuel (coal) is stocked up for approximately 28–30 days in spite of the current situation [4].

The majorly hit sector was renewable generation in terms of fuel suppliers. Firstly, they were in must-run condition so for the continuous operation stock availability is must. But due the ongoing pandemic, the supply chains were adversely affected. The goods and services supply came to a halt for approximately four weeks contributing a major loss in that sector [2]. The production was also a major concern since the labor was not available at the site. These logistical delays affected the solar power industry up to a large extent.

3 Impact on Installed Capacity of Generation

Installed Capacity of Generation refers to as the intended full load sustained generation of power plants that can be produced under certain conditions. A recent study shows that the installed capacity of the coal-based power plants may saw a decline for the first time as a result of this pandemic in 2020 [7]. This is because of the decommissioning of the coal fired power stations in the first half of this year. These disruptions may also contribute to heavy loss in the power sector.

As of July 2020, India has a total thermal installed capacity of about 2,31,456 MW contributing almost 62.2% of the total installed capacity of India [8].

With this data, we can estimate the contribution of thermal sector in power generation. So, the decommissioning of the plants and generators with the reducing demand made a significant impact on the Indian Power Sector.

The Renewable Energy Sources (RES) contributes about 23.7% of the total installed capacity and since they were on "must-run" condition, the impact was not much in this aspect. However, the deployment of RES in the grid should follow the grid code provided by transmission operators [20] (Table 2).

Table 2 All India Region wise Installed Capacity as on April 30, 2020 [20]

S. No	Region	Thermal					Nuclear	Hydro	RES@MNRE	Grand total
		Coal	Lignite	Gas	Diesel	Total				
1	NR	53,440	1580	5781	0	60,801	1620	20,086	16,870	99,377
2	WR	72,935	1540	10,806	0	85,282	1840	7623	26,043	120,787
3	SR	44,095	3490	6492	434	54,510	3320	11,775	42,474	112,078
4	ER	27,285	0	100	0	27,385	0	4639	1499	33,523
5	NER	770	0	1776	36	2582	0	1577	365	4523
6	Islands	0	0	0	40	40	0	0	18	58
All India		198,525	6610	24,955	510	230,600	6780	45,699	87,269	370,348

Table 3 Energy met in MU during adversely affected months

Generation type	Energy met (MU)			Percentage contributed		
	February	March	April	February	March	April
Thermal	82,602	76,312	60,883	79	76	72
Nuclear	3098	3960	4140	3	4	5
Hydro	8424	9100	9660	8	9	11
Renewable	10,706	10,830	10,374	10	11	12

The table above gives a clear picture of the region wise overall installed capacity of India. The major contributor is thermal with a capacity of 230,600 MW followed by renewable energy sources with a capacity of 87,269 MW [5]. The renewable sector is playing a vital role during this pandemic but may be affected due to the unavailability of the raw materials (Table 3).

4 Impact on Financial Situation of Discoms

The Distribution Companies are responsible for the distribution of the electricity to the customers [21]. They are further classified as state-owned or government owned distribution companies and private owned distribution companies. As per the recent update, the Indian Electricity Distribution will be completely privatized [19]. The major reason behind this is to increase the efficiency in the power sector and hence will be more profitable. Hence, a secured new grid connection can be achieved.

Due to these factors, the distribution companies were so far under huge financial distress and this pandemic has made the situation more badly. The domestic demand saw a rise since major population was staying indoor. The electricity tariff from the consumer saw an increase but the DISCOMs were unable to collect the cost of power from the consumer end because of the movement restriction of individuals as well [22]. This might not contribute much in the financial situation of the distribution companies. This is because the cost of the industrial electricity bills is quite high.

Along with this, some of the state utilities gave more subsidy to the domestic consumers for the cost of power for some period of time to combat the effects of this corona virus pandemic [23]. Also, the consumers were provided with a moratorium for payment of electricity bills in some states. The major profit earned by the distribution companies is from the industrial consumption and that was significantly turn down during the nationwide lockdown resulting in lesser cash in hand for the distribution companies.

The Distribution Companies were still paying for the power they purchased bearing high debts and huge financial losses. As a matter of fact, before the pandemic situation, in the end of February 2020, the total dues of distribution companies to generation companies came approximately Rs. 92,602 Crores and must have saw an upswing till date [12]. The revenue deficit on all India basis is estimated to be about

Rs. 130 billion per month [24]. This sector could be made profitable by reducing cross-subsidies and which is itself a challenging task.

5 Impact on Capacity Addition

Capacity Addition may refer to as to the addition of new projects and improving the commissioned projects in subject to the increased production or the increased capacity. As planned, India is expected to add 131.31 GW of generation capacity by 2022 [1]. The major contribution will be of the renewable power capacity. But this global widespread might delay as expected.

The major reason behind the delay in the project is the less availability of goods and services and the migrant labor. The construction activities came to a halt, the global supply chains were disrupted and the migrant labor left the workplace since they were not paid enough wages. All these factors lead to the delay in the execution of the projects that would increase the pre-operative expenses, the overall cost of the project and so on.

The restriction of movement during the nationwide lockdown affected the transportation of raw materials or the key components that were required for setting up new projects. These logistical delays will affect other planning's as well which in turn would impact the expected returns from the projects. April 30 onwards, all these operations were declared as essential services but to for the efficient and smooth running of the projects with meager staff was a big challenge [6]. Also, because of the current scenario, the companies are hesitating in investing for future projects since they are unable to collect their revenues from the earlier projects.

With COVID-19, delays in the solar power projects were affected because the shipment of solar panel were stopped. On the other hand, India is expected to exceed the solar installed capacity by 8000 Gigawatt (GW) till 2050. Hence, it is becoming a challenging task now [12] (Fig. 7).

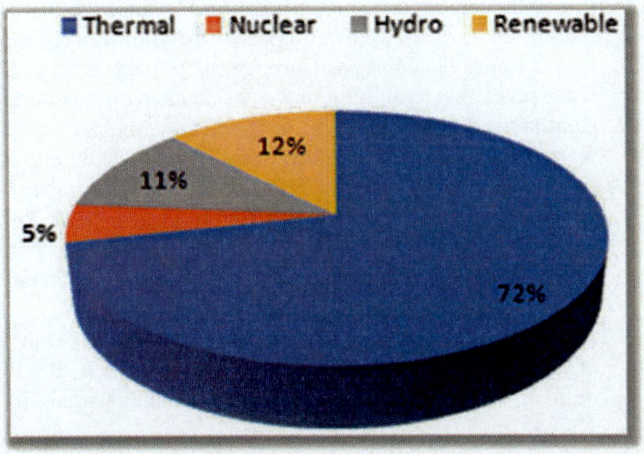

Fig. 7 Contribution of different sectors in All India Generation

6 Conclusion

The global widespread has impacted the power sector adversely. The impact might not be significant till date but in coming times the effect of this pandemic would be huge. The real-market of electricity that was estimated to be in operation since April 2020. But due to trial runs, we were unable to accomplish the task. Also, a new scheme that was about to be launched this year for the benefit of Discoms who are bearing huge loss providing relief by installing smart meters and incentives for rationalization of the tariff will be delayed making the situation more critical. These factors might worsen the situation resulting in anticipated economic slowdown. Hence, need of resilient power sector is much more increased to combat the spread of this global widespread safeguarding the future of power sector.

References

1. POSOCO—National Load Despatch Center. (2020). Retrieved April 15, 2020, from https://posoco.in/
2. CEA. (2020). Retrieved April 15, 2021, from http://cea.nic.in/
3. Saini, D. K., Nair, R., Yadav, M., & Prasad, R. (2018). Modified algorithm for fault detection on AC electrical traction line: An Indian climate case study. *IEEE Transactions on Transportation Electrification, 4*(4), 936–950. https://doi.org/10.1109/TTE.2018.2859379
4. D. K. Saini, R. K. Nair C, B. V. Venkatasubramanian and M. Yadav, "Effect of Unaccounted Parameters on Reactive Power Compensation in Indian Electric Traction Line," in IEEE Access, vol. 8, pp. 182679–182692, 2020, doi: https://doi.org/10.1109/ACCESS.2020.3028582
5. Welcome to Government of India | Ministry of Power. (2021). Retrieved April 15, 2021, from https://powermin.nic.in/
6. Yadav, M., Pal, N., & Saini, D. K. (2021). Resilient electrical distribution grid planning against seismic waves using distributed energy resources and sectionalizers: An Indian's urban grid case study. *Renewable Energy, 178*, 241–259.
7. Yadav, M., Pal, N., & Saini, D. K. (2020). Microgrid Control, Storage, and Communication Strategies to Enhance Resiliency for Survival of Critical Load. *IEEE Access, 8*, 169047–169069. https://doi.org/10.1109/ACCESS.2020.3023087
8. Power System Operation Corporation Limited. (2020). *Pan India lights off event-impact on Indian power system.* Power System Operation Corporation Limited.
9. Power Loads Changing as Coronavirus Impacts Energy Sector. Retrieved 2021, from https://www.powermag.com/blog/power-loads-changing-as-coronavirus-impacts-energy-sector/
10. Coronavirus expected to have major impact on private power. Retrieved 2021, from https://www.thehindubusinessline.com/economy/coronavirus-expected-to-have-major-impact-on-private-power-sector-players-indian-chamber-of-commerce/article31229909.ece
11. Covid-19 Impact: Energy Sector Year in Review 2020. Retrieved 2021, from https://oxfordbusinessgroup.com/news/covid-19-impact-energy-sector-year-review-2020
12. COVID-19: Potential impacts on the electric power sector. Retrieved 2021, from https://crsreports.congress.gov/product/pdf/IN/IN11300
13. A. Yadav, N. Pal, J. Patra, and M. Yadav "Strategic planning and challenges to the deployment of renewable energy technologies in the world scenario: its impact on global sustainable development," in Environment, Development and Sustainability 22(1), 297–315. https://doi.org/10.1007/s10668-018-0202-3

14. D. Bose, D. Saini, M. Yadav, S. Shrivastava, N. P, Decentralized solar energy access and assessment of performance parameters for rural communities in India, Sustain. Clim. Chang. 14 (2021) 103e114.
15. D. Bose, D. Saini, M. Yadav, S. Shrivastava, N. Parashar, "Review of sustainable grid-independent renewable energy access in remote communities of India." Integr, Environ. Assess. Manag. 17 (2021) 364e375
16. IEA. Covid-19 impact on electricity—Analysis. Retrieved 2021, from https://www.iea.org/reports/covid-19-impact-on-electricity
17. Covid-19 update: Power sector situation in lockdown countries. Retrieved 2021, from https://www.power-technology.com/comment/covid-19-power-sector-lockdown-countries/
18. The Himalayan Times. COVID-19 in Nepal's power sector. Retrieved 2021, from https://thehimalayantimes.com/opinion/covid-19-in-nepals-power-sector/
19. McKinsey. COVID-19's potential impact on healthcare. Retrieved 2021, from https://www.mckinsey.com/industries/healthcare-systems-and-services/our-insights/understanding-the-hidden-costs-of-covid-19s-potential-impact-on-us-healthcare
20. M. Yadav, D. K. Saini, and N. Pal. (2017). Review and compliances of grid code with renewable energy (RE) integration. In Proceedings of the International Conferences on Large Scale Grid Integration of Renewable. Energy in India (pp. 1–9).
21. COVID-19's impact on the global energy and power industry. Retrieved 2021, from https://coronavirus.marsh.com/us/en/insights/research-and-briefings/covid-19-impact-on-the-global-energy-and-power-industry.html
22. Covid 19—Power sector legal preparedness—Coronavirus. Retrieved 2021, from https://www.mondaq.com/india/litigation-contracts-and-force-majeure/948818/covid-19-power-sector-legal-preparedness
23. Economic impact of coronavirus Google Slides & PPT template. Retrieved 2021, from https://slidesgo.com/theme/economic-impact-of-coronavirus
24. Impact of Covid-19 on the global energy sector – pv. Retrieved 2021, from https://www.pv-magazine.com/2020/04/24/impact-of-covid-19-on-the-global-energy-sector/

Comparative Study of Ergonomic Assessment Tools on the Blue Collar Workers in a University

T. Ahmed Thanveer, Surendar Varadharajan, and P. A. Arun

1 Introduction

Due to the lack of time and the primary work entrusted, it is observed most of the workplaces are having reactive approaches [1]. It is disappointing to see that only a very few workplaces are taking proactive measure to prevent injuries. Work related musculoskeletal disorders can affect the smooth running of business due to the worker compensation, reduced productivity, absenteeism, worker turnover and poor quality [2]. Hence the industries should aim in reducing the musculoskeletal disorder and increase profit. While comparing with occupational diseases which are caused due to exposure of a particular agent, musculoskeletal disorders are mostly multi-factorial [3]. It's always a challenge for the epidemiologists and ergonomists who are conducting research to find the factors that contribute to WMSDs as musculoskeletal disorder are multi-factorial.

Different ergonomic tools are used to access the risk due to the exposure with WMSDs. The ergonomic hazards can be analyzed through different approaches, quantitative, qualitative and semi quantitative. The quantitative method requires numerical computation and expertise to do the study. NIOSH can be included in quantitative analysis and it requires more data and time for conducting the study [4]. The qualitative tool uses basic observation data and makes the work easy for the observer. Studies showed that certain ergonomic tools are flexible and easy to use but at the same time some require more resources and time.

The techniques to quantify the discomfort and stress due to awkward postures can be broadly classified to observational and instrument based. Instrument based technique uses devices to continuously monitor and record the posture of the worker during the work. Observation techniques are more preferred due to their low cost,

T. Ahmed Thanveer · S. Varadharajan · P. A. Arun (✉)
Department of HSE and Civil Engineering, School of Engineering, Unviersity of Petroleum and Energy Studies, Dehradun, Uttrakhand, India
e-mail: arun.pa@ddn.upes.ac.in

N. A. Siddiqui et al. (eds.), *Advances in Behavioral Based Safety*,
https://doi.org/10.1007/978-981-16-8270-4_17

ease of conducting and the non-interference with the work. It's unfortunate to say that despite of all these tools ergonomic hazards still exist uncontrolled. One of the reason being the analyst finds it difficult to choose between the tools for performing ergonomic analysis from the different options available.

The factors that have to be looked into during the study are, weight lifted, frequency, age, duration, movement and body part affected. Hence through this paper, the selection process is made easier by conducting study and suggesting the appropriate tool.

2 Methodology

For conducting the study on ergonomics, there are different tools which are already developed and in use. The techniques available were studied and foremost preferred techniques in use were selected for the project. The tools selected were RULA, REBA, NIOSH lifting equation and WISHA. The ergonomic assessment tools were applied on the blue-collar workers in the university to study the risk possessed by them. The workers are involved in different activities like cleaning floor, cutting bushes, carrying waste, watering plants etc. Observation and interview of the workers were done to get a clear idea of the work carried out.

Age of the worker and the duration of the work were noted to incorporate that into the study. An activity was chosen and the workers were observed for a considerable amount of time, and the most common posture was captured. The photographs captured were studied with the abovementioned tools. The observation included different activities; each activity was studied with the selected tools. The observations were compared in different aspect to suggest the pros and cons of the specific tools. The studies in the literature mainly focused on a specific work which was analyzed by different tools. Here more focus was given in suggesting the best suited tool for the activities selected. The factors which effect WMSD, which were looked upon by the individual tools were also found and compared (Fig. 1).

3 Ergonomic Risk Assessment Tools

Ergonomics is gaining widespread emphasis in recent times and we have different tools to study and measure ergonomic risk in the workplace. The increase in the long-term effects of repetitive motion injuries has brought light in the authorities to study the problem and find the solution to them [5]. The selected techniques for the study are the following.

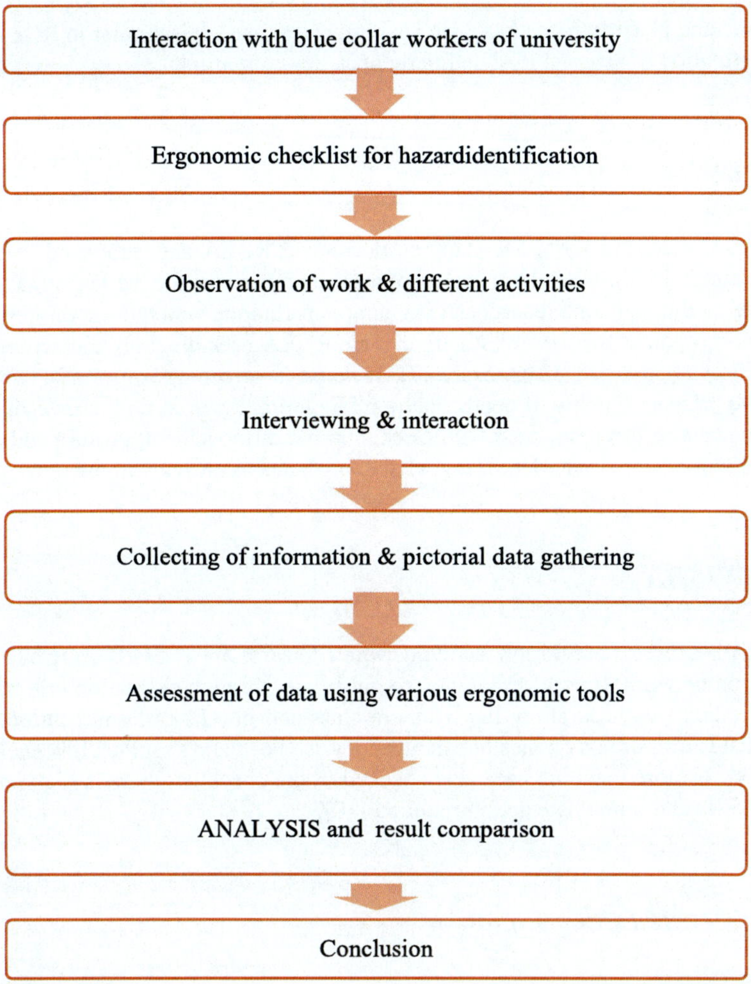

Fig. 1 Methodology

3.1 REBA

REBA helps to analyze various tasks in which systems are sensitive to musculoskeletal risks. REBA is logical and quick risk estimation for the whole body WMSDs of the workers [6]. A worksheet is used for the evaluation of the chosen posture in the particular job. It divides body to segments which are coded according to movement planes. Neck, legs and trunk come in group A. Wrist, upper and lower arm comes in group B. The scoring is done in a manner it increases with the deviation from neutral position by observing each body part. It is proved best for the static, dynamic and places where change in position is frequent. It is similar to RULA and gives attention to external load acting on neck, legs and trunk.

3.2 RULA

E Nigel Corlett and Lynn McAtamney developed RULA and published in 1993. Rapid upper limb assessment is an ergonomic analysis tool for the investigation of work related upper limb disorder. It is a simple technique which does not need any special equipment for the investigation. This RULA uses diagram and score table for evaluating the risk of exposure [7]. It helps in screening a group of workers who are affected quickly. It needs only a clipboard and pen to do the investigation [8]. Muscular effort associated with force, number of movement, posture and static muscle work can help in identifying what contributes to muscle fatigue.

3.3 WISHA

WISHA is a direct observation technique where hazards are assessed using checklist based on the work. It uses verbal cues to guide user through evaluation criteria [9]. In accordance with duration, the risks are classified into hazard and caution. The checklist is of four body zones like hand and wrist, neck and shoulder, low back and knee [9]. If more than one hazard is checked in the checklist, there is high chance for WMSD and immediate action required [10].

3.4 NIOSH Lifting Equation

To use NIOSH lifting tool first we need to collect measures and information about the task performed and record the information for calculating the RWL. The data should be collected by direct observation and interviewing the employees involved in the work. The NIOSH equation is used to access manual handling in workplaces and handling risk in working places associated with tasks like lifting and lowering [11]. The Recommended Weight Limit (RWI) is calculated by multiplying different task variables.

Task variables include

H-Horizontal distance between the object and the body

V-Vertical distance between the object to the floor

D-Distance vertically moved by the object

A-Twisting or asymmetric angle

F-Frequency

C-Coupling

The NIOSH Lifting Equation,
(1) $RWL = LC \times HM \times VM \times DM \times FM \times CM \times AM$ (1)
(2) Lifting Index (LI) $LI = Weight/RWL$ (2)
The RWI gives the acceptable load that an employee who is fit and healthy could lift (8-h shift). To find out relative level of MSDS risk and physical stress. We calculate the lifting index (LI).

The LI if < 1 shows low risk to employees who are healthy. If $LI \geq 1$ it denotes high risk to some amount of population [12]. Hence higher the LI higher the risk for back injury and MSDS. Along with RWL, Lifting Index (LI) is also calculated to provide a relative level of MSD risk and physical stress. The risk levels are categorized into three; (a) $LI < 1$, as Safe (b) $1 < LI < 3$, as Increased risk and (c) $LI > 3$, as Not safe.

4 Result and Discussion

4.1 Tools Applicable for Activities

The blue-collar works in the university were observed and 20 work activities were selected for the study. RULA and REBA were used to evaluate all the activities because they were primarily based on posture evaluation. NIOSH was applied to only works which involved lifting weight and hence could only be applied to 5 activities. The WISHA checklist included different work situations, so the selected activities were compared with them and only 11 activities could be evaluated. By analyzing the results, it showed RULA and REBA are universal tools. NIOSH is confined to a particular activity lifting. WISHA checklist can be improved by adding more work conditions to increase the range of applicability (Fig. 2).

4.2 The Techniques and the Factors Considered

It was found that the exposures like awkward posture, load or force, movement frequency, duration and vibration have to be considered for measuring, the risk possess due to ergonomic hazards. All the factors were considered in WISHA checklist risk assessment. Duration and Vibration were not included in RULA and REBA. WISHA is a preliminary risk assessment tool and gave only qualitative output, it couldn't be relied completely as it doesn't generate a numeric value. Duration is a major component missing in RULA and REBA because most of the ergonomic risk arise due to long repeated work activities (Table 1).

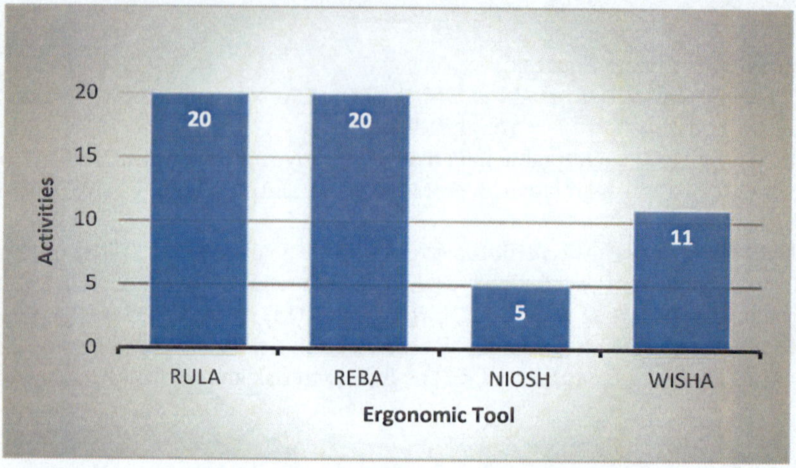

Fig. 2 Ergonomic tools applicable

Table 1 Exposure considered in each technique

SL. NO	Technique	Posture	Load/force	Movement frequency	Duration	Vibration
1	RULA	Y	Y	Y	–	–
2	REBA	Y	Y	Y	–	–
3	NIOSH	Y	Y	Y	Y	–
4	WISHA	Y	Y	Y	Y	Y

RULA

Rapid upper limb assessment is an ergonomic analysis tool for the investigation of work related upper limb disorder. The risk was categorized to four namely negligible, low, medium and high. The results showed all the activities contained at least a small amount of risk. The result showed 25% of the work activities under high risk, 35% under medium risk and 40% in low risk. Most of the activities which were having high risk came under gardening works. Both hands cannot be examined separately and lack of means to combine them is a weakness of RULA.

REBA

REBA is logical and quick risk estimation for the whole body WMSDs of the workers. REBA classified risk into 5 categories namely negligible, low, medium, high and very high. The results showed all the activities contained at least a small amount of risk. The result showed 30% of the work activities under high risk, 60% under medium risk and 10% in low risk. Most of the activities which were having high risk came under weight lifting and gardening works.

WISHA

WISHA is a preliminary risk identification technique that of qualitative in nature. The activities were classified into caution and hazard according to the nature of the work. Caution indicates risk exists which is not of serious nature but can be improved and Hazard indicates fast action has to be taken to protect the workers from ergonomic hazards. 64% of the activities came under caution and 36% under hazard. Due to the limitation of number of activities that could be included in the checklist, all observations could not be analyzed using the tool.

NIOSH Lifting Equation

The NIOSH is used to assess the risk in lifting and lowering the materials. The Recommended Weight Limit (RWL) is the weight an employee can lift for 8 h without the risk of musculoskeletal disorders. NIOSH equation is tested in many laboratories and well-documented tool. It is having a sound scientific background and gives good result for the health risk related to back. The demerit of the technique increases skill and knowledge required [1]. Practical limitations also cannot be neglected. It is also observed that it is most useful for OHS ergonomists/Practitioners and researchers.

Out of 20 activities considered, NIOSH was applicable to only 5 activities that involved lifting or lowering operations. 60% of these activities were found safe and the remaining 40% were under the increased Risk category.

5 Discussion

By the comparative study of the results (Table 3) of 20 different activities with the four tools RULA, REBA, NIOSH Lifting Equation and WISHA results were gathered. The results of RULA and REBA agreed with each other for around 50% of activities. But for the other activities when compared it was found RULA underestimated the risk in many cases. As REBA considered whole body assessment it has a better understanding of the risk involved [13]. For two activities waste collecting and cleaning seats, RULA showed high risk and REBA medium risk, this indicates the lack of scope in RULA.

The lifting activities which showed safe using NIOSH lifting technique came under medium risk in REBA. And no common trend could be found among the techniques which involved lifting activity. It indicates the inefficiency of the other tools in finding risk involved in lifting activities. The factors like horizontal distance, vertical height, Asymmetric angle are not considered RULA and REBA. 90% of the results of WISHA agreed with that of REBA. Even though WISHA checklist is a qualitative tool, it was proved to be reliable and can be used as a good preliminary study tool. During various activities, the discomforts like low back pain, neck pain

Table 2 Risk chart

Color Indication	Risk
Green	Negligible
Yellow	Low
Orange	Medium
Red	High

and wrist aches were identified with the workers. Aged workers were found to be more affected by ergonomic hazards.

Some of the general recommendations while selecting a tool include, the persons, who carry out the ergonomic studies, should have a clear idea about the aim and objective of study. If it is found, that one technique is insufficient, it is recommended to apply multiple techniques to achieve the goal. The hazards which have to be emphasized such techniques have to be adapted which gives quantitative outputs, which are like RULA, REBA and NIOSH. The observer has to be trained to avoid the methodological errors that can affect the result [14]. Works which involve bending, reaching, stooping and grasping for long time are causing wear out in the body. It is very clear to avoid body aches, workers need to maintain proper posture and use right tool. The work has to be maintained in the comfort zone by eliminating unwanted movements and following correct procedure. The risks are color coded according to Table 2 and are tabulated in the Comparative table of results (Table 3).

6 Conclusion

The biggest challenge faced by the professional would be the selection of the ergonomic assessment tool according to their need. General observation techniques appeared more helpful for the HSE practitioners to evaluate the risk with the limited time and resources. From this project, it was very clear that there is no universal tool to evaluate every work. Every tool had its own strength and weakness. It was also found that all the factors that affect the MSDs are not considered in every tool. So it comes to the responsibility of the user to select the most appropriate one by considering the data available, nature of work, the resource available, time duration, cost consideration and accuracy. The development of task-based specific ergonomic tool would help in cutting down using of multiple tools. In Korea, the tool ALLA was developed to assess the risk in agriculture when they were unable to identify the increasing skeletal diseases in the country [15]. Understanding other tools did not focus on the lower limb risks they developed a specific tool to address the problem.

Works which involve bending, reaching, stooping and grasping for long time were found to be the main cause of wear out in the body. To avoid body aches and various MSDs diseases, the workers need to maintain proper posture and use the suited tool. And thereby maintaining work in the comfort zone without much deviation from the neutral position.

Table 3 Comparative table of results

SL.NO	ACTIVITY	RULA	REBA	NIOSH	WISHA
1	Waste collecting	7	7	N/A	Caution
2	Cutting Branches	7	8	N/A	Caution
3	Sweeping Outdoor	4	5	N/A	N/A
4	Collecting Waste	6	8	N/A	Hazard
5	Lawnmower1	4	5	1.5	N/A
6	Waste Handling	5	8	N/A	N/A
7	Weed Pulling	7	8	N/A	Hazard
8	Lawnmower pushing	4	3	N/A	N/A
9	Grass Cutting	7	9	N/A	Hazard
10	Weight Carrying	6	8	0.5	N/A
11	Lawnmower pulling	5	6	N/A	N/A
12	Bush Cutting	4	5	N/A	Caution
13	Arranging Chair	4	5	0.5	Caution
14	Cleaning Seats	7	6	N/A	Caution
15	Sweeping indoor	3	4	N/A	Hazard
16	Carrying	4	4	0.5	N/A
17	Waste collecting	6	7	1.1	N/A
18	Watering	4	3	N/A	N/A
19	Mopping	5	5	N/A	Caution
20	Table Cleaning	6	5	N/A	Caution

References

1. Pascual, S. A., & Naqvi, S. (2008). An investigation of ergonomics analysis tools used in industry in the identification of work-related musculoskeletal disorders. *International Journal of Occupational Safety and Ergonomics, 14*(2), 237–245.
2. Ross, P. (1994). Ergonomic hazards in the workplace: Assessment and prevention. *AAOHN Journal, 42*(4), 171–176.
3. David, G. (2005). Ergonomic methods for assessing exposure to risk factors for work-related musculoskeletal disorders. *Occupational Medicine, 55*(3), 190–199.
4. Dempsey, P. G. (2002). Usability of the revised NIOSH lifting equation. *Ergonomics, 45*(12), 817–828.
5. Schneider, S., & Susi, P. (1994). Ergonomics and construction: A review of potential hazards in new construction. *American Industrial Hygiene Association Journal, 55*(7), 635–649.
6. Ansari, N., & Sheikh, M. (2014). Evaluation of work Posture by RULA and REBA: A Case Study. *IOSR Journal of Mechanical and Civil Engineering, 11*(4), 18–23.
7. McAtamney, L., & Corlett, E. N. (1993). RULA: A survey method for the investigation of work-related upper limb disorders. *Applied Ergonomics, 24*(2), 91–99.

8. Lowe, B., et al. (2018). Assessment methods used by certified ergonomics professionals. In *Proceedings of the Human Factors and Ergonomics Society Annual Meeting*. Sage CA: SAGE Publications.
9. Rahman, M. N. A., and Zuhaidi, M. F. A. (2017). Musculoskeletal symptoms and ergonomic hazards among material handlers in grocery retail industries. In *IOP Conference Series: Materials Science and Engineering*. IOP Publishing.
10. McGaha, J., et al. (2014). Exploring physical exposures and identifying high-risk work tasks within the floor layer trade. *Applied Ergonomics, 45*(4), 857–864.
11. Asadi, N., et al. (2015). A comparative assessment of manual load lifting using NIOSH equation and WISHA index methods in industrial workers of Shiraz City. *Journal of Health Sciences and Surveillance System, 3*(1), 8–12.
12. Takala, E.-P., et al. (2010). Systematic evaluation of observational methods assessing biomechanical exposures at work. *Scandinavian Journal of Work, Environment & Health, 36*(1), 3–24.
13. Manavakun, N. (2004). A comparison of OWAS and REBA observational techniques for assessing postural loads in tree felling and processing. In *FEC–FORMEC–2004 conference.*
14. Kale, P., & Vyavahare, R. (2016). Ergonomic Analysis Tools: A Review. *International Journal of Current Engineering and Technology, 6*(4), 1271–1280.
15. Kong, Y.-K., et al. (2018). Comparisons of ergonomic evaluation tools (ALLA, RULA, REBA and OWAS) for farm work. *International Journal of Occupational Safety and Ergonomics, 24*(2), 218–223.

Biomonitoring of Atmospheric Heavy Metal Deposition by using Moss species *Bryum Argenteum*

Suresh Kumar Kolli, Srinivas Namuduri, R. Uma Devi, Vishnupriya Sowjanya Indukuri, M. Satyanarayanan, and Akbar Ziauddin

1 Introduction

Air pollution has become a major contributor to environmental degradation. Air pollution is caused by emissions from industrial chimneys and automobile exhausts. Rapid developments in the industrial and transportation sectors, combined with rising human population demand, have become significant sources of air pollution [1]. Human activities are responsible for the release of chemical compounds or metallic elements such as heavy metals, sulfur, and nitrogen compounds into the atmosphere, which causes damage to the health of human beings and the environment [2–4]. The intensity with which emissions are distributed is determined by the type of emission source, the composition of emissions, and the weather conditions [5, 6]. The majority of emissions occur very close to the source, but some emissions can travel thousands of kilometers [7–9].

Air quality can be monitored by directly measuring pollutants in the air or deposition, building models depicting the spread of contaminants, or using biomonitors

S. K. Kolli (✉) · S. Namuduri
Faculty, Department of Environmental Sicence, GIS, GITAM Deemed to be University, Rushikonda, Visakhapatnam, India
e-mail: skolli@gitam.edu

R. U. Devi
School of Science, GITAM (Deemed to be University), Hyderabad 502329, India

V. S. Indukuri
Scholar, Department of Environmental Sicence, GIS, GITAM Deemed to be University, Rushikonda, Visakhapatnam, India

M. Satyanarayanan
Geochemistry Division, HR-ICP-MS Lab, National Geophysical Research Institute, Hyderabad, Telangana, India

A. Ziauddin
Advance Career Training & Consulting, Abu Dhabi, United Arab Emirates

N. A. Siddiqui et al. (eds.), *Advances in Behavioral Based Safety*,
https://doi.org/10.1007/978-981-16-8270-4_18

[10–13]. Direct air quality measurements provide information about pollutant levels, but they are very expensive, and there is a risk of contamination when determining low concentrations [14].

Biomonitors provide data on the number of pollutants as well as their impact on the occurrence and condition of biomonitors [14]. Despite the fact that the methods are quick and inexpensive, they may only provide a rough picture of air quality and pollutant deposition.

Biomonitors/bioindicators are organisms or parts of organisms that depict the presence of pollutants based on specific symptoms, reactions, morphological changes, or concentrations [15–19]. There is a lot of variation in the terms biomonitor and bioindicator. However, bioindicator refers to all organisms that provide information on the quality of environmental changes, whereas biomonitor refers to organisms that provide quantitative data on the domain's quality [20–22].

In general, organisms are classified based on their origin into passive biomonitors, which monitor organisms that naturally occur in the study area. Active biomonitors are organisms that are brought into the research area for a specific period of time under controlled conditions [17, 20, 23, 24].

Visakhapatnam's industrial development is conspicuous to urban agglomeration, and the city is situated in a topographical bowl formed by two hill ranges. A significant portion of the town is located within the bowl area, where the majority of the industrial and commercial activities are concentrated within a 10-km radius of the Bay of Bengal's shore. Because of the city's unusual geographical location, wind movement is either eastern or western and is engulfed within the hill ranges. As a result, there is a chance that air pollution levels will rise within the city. The town has witnessed an influx of people from neighboring places for their livelihood. It has increased the concentration of industries, traffic over time. Therefore, this work aimed to appraise the concentrations of trace metals in Bryumargenteum, which is grown in different areas of Visakhapatnam, to ascertain the metal pollution levels.

2 Materials and Methods

2.1 Study Area

Visakhapatnam city was selected as a study area for estimation of heavy metal concentrations by using mosses as bioindicators. Visakhapatnam city is situated in North coastal Andhra Pradesh. It lies between latitude 17,040′30″–170 45′N and longitudes 83,010′–820 21′E, covers above 160km^2 in the Survey of India Topographic Map of 65 O/1 and 65 O/2, and is located in Andhra Pradesh, India. The climate in Visakhapatnam is generally tropical humid with mean daily maximum temperatures is in the range of 27–34 degrees centigrade and mean daily minimum temperatures varies between 140 and 280 C. The annual mean humidity is 77%, and the wind direction is generally from the Southwest toward North West. And it is 0.8 m below mean sea level (MSL). Due to the peculiar geographical location of the city, wind movement is either eastern or western and is engulfed within the hill ranges. Hence, there is a possibility of the buildup of air pollution levels within the city. Visakhapatnam is selected for the case study since numerous sources emit air pollutants, including several major and minor industries like steel, refinery, fertilizers, port activities, etc., located within the city study area. The study was carried out in industrial and urban areas characterized by heavy industrial activity and traffic density.

2.2 Sampling

Both moss and air samples were collected from three industrial areas, like Autonagar area (BHPV), Parawada industrial area (NTPC), Gajuwaka area. And a control sample was collected from Rushikonda (GITAM) area. The sampling region from Autonagar is 13 km, Parawada is 10 km, Gajuwaka is 12 km, and Rushikonda (control area) is 2 km away from the seafloor. The primary objective of establishing the sample plots was to monitor concentration changes in different industrial areas concerning the control area. Moss samples were collected from four regions and mixed to get a composite and representative sample. Moss samples were taken from old walls of the sampling regions. The principal moss species sample was Bryumargenteum. Air samples were collected by using a Respirable Dust Sampler on filter papers in respective sampling areas.

2.3 Sample Preparation

Samples were collected manually, free from solid litter, dust, and other unwanted material. The specimens dried in a hot air oven for 24 h at 400C. Then the material was taken and crushed with a mortar and sieved (UNECE 2001). As per Rühling [25], 0.5 g of sieved moss material was taken in Teflon beakers, and 30 ml of concentrated Nitric acid (HNO3) was added in each beaker and kept for overnight digestion by closing the lids. The next day, these beakers were heated on a hot plate at 130–1400C for 2 h by keeping the tops. The acid content should be maintained at 10 ml in the beakers. Toward the end, about 4–5 ml of hydrogen peroxide (H_2O_2) must be added drop-wise and heated further, and the volume was reduced to about 10 ml. All organic material gets oxidized during this entire process, and the inorganic contents are extracted into the solution. The digested solutions were transferred to a 250 ml volumetric flask and diluted to 250 ml with deionized water [26, 27, 28].

For air samples, filter papers were collected from the different areas were taken and cut into small pieces. These pieces were taken in Teflon beakers, and 30 ml of concentrated Nitric acid (HNO_3) was added to each beaker and kept for overnight digestion by closing the lids. The next day, these beakers were heated on a hot plate at 130–1400C for two hours, keeping the lids. When the solution comes to about

10 ml, heating was stopped and transferred to a 250 ml volumetric flask and diluted to 250 ml with deionized water.

2.4 Analysis

The concentrations of different metals Chromium (Cr), Manganese (Mn), Iron (Fe), Nickel (Ni), Copper (Cu), Zinc (Zn), Arsenic (As), Selenium (Se), Cadmium (Cd), and Lead (Pb) were analyzed by ICP-MS (Inductive Coupled Plasma Mass Spectroscopy) (Perkin-Elmer Sciex Instrument, Model ELAN DRAC II, USA). For calibration and to check the accuracy of the analysis of filter paper and moss material, NIST 1643e reference material is used for quality control. Reagent blanks were used wherever appropriate to ensure accuracy and precision. Further details of ICP-MS are presented elsewhere [29].

3 Results and Discussion

Several works have been proved the significant role of mosses as bioindicators for monitoring heavy metals [30, 31]. In the present study, the significant uptake of elements by mosses is relatively straightforward for heavy metals. The heavy metal concentrations such as Chromium (Cr), Manganese (Mn), Iron (Fe), Zinc (Zn), Arsenic (As), Selenium (Se), Nickel (Ni), Copper (Cu), Cadmium (Cd), and Lead (Pb) determined in mosses collected from sampling sites are higher than air concentrations in respective sampling sites. The assessment of toxic elements deposition at different areas by bryophytes shows variation in content. The increased amount of metal accumulation in moss Bryumargenteum resulted from its lifetime deposition [32]. A significant source of heavy metal in the urban area is metallurgical process, automobile exhaust emission, oil combustion, and processing of crustal material [33, 34]. The trace metal concentration at different sampling sites was shown in Fig. 1. The trace metal concentration in moss samples at various sampling sites was shown in Fig. 2.

In the discussion, described each trace metal concentration at different sampling sites.

3.1 Chromium

Chromium concentrations in moss samples were showed in Tables 1 and 2. The chromium concentrations in air samples were 14.573 ng/g–19.078 ng/g, whereas Cr concentrations in the moss samples range are 646.4 ng/g–787.76 ng/g. High chromium concentrations were observed in the air sample of the control area

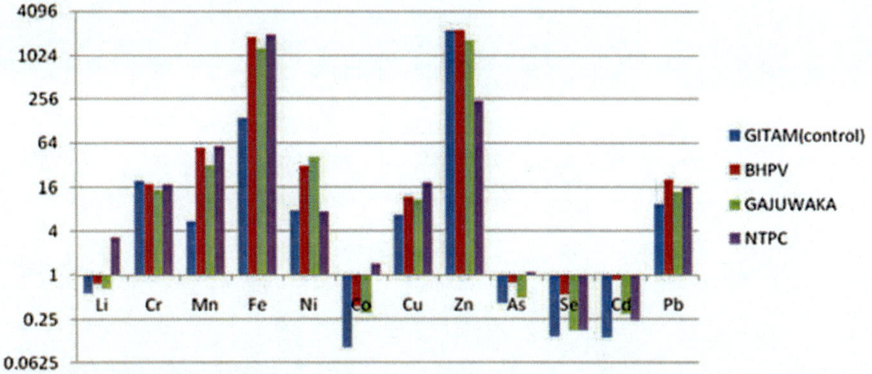

Fig. 1 Trace metal concentrations in air samples at different sampling sites of Visakhapatnam

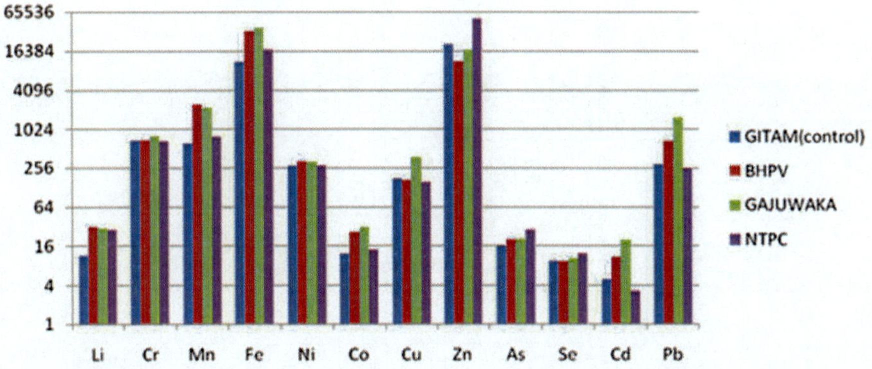

Fig. 2 Trace metal concentrations in moss (*Bryum argenteum*) samples at different sampling sites of Visakhapatnam

Table 1 The Bio concentration factor values for heavy metals in the study areas

Location	Cr	Mn	Fe	Ni	Cu	Zn	As	Se	Cd	Pb
Rushikonda	34.97	113.6	78.69	34.74	26.16	9.11	37.59	65.35	34.85	31.3
Autonagar	39.85	45.11	18.16	10.34	13.88	4.88	25.02	17.01	12.84	33.11
Industrial area	54.05	69.00	28.86	7.95	35.08	10.06	41.97	58.28	66.21	111.47
Parawada	37.4	13.45	8.61	35.67	8.5	206.47	26.03	72.23	13.91	15.7

(GITAM) (19.078 ng/g). Accumulation of Cr metal in moss species Bryumargenteum was highly regarded in Gajuwaka (787.76 ng/g). The lowest absorption was

Table 2 Trace elemental concentration (in ng/g dry wt.) of moss samples collected from sampling sites of Visakhapatnam

Location	Sample	Cr	Mn	Fe	Ni	Cu	Zn	As	Se	Cd	Pb
Rushikonda (control area)	Air	19.078	5.453	137.981	7.776	6.593	2233.659	0.415	0.142	0.14	9.407
	Moss	667	619.16	10,858.48	270.2	172.48	20,361.68	15.6	9.28	4.88	294.44
Autonagar (industrial area)	Air	17.256	53.622	1791.55	30.784	11.808	2240.132	0.788	0.529	0.841	19.871
	Moss	687.68	2419.24	32,548.84	318.44	164	10,942.28	19.72	9	10.8	658.08
Gajuwaka (industrial area)	Air	14.573	31.27	1272.492	40.187	10.859	1650.459	0.486	0.175	0.296	13.573
	Moss	787.76	2157.72	36,730.08	319.68	380.96	16,616.8	20.4	10.2	19.6	1513
Parawada industrial area	Air	17.28	57.02	1958.86	7.59	17.99	241.23	1.10	0.17	0.23	15.55
	Moss	646.4	767.48	16,871.92	270.8	153.08	49,807.56	28.64	12.28	3.2	244.24

observed in the sampling site of the Parawada industrial area (646.4 ng/g). The significant sources of chromium emissions are from the steel industry, combustion of tires [35–37].

3.2 Manganese

In the moss sample, concentrations of Manganese were represented in Tables 1 & 2. The Manganese concentrations in air samples ranged from 5.5 ng/g to 57.02 ng/g. The significant sources of Manganese are vehicular movement and combustion of coal [37, 38].

Parawada industrial area moss samples show a lower accumulation of Mn. The maximum and minimum Manganese concentrations in moss samples found were 619.16 ng/g and 2419.24 ng/g.

3.3 Iron

The observed iron concentrations in air samples range is from 137.981 ng/g to 1958.863 ng/g. A substantial amount of iron was found in the air sample collected in Parawada industrial area (1958.863 ng/g). The familiar sources for the iron emissions are the steel industries, traffic emissions, and combustion [16].

The maximum and minimum concentrations of Iron in moss samples found were 36,730.08 ng/g and 10,858.48 ng/g. Maximum accumulation of Iron in Bryumargenteum was found in the area of Gajuwaka. Minimum accretion of Iron in moss species was found in the control area due to low atmospheric concentration.

3.4 Nickel

Nickel concentrations in air samples in all sampling sites range from 7.594 ng/g to 40.187 ng/g. The maximum value of air samples was found in the sampling site of Gajuwaka, whereas the minimum value was observed at the site Parawada industrial area, which was less than the control area air concentration. Nickel was emitted as an air pollutant from thermal power stations, where the combustion of coal was significant activity [31].

Nickel's maximum and minimum concentrations in moss samples found were 319.6 ng/g and 270.2 ng/g. Accumulation of Nickel in moss Bryumargenteum was highest in the Gajuwaka sampling site due to increased atmospheric concentration. Minimum nickel concentration in moss samples was found in the control sampling site.

3.5 Copper

The observed copper concentrations in air samples range is 6.593 ng/g to 17.994 ng/g. Minimum copper concentration was found in the control area GITAM. Maximum copper concentration was observed in the Parawada industrial area. The copper concentration in the air observed was not at an objectionable level. The atmospheric deposition of copper may be due to coal combustion and from smelters and metallurgic industries [30].

The long-term accumulation of moss for copper metal ranges from 153.08 ng/g to 380.96 ng/g. Minimum accumulation of copper metal was observed in the Parawada site, which has the highest atmospheric copper concentration. Maximum copper accumulation was observed in the Gajuwaka sampling site.

3.6 Zinc

Zinc levels in air samples were found between 241.237 and 2233.659 ng/g. Minimum zinc concentration in air was found in the Parawada industrial area site. The maximum concentration was observed in the Control area. The primary sources for zinc emission are smelters and the combustion of fossil fuels [30].

The moss zinc accumulations range between 10,942.28 ng/g and 49,807.56 ng/g. The highest accumulation of zinc element was found in Parawada industrial area, where the lowest atmospheric zinc concentration was found. This is due to the impairment of the accumulation capacity of moss due to high concentrations. The lowest accumulation of zinc in the moss sample was observed in the Autonagar area.

3.7 Arsenic

Arsenic was highly toxic to the environment. The air concentrations of arsenic range between 0.415 ng/g and 1.104 ng/g. The lowest air concentration of arsenic was found in Control (0.415 ng/g), a control site. Maximum arsenic concentration in air samples was observed in the Parawad industrial area. Arsenic emissions may be from coal and tires, and glass manufacturing industries [39].

The Arsenic concentration of the moss sample was recorded between 15.6 ng/g and 28.64 ng/g. The maximum Arsenic concentration in Bryumargenteum was observed in Parawada industrial area, which has a high atmospheric arsenic concentration. And the minimum Arsenic concentration in moss was found in the control site due to low arsenic concentration in the atmosphere.

3.8 Selenium

Selenium concentration in air samples of different sampling sites ranges between 0.142 ng/g and 0.529 ng/g and toxic. Minimum air selenium concentration was found in the area 0.142 ng/g. Maximum air selenium concentrations were found in the Autonagar area. Minor emissions of Selenium were due to thermal power stations and coke oven batteries [31].

Accumulation of Selenium was dependent on the atmospheric selenium concentrations and climatic factors. The maximum deposition in moss was found in Parawada industrial area, and the minimum deposit was observed in the Autonagar area. High concentration in the atmosphere may retard the uptake of Selenium by moss species Bryumargenteum. The moss selenium concentration ranges between 9 ng/g and 12.28 ng/g.

3.9 Cadmium

Cadmium is the most toxic metal, and it is primarily present as industrial cadmium dust or fume. The Cd concentrations in the air range between 0.14 ng/g and 0.841 ng/g. Maximum atmospheric concentration was found in the Autonagar site, whereas minimum concentration was found in Rushikonda (GITAM), i.e., the control area. The familiar sources of Cadmium emissions are fertilizer industry, combustion, and vehicular emissions [40].

The accumulation of cadmium in moss species Bryumargenteum was ranging between 3.2 ng/g and 19.6 ng/g. the maximum accretion was found in the Gajuwaka area, whereas minimum accumulation was found in Parawada industrial area.

Lead

Lead concentration in different sampling sites was ranging from 9.407 ng/g to 19.871 ng/g. The lowest atmospheric concentration was found in the Control area, whereas the highest concentration was found in the Autonagar area. The primary emission sources for lead elements were smelters, traffic, and combustion of fossil fuels and tires [30].

The accumulation concentration of lead was between 244.24 ng/g and 1513 ng/g. The maximum accumulation concentration in moss was found in the Gajuwaka area, while the minimum was observed in Parawada industrial area.

The Bioconcentration factor (BCF) of heavy metal(s) was calculated as shown below.

BCF = Metal concentration in moss (mgkg-1)/Metal concentration in air environment (mgL-1 or mgkg-1).

It is clearly observed that the accumulation of trace metals in moss samples was comparatively high in all the industrial areas than the control site. Compared to other moss samples collected from industrial areas, Autonagar moss samples show high levels of heavy metals. In Parawada industrial area, where high atmospheric concentrations were found, significant accumulation in moss (Bryumargenteum) was not observed. The concentrations of heavy metal in the air are in the order of Fe > Zn > Mn > Ni > Pb > Cr > Cu > Cd > As > Se. The uptake efficiency of studied heavy metals in the moss species Bryumargenteum is in the order Fe > Zn > Mn > Pb > Cr > Ni > Cu > As > Cd > Se.

Bryumargenteumcan is used as a bioindicator for atmospheric pollution [41]. The uptake efficiency of the majority heavy metals follows the order Pb > Co, Cr > Cu, Cd, Mo, Ni, V > Zn > As in most studies [42].

Analyte	Cr	Mn	Fe	Ni	Cu	Zn	As	Se	Cd	Pb
% increase in industrial area to control area	15	74	70	15	54	59	45	24	75	80

The amount, duration, and intensity of heavy metals in the atmosphere influence moss accumulation [22]. If there are large levels of certain cations in the depositions, these affect the uptake of other cations due to ion competition.

4 Conclusion

The ease of sampling, the lack of complex and expensive elevated equipment and the consolidated and time-integrative behavior patterns of the moss biomonitor provides the advantages for future biomonitoring of atmospheric trace elements, particularly in large-scale surveys.

Biomonitoring is a critical method for determining the source. It is a simple and straightforward process to collect mosses in specific areas ranging from pollution-free background regions to highly polluted areas. By collecting mosses during the pre-monsoon, monsoon, and post-monsoon seasons, it is possible to determine the specific trace element pollution area. This research can also predict the appropriate moss species, which can be used as a biomonitor for a single trace element or a group of trace elements.

The present study reveals the content of heavy metals (Cr, Mn, Zn, As, Fe, Ni, Cu, Se, Cd, and Pb) in moss Bryumargenteum collected in the industrial zones of Visakhapatnam. The species grows prominently in the urban areas and forms dense carpet in moist and shady places, particularly in the winter months. The results show that these species can accumulate high amounts of trace metals present in particulate matter of its surrounding atmosphere. Thus, the moss Bryumargenteum can be used as a heavy metal indicator in regional studies as follows.

Other approaches have a great difficulty to obtain such detailed and accurate variations in time and space at a reasonable cost. Since the current study was preliminary, and a systematic protocol needs to be developed.

Acknowledgements We are thankful to DST–FIST facilities at the department, National Geophysical Research Institute (NGRI) Hyderabad for ICP-MS analysis and GITAM Deemed to be University for providing the indispensable facilities.

References

1. Ahuja, D., & Marika, T. D. (2009). Sustainable energy for developing countries ,S.A.P.I.EN.S. Retrieved from http://sapiens.revues.org/823
2. CPCB. (2009). *Hazardous air pollutants*. Delhi, India: Central Pollution Control Board, Ministry of Environment & Forests.
3. Florentina, I., and Ion, B. (2011). The effects of air pollutants on vegetation and the role of vegetation in reducing atmospheric pollution, the impact of air pollution on health, economy, environment and agricultural sources. Retrieved from http://www.intechopen.com/books/the-impact-of-air-pollution-on-health-economy-environment-andagricultural-sources/the-effects-of-air-pollutants-on-vegetation-and-the-role-of-vegetation-in-reducingatmospheric-pollu
4. Kampa, M., & Castanas, E. (2008). Human health effects of air pollution. *Environmental Pollution, 151*, 362–367.
5. Briggs, D. (2003). Environmental pollution and the global burden of disease. *British Medical Bulletin, 68*, 1–24.
6. Seppänen, O., and Kurnitski, J. (2009). Moisture control and ventilation. In: *WHO guidelines for indoor air quality: Dampness and Mould*. Geneva: World Health Organization.
7. Haag, A. L. (2007). The even darker side of brown clouds. *Nature climate Change, 1*, 52–53.
8. Kvetoslav, R. (Ed.). (2000). *Aerosol Chemical Processes in the Environment*. London: CRC Press.
9. Self, S. (2006). The effects and consequences of very large explosive volcanic eruptions. *Philosophical Transactions of Royal Society A, 364*, 2073–2097.
10. Taddesse, A., Chaubey, A. K., Tesfaye, T., & Yehunie, A. (2012). Application of instrumental neutron activation analysis technique for investigation of air pollution in Addis. *International Journal of Scientific & Engineering Research, 3*(10), 8–16.
11. Ataabadi, M., Hoodaji, M., & Najafi, P. (2010). Heavymetals biomonitoring by plants grown in an industrial area of Isfahan' Mobarakeh steel company. *Journal of Environmental Studies, 35*(52), 25–27.
12. Vučković, I., Špirić, Z., Stafilov, T., & Kušan, V. (2012). Biomonitoring of air pollution with zinc in croatia studied by moss samples and ICP-AES. *Contributions, Section of Natural, Mathematical and Biotechnical Sciences, 33*(1–2), 47–60.
13. Poikolainen, J. (2004). Mosses, epiphytic lichens and tree bark as biomonitors for air pollutants – specifically for heavy metals in regional surveys. Academic Dissertation, Faculty of Science, University of Oulu.
14. Fatoba, P. O., Ogunkunle, C. O., & Olawepo, G. K. (2012). Assessment of atmospheric metal depositions in the industrial areas of the southwest of Nigeria. *Ethiopian Journal of Environmental Studies and Management EJESM, 5*(3), 260–267.
15. Govindapyari, H., Leleeka, M., Nivedita, M., & Uniyal, P. L. (2010). 2010, Bryophytes: Indicators and monitoring agents of pollution. *NeBIO, 1*(1), 35–41.
16. Markert, B., Wappelhorst, O., Weckert, V., Herpin, U., Siewers, U., Friese, K., & Breulmann, G. (1999). The use of bioindicators for monitoring the heavy-metal status of the environment. *Journal of Radioanalytical and Nuclear Chemistry, 240*(2), 425–429.

17. Hoodaji, M., Ataabadi, M., and Najafi, P. (2012). Biomonitoring of airborne heavy metal contamination, air pollution: Monitoring, modelling, health and control. Retrieved from http://www.intechopen.com/books/air-pollution-monitoring-modellinghealth-and-control/biomonitoring-of-airborne-heavy-metal-contamination
18. Pöykiö, R., Hietala, J., & Nurmesniemi, H. (2010). Scots pine needles as bioindicators in determining the aerial distribution pattern of sulphur emissions around industrial plants. *International Scholarly and Scientific Research & Innovation, 4*(8), 92–95.
19. Blagnytė, R., & Paliulis, D. (2010). Research into heavy metals pollution of atmosphere applying moss as bioindicator: A literature review. *Environmental Research, Engineering and Management, 4*(54), 26–33.
20. Markert, B., Vazquez, M. D., & Wappelhorst, O. (2004). Determination of 28 elements in aquatic moss Fontinalis antipyreticahedw and water from the upper reaches of the river Nysa, by ICP-MS, ICP-OES and AAS. *Water, Air, and Soil Pollution, 152*, 153–172.
21. Zhou, Q., Zhang, J., Jianjie, F., Shin, J., & Jiang, G. (2008). Biomonitoring: An appealing tool for assessment of metal pollution in the aquatic ecosystem. *Analytica chimica acta, 606*, 135–150.
22. Chakrabortty, S., & Paratkar, G. T. (2006). Biomonitoring of trace element air pollution using mosses. *Aerosol and Air Quality Research, 6*(3), 247–258.
23. IAEA-TECDOC-1338. (2003). Biomonitoring of atmospheric pollution (with emphasis on trace elements)—BioMAP II. In *Proceedings of an international workshop organized by the International Atomic Energy Agency*. Instituto Tecnológico e Nuclear and the Universidade dos Açores.
24. Namieśnik, J. (2001). Modern trends in monitoring and analysis of environmental pollutants. *Polish Journal of Environmental Studies, 10*(3), 127–140.
25. Rhüling, A. (1998). *Atmospheric heavy metal deposition in Europe 1995–1996*. Copenhagen: Nordic Council of Ministers.
26. Balaram, V. (2005). Recent developments in analytical techniques for characterization of ultra pure materials—An overview. *Bulletin of Materials Science, 28*(4), 345–348.
27. Satyanarayanan, M., Balaram, V., Roy, P., & Gnaneshwar, Rao T. (2006). Inter laboratory collaborative analysis of rainwater reference material. *Current Science, 91*(2), 235–238.
28. Marin, Ş., Lăcrimioara, Ş., & Cecilia, R. (2011). Evaluation of performance parameters for trace elements analysis in perennial plants using ICP-OES technique. *Journal of Plant Development, 18*, 87–93.
29. Reddy, S. L., Fayazuddin, M., Gangi Reddy, N. C., Ahmad, A., Reddy, G. S., Rao, P. S., Reddy, B. J., & Frost, R. L. (2008). Characterisation of bhringaraj and guduchi herb by ICP-MS analysis, optical absorption, infrared and EPR spectroscopic methods. *Spectrochimica Acta Part A: Molecular and Biomolecular Spectroscopy, 71*(1), 31–38.
30. Ana-Maria, R. U. S. U., Bartok, K., & Nemeth, R. (2003). Lead, Copper and Zinc content in native lichens and mosses collected from Baia Mare smelter area. *Contribution Botanice, 38*(1), 113–120.
31. Grodzinska, K., Frontasyeva, M., Szarek-Lukaszewska, G., Klich, M., Kucharska-Fabis, A., Gundorina, S. F., & Ostrovnaya, T. N. (2003). Trace element contamination in industrial regions of Poland studied by moss monitoring. *Environmental Monitoring and Assessment, 87*, 255–270.
32. Sabovljevi, M., VanjaVukojevi, A. S., & Vujii, M. (2009). Deposition of heavy metals (Pb, Sr and Zn) in the county of Obrenovac (Serbia) using mosses as bioindicators. *Journal of Ecology and The Natural Environment, 1*(6), 147–155.
33. Pal, R., Kumar, A., Gupta, A., & Mahima, T. A. (2014). Source identification and distribution of toxic trace metals in respirable dust (PM10) in Brasscity of India. *Global Journal of Human-Social Science: B Geography, Geo-Sciences, Environmental Disaster Management, 14*(5), 1.
34. Vats, S. K., Singh, A., Koul, M., & Uniya, P. L. (2010). Study of metal absorption by two mosses in Delhi region, India. *Journal of American Science, 6*(3), 176–181.
35. Vučković, I., Špirić, Z., Stafilov, T., & Kušan, V. (2013). Moss biomonitoring of air pollution with chromium in Croatia. *Journal of Environmental Science and Health, Part A: Toxic/Hazardous Substances and Environmental Engineering, 48*(7), 829–834.

36. Bačeva, K., Stafilov, T., Šajn, R., & Tănăselia, C. (2012). Moss biomonitoring of air pollution with heavy metals in the vicinity of a ferronickel smelter plant. *Journal of Environmental Science and Health, Part A: Toxic/Hazardous Substances and Environmental Engineering, 47*(4), 645–656.
37. Shakya, K., Chettri, M. K., & Sawidis, T. (2001). Appraisal of some mosses for biomonitoring airborne heavy metals in Kathmandu valley. *Eco Print, 8*(1), 69–75.
38. Marinova, S., Frontasyeva, M., Yurukova, L., Strelkova, L., Steinnes, E., & Marinov, A. (2010).The moss techniques for air pollution study in Bulgaria. In *AIP Conference Proceedings, 1203*(1), 777–782.
39. Blum, O. (2007). Atmospheric heavy metal deposition in Romania and neighboring countries: Comparative evaluation on the basis of all European moss monitoring. In R. N. Hull, C.-H. Barbu, & N. Goncharova (Eds.), *Strategies to enhance environmental security in transition countries* (pp. 369–385). Dordrecht: Springer.
40. Otvos, E., Pazmandi, T., & Tuba, Z. (2003). First national survey of atmospheric heavy metal deposition in Hungary by the analysis of mosses. *The Science of the total Environment, 309*(1–3), 151–160.
41. Pradhan, N., & Joshi, S. D. (2008). A diversity account of Bryaceae (Bryophyta: Musci) of Nepal. *Journal of Natural History Museum*,19–26.
42. Zechmeister, H. G., Dullinger, S., Hohenwallner, D., Riss, A., Hanus-Illnar, A., & Scharf, S. (2006). Pilot study of road traffic emissions (PAHs, Heavy metals) measured by using mosses in a tunnel experiment in Vienna. *Austria. Environmental Science and Pollution Research, 13*(6), 398–405.

Glacier Melting: Drastic Future

Amit Awasthi, Rishi Dewan, and Geetesh Singh

1 Introduction

Glaciers are the bulky mass of snow and ice forms in the area where the rate of snowfall frequently surpasses the rate at which the snow melts. When snowfall, and remains in one location over a many years compress into large, thickened ices masses give rise to formation of the Glaciers. Glaciers are sensitive indicators of changing climate. Movement is one of the unique property of Glacier which makes it different from mountain or ice. Glaciers survive, if mean annual temperatures are close to the freezing point, winter precipitation yields substantial accumulations of snow, temperatures throughout the year are so high that it'll not completely loss the previous winter's snow accumulation. Before understanding about the Glaciers and its melting, first let us understand about the difference between Glaciers and Dead Ice.

As discussed, Glacier is a term which in general or broadly represents "the huge mass of ice formed by depression and re-crystallization of snow which are formed year after year process. Its formation is by periods of cold climate, the accumulation and degeneration of ice several times. Not all the huge masses of ice are known as Glaciers. Glaciers differ from dead ice on the basis of its motion. The maximum speed of glacier is up to 40 m per day [1]. The glacier with zero speed is known as dead ice. The movement of Glacier is slow and due to its own weight. These are formed when snow remains in one place for a long time to transform into ice.

For the formation of glacier, it is imperative that the snow accumulates in winter is more than snow melts in summer. As per the researches done, it is believed that

A. Awasthi · R. Dewan (✉)
Department of Physics, University of Petroleum & Energy Studies, Dehradun 248001, India
e-mail: rishi@ddn.upes.ac.in

G. Singh
Department of Petroleum & Earth Science, University of Petroleum & Energy Studies, Dehradun 248001, India

N. A. Siddiqui et al. (eds.), *Advances in Behavioral Based Safety*,
https://doi.org/10.1007/978-981-16-8270-4_19

there may be 21 glacial cycle during last ice age alternating with number of warm glacial cycle. 30 million cubic km of earth is covered with ice. Pleistocene, Ice age which results the formation of glaciers in Himalayas [2].

1.1 Colour of Glacier

Glacier appears to be blue in color. The reason for blue color is water present in it. Actually compressed ice absorbs other color except blue color thus it appears to be blue. While another reason is absence of air bubbles. Air bubbles exhibit white color, but as per high pressure there are no air bubbles thus blue color is seen.

1.2 Movement of Glaciers

One of the unique property of Glaciers is their movement. It is also important to understand the phenomenon behind the movement of these glaciers. Gravitational force on the ice mass or sheer weight of a thick layer of ice is supposed to be main reason behind the gradual flow of these glaciers. Due to softer behavior of Ice in comparison to rock, Ice deformed easily by the persistent pressure of its own weight. Depending upon the amount of snow evaporate, accumulate or melt, Glaciers periodically retreat or retreat itself. Irrespective of the retreatment, glaciers like a conveyor belt deform and move downslope. In general, advancing and retreating require years or decades to show substantial effect. But if due to some reasons, when glaciers retreat rapidly, movement may be noticeable within few months or years. Movement of glaciers depends upon its advance and retreat, i.e., if ice and snow added are more than loss during evaporation, calving or melting then glaciers will advance and vice versa. This can be clearer if we understand the type of zones formation in glacier. Accumulation Zone is the zone where snow is added to the glacier and creates to turn to ice also termed as input zone. This zone is generally upper region of the glacier through which water seeps and slowly forms vertical lands and horizontal ice lenses. This is the zone where the ice flows like a conveyor belt which is driven by gravity. Other zone is ablation zone where the glacier loses ice through evaporation, calving and melting and is also called as output zone. It includes the lower region of glacier. More about the zone exhibited by Glaciers is discussed in the coming section.

1.3 Zones Exhibited by Glacier

The normal valley glacier exhibits two zones, those are accumulation zone and ablation zone. Firstly, the accumulation zone is addition of snow to glacier by means of snowfall, hail, rain, drifting of snow etc. In this a new layer with respect to previous year is added to glacier. The new layer is easily distinguishable as it is clear. The ablation zone is the melting of ice from glacier by means of melting, evaporation etc. The ablated or melted ice is highly dirty and melted water results in formation of ponds or rivers. Theoretically, the lines which separate the accumulation zone and ablation zone are known as equilibrium line. To determine the health of Glacier, glaciologists calculate the ablation of ice and accumulation of ice. If ablation process is higher than accumulation zone then the Glacier health is in danger. It has been estimated that some of the glaciers are undergoing an abnormal retreat, due to global warming and were likely to vanish in next 50 odd years and so on. For instance, Siachen Glacier, the second largest Glacier is about 74 km long showed a great retreat in it. Snout of glacier is observed and predicted that it is likely to vanish in next 50 years [3, 4].

1.4 Classification of Glaciers

Glaciers are not only found in Antarctica and Green land but they are found in almost every continent of the world, because some places in these continents have specific geographic and climatic condition that support the formation of galactic ice in these areas. Major conditions for the existence/formation of glaciers are that more snow should be accumulate in winter than melt in the summer, this is the reason why glaciers found mostly in mountainous or the polar region. As per the information compiled by the world Glacier Monitoring Services, glaciers covered area of 352,500 square kilometers of Arctic islands and Antarctica without including the main ice sheet of Antarctica and Greenland Ice Sheet of Arctic. Glaciers are classified by their size (i.e., ice sheet, ice cap, valley glacier, cirque glacier), location and thermal regime (i.e., polar vs. temperate).

On the basis of location, Glaciers are mainly classified into two category Alpine glaciers and Continental glaciers. Alpine glaciers located only on the top of mountain whereas continental glaciers are located on continental landmasses like Greenland or earth's poles irrespective of elevation. In general, Continental glaciers are greater in size in comparison to Alpine glaciers. Both types of glaciers create landforms due to weathering and deposition. On the basis of different type of landform created during the movements of glaciers can be named as Mountain Glacier, Valley glaciers, Tidewater glaciers, Piedmont glaciers, Hanging glaciers, Cirque glaciers, Ice Aprons, Rock Glaciers, Ice shelves, Icefields, Ice caps, Ice streams, Ice sheets, etc. These different type of Glacier by its melting property are one of the reason which is responsible for existence of life on earth. Many living things are survive on the earth because of Glacier melting.

2 Significance and Insignificance Prospective of Glacier Melting

It is important to understand that, no doubt melting of Glaciers is essential for the survival of life on earth. But if the melting of Glacier is not supported by the accumulation of ice and speed of Glacier melting increase day by day then it will create direct and indirect short and long-term drastic effect on the ecological system. Glacier melting is the natural process and it is essential to glacier to be melted. As glacier melting supports agricultural irrigation. It helps in production of hydroelectricity. It also provides an environment to aquatic animals. Glacier is an important source of water across whole world. Many source of water in desert is downstream of mountainous water originated from glacier. Many of ponds, lakes, rivers, wells and other many source are constructed due to melting of Glacier. Glacier ice is also been used commercially. Its source is used as drinking purpose for cold water before invention of refrigerators. Ice is used in drinks for people of Norway's [5]. In many regions of world, melt water is used as electric power generation because very high pressure generated due to flow of water. So many stations works on power generated from glaciers.

Glaciers also benefit in terms of tourism provide to the country. As the scenery of glacier is very attractive, people fond to visit these places and this visiting economically benefits the countries. Other than homo-sapiens, Glacier also benefits other species. Many species of plants and animals have their origin because of environment provided by the Glacier.

Glacier retreats affect aquatic eco system on basis of water temperature and stream flow. Glacier works as a planet air conditioner, which regulates the climate and have significant effect on climate. Glacier acts as a mirror, which reflects UV rays in atmosphere and hence regulates temperature on Earth. But due to ozone depletion, intense rays from sun hit glacier and more refection takes place which leads to global warming and then, these global warming melts glaciers. Thus, the increase in melting impacts our planet adversely. Its effect can be seen directly or indirectly on agriculture, coastal areas, electricity demand, irrigation, etc.

2.1 Effect of Glacier's Melting on Agricultural Practices

Melting of glacier will directly and directly influence agricultural output. Irrigation is supply of fresh water to plants for their growth. It is known that plants need water, air and sunlight for their growth. Due to rising of sea level and rapid melting of Glaciers, various contaminations that have been freezed in glaciers will melt and will influence the health of earth. Increase in melting speed of glacier is a serious note especially for agriculture output. More melting of Glacier will increase the flow of water in rivers which affect the irrigation process of river and related canals. Except some of few areas on earth which depend on rainfall will remain unaffected as they are not directly or indirectly depended on glacier water [6]. If glacier will extinct, no rivers will flow, thus lacked freshwater will negatively affect the agriculture and during the dry season agricultural land will face the tremendous shortage of water. Practice of burning of agriculture reside is also increases which is support indirectly by over melting of Glaciers m which have significant effect on the respiratory system [7]. This shortage causes the significant negative effect on agricultural output. In the beginning, when rapid melting of Glacier occurred, it cause water level high in many rivers and lakes will cause flood in river and may destroy the planted crops. Sea level rise will cause floods in many areas which are near sea or oceans, so this leads to destruction of land permanently. Thus, melting of glaciers affects agricultural output in many ways. If precautions will not be taken, then in upcoming years, permanent disappearance of agricultural land will take place. As the vanishing of agricultural land will create the appetite of food in many areas. Therefore, it affects the livelihood on earth and it will influence various species on planet. Although future conditions are worse due to melting of glacier on irrigation but in present it is considered that there may be positive effect on agriculture.

2.2 *Effect of Glaciers Melting on Electricity Production*

The increment of greenhouse gases in atmosphere leads to global warming and melting of glaciers. As hydroelectricity is produced by glacier melt water, when melt water falls on turbine, electricity is generated. Therefore, melted glacier water is essential for production of electricity. Researchers concluded that effect of electricity on glacier and effect of glacier on electricity is dynamic in nature. They both are interrelated to each other as by the consumption of electricity excessive amount of carbon dioxide is released which leads to warming of globe and later on melting of glaciers. If glacier will be extinct, then generation of hydroelectricity will also be affected drastically [8].

2.3 *Effect of Glaciers Melting on Coastal Areas*

Coastal areas are considered to be as most ecological and economical areas on the earth. As in the coastal areas, the country invests much and gets much more from tourism, forests etc. Coastal areas are threatened by global climatic change. Recently there are only minor effects on coastal areas due to glacier melting. But its predicted magnitude in coming centuries will trigger a worst effect on coastal areas. As these changes continue, we risk serious degradation of marine ecosystem which will show adverse effects on human health and welfare. To decline the problem of coastal areas one must show major focus of concern regarding climate change. It is known that over different places on Earth the temperature (atmosphere) is different, somewhere the temperature is high and another there is low. There is variability over Earth is natural variability while it has been noticed that beside of this natural variability the invariable temperature risen has been taking place. Climatic changes due to human activities will lead to distinction of coastal areas. Due to these activities, the greenhouse gases such as carbon dioxide, nitrogen dioxide etcis increasing which leads to global warming. The warming of globe results in melting of glacier thus rise of sea level at coastal areas. And the physical and chemical changes in marine system are clearly observed.

According to IPCC (2001) that the temperature at sea surfaces has risen 0.4–0.8 °C over past century [9]. Due to human activities, the greenhouse gases are released in atmosphere, which warm the planet and depletes the ozone layer and ultraviolet rays directly hit the earth surface which rises temperature of atmosphere. Different physical changes can be easily seen on coastal areas are the warming of ocean water, expansion of ocean, rise in sea level etc. According to IPCC (2001), the sea level rises 2 mm per year because of melting of Glacier [9]. Chemical changes are also observed because of increase in the concentration of carbon dioxide in atmosphere. According to IPCC (2001), the carbon dioxide concentrations are to be risen from 280 to 540–970 ppm by 2100 year [9]. It has been estimated that from year 1894 to 1994, the half of carbon dioxide released is now stored in oceans [10]. This is the probable reason that the chemical changes have been observed in oceans and continuing uptake of atmospheric carbon dioxide from oceans decreases oceanic pH over the next few centuries. It has been estimated that the pH decreases from 0.3 to 0.5 units over the next 100 years and from 0.3 to 1.4 units over next 300 years which in turn depends on carbon dioxide emission in atmosphere [11]. Increasing in carbon dioxide level in atmosphere leads to depletion of ozone layer and leads to enhanced level of ultraviolet rays at earth's surface [12].

3 Depressing Truth

It was around the start of twentieth century that the global network first time understood the conceivable effect of the rise in environmental temperature on the mountain ice sheets. It was trusted that such an examination over next 100 odd years or so would empower the researchers to build up the connection between the environmental change and the icy mass vacillations. Glaciologists of mid-twentieth century were impaired by the way that they almost no data was at the hand about amassing zone of ice sheets, when contrasted with exhibit time [5, 6]. Softening of Icy mass is normal phenomenon however because of human errors, the rate of withdraw is exceptionally quick. Learning all about melting, a question here arises that why and how are glaciers melting? Glaciologists assess the planet and found earth to be warmer irrespective to past. This warming of globe is termed as Global Warming. Global warming, a short term, contains a giant and horrible meaning is the main reason for glacier melt down. From last three decades melting of Glacier had been speeded up [13]. There are many reasons for global warming such as the excessive use of chemical fertilizers in croplands, the deforestation causes our planet to be warmer.

Plants help to circulate the temperature on earth and avoid the risk of warming but if the quantity of plants will be reduced then automatically the temperature will rise. Another main reason is Greenhouse gases causing pollution and creating hole in ozone layer (which protects our earth from harmful ultraviolet rays), when sun rays fall directly on earth it increases temperature rapidly and causes glacier to melt [14].

4 Studies Done on Indian Himalyas

Different studies on Indian Himalayas can be broadly understood into three phases as per the MOEF discussion paper [15].

First phase: In view of distributing information, the primary stage involved the studies done from early twentieth century to 1950AD. In this stage, main focus was done to understand the complexities in social part of the Glacier, local ice cover expansion, geomorphology and component of their transient and procedure of latest ice age. The short history of first stage is started in 1905 with Ice sheet snout formation, which was first time quoted by the scientist. Noses of ice sheets in Himalayas were first time conveyed by Geological study of India in 1906. Geologists reported different parameters from Jammu and Kashmir to Sikkim by drawing huge scale maps via plane table and telescope. Photos were taken from different plots for future relationship by inspecting more than 19 ice sheets. The icy mass snout was gently balanced between two primary parameters-temperature and snow precipitation, and would experience change in both elevations and appearance [16].

Second phase: The second stage consider the observation and studies done on ice sheets from 1957 to 1970AD. The primary field of research in this period of study was the stretched examinations to expand mindfulness toward recording the perpetual ice masses of the Himalayas. This movement was expanded nearly till seventies. In this period, main focus on the icy masses available at Siachen, Mamostong, Kumdan, Machoi in J&K, Barashigri, Sonapanii, Guglu in H.P, Gangotri, Arwa, Poting, Milam, Pindari, Shankalpa, Kalganga, Bamlas, Safed, Bhilmagwar, Pachu, Burphu in Uttarakhand and Zemu in Sikkhim. In year 1965, geophysical techniques of seismic and resistivity were started at the Zemu glacier (Sikkim) for understanding the slight deviation in evaluation of ice thickness. The investigations involve the comparison of nose position of the individual ice sheet with its prior mapped position [17]. During these studies, it is observed that Saichen ice sheet withdrew by around 400 m between 1929 and 1958 AD. Kumdan ice sheets of the Upper Shyok valley observed to experience diverse phases in year 1958 [18].

Third phase: Third phase represent the post-1970 duration in which international hydrological program was introduced. In this phase main focus were done on 1. Glacier Inventory, 2. Glacier Regimen or Mass Balance, 3. Glacier Hydrometry, 4. Suspended Sediment Transport, 5. Glacier Flow Movement, 6. Monitoring of the glacier snout, 7. Geo-morphological studies of the peri-glacial zone, including the identification and mapping of the glacier landforms, to establish the extent of ice cover during the various phases of the last Ice Age, 8. Special studies like Thermal profiling of the glacier ice and artificial augmentation of glacier ice melting. The investigations were done mainly 10 Glaciers and those were: Nehnar (J&K), Gara, Gorgarang, Shanegarang, Baspa(HP), Gangotri, Dunagiri, Tipra Bamak, Chorabari (Uttarakhand) and Zemu & Changme Khangpu (Sikkim) [19].

From these studies, no doubt it is clear that researcher eagerly focuses and watches the behavior of glaciers for the better and safe future of Earth. But for better strategies, more number of studies will be required to understand and control the melting of Glaciers. In the coming section, studies will be discussed in which the focus of measurement is Indian Himalayas.

4.1 National Status

Many Indian scientist studies the behavior of Glaciers and some of the studies will be discussed in the coming section. Main aim to discuss the national status here to highlight that many studies are already done but still many more required to understand the behavior of Glaciers so that necessary steps will be taken accordingly for the better health of Glacier. Because if our Glaciers are healthy then only life will sustain safely for a long time on Earth. Berthier et al. [20] used Remote Sensing data to monitor glacier elevation changes and mass balances in the Spiti/Lahaul region (32.2°N, 77.6°E, Himachal Pradesh, Western Himalaya, India). And in 2004, the digital elevation model is derived from two SPOT5 satellite optical images without using any ground control points. And it has been concluded that the rate of ice loss is twice higher in 1994–2004 than 1997–1999.

Bhambri et al. [21] studied the glacier changes in Garhwal Himalayas, India from 1968 to 2006 on the basis of remote sensing. For his studies, he used Corona and Advanced Spaceborne Thermal Emission and Reflection Radiometer (ASTER) satellite images. He concludes that in Garhwal region of Himalayas, the retreatment of Glacier is low in comparison to previous studies.

Kulkarni et al. [22] estimated glacier retreat for more than 460 glaciers from 1962 for Baspa, Chenab and Prabhati basins. He used Indian Remote Sensing satellite data for his studies. It has been concluded that higher retreat of smaller glacier and climate is influencing the sustainability of Indian Himalayas.

Dobhal et al. [23] had conducted the mass balance studies from 1992 to 2000 on the Dokriani Glacier of Garhwal Himalaya, India. He concluded that there are continuous negative trends in mass balance of Dokriani Glacier which clearly indicates that this Glacier is also rapidly melting as other glacier are melting.

Wagnon et al. [24] studied Chhota Shigri Glacier (32.2°N, 77.5°E; 9 km long) located in Lahaul and Spiti Valley, Himachal Pradesh. After the long time studies on mass balance, he also verified the insignificant impacts on Glaciers. These studies clearly showed that Health of Glaciers are not good and speed of melting of glacier increases and if it is not controlled or reduced then it will pose serious effect on the coming generation.

4.2 International Status

Himalayas is the source of 10 largest rivers in Asia. So, due to rapid melting of glaciers, uncertainty in river is observed. The climatic changes in greater Himalayas can only be addressed through increased regional collaboration in scientific research and policy making. So due to this reason Himalayas is closely observed by international scientific community. International status includes the studies done by international authors on the Glaciers of Indian Himalayan.

Kenneth Hewitt et al. [25] studied the Glacier expansion and the 'Elevation Effect' of Karakoram Himalayas. The traces for glacier expansion was founded in the central Karakoram region, in contrast to worldwide decline of mountain glacier.

Kaab et al. [26] used advanced Spaceborne Thermal Emission and Reflection Radiometer (ASTER) instrument and the Shuttle Radar Topography Mission (SRTM) for Glacier's studies in Indian Himalayas. The study signifies the space based remote sensing and provides new insights for studying glaciers at high elevation.

Barh et al. [27] studied the physical basis of Glacier volume area scaling. The scaling analysis of mass and momentum conversation equation showed that glacier volume can be related to power law more easily in comparison to observed glacier surface area.

Scherler et al. [28] has studied spatially variable response of Himalayan glacier to climatic change affected by debris cover. He concludes that no direct uniform response of Himalayan glaciers is observed with respect to climatic change. And hence more studies will be required for authenticate conclusive remarks.

5 Discussion

According to IPCC 2007 reports, sea levels will rise by 8–23 in. till the end of this century [29]. The effect of glacier melting is drastic, and arctic is one of the worst places affected by global warming. By 2040, the region is expected to have a completely ice free summer or even earlier.

As per studies done it has been concluded that there will be drastic effect of glacier melting on planet. As glacier water is the fresh source of water and used for drinking purposes, thus by rapid melting of Glaciers, the fresh water is merging in oceans that are salty water. Thus, the fresh water is spoiling rapidly. Another major issue arises due to the melting of glacier is rise in sea level. This rise is responsible for flood in many coastal areas and even disposal of some areas. In India, except one or two, all glaciers are showing negative mass balance which clearly indicates the retreatment of Glaciers. The Maldives are at great risk due to glacier melting. It also leads in reduced agricultural output. The agricultural field which depends on glacier water will be greatly affected [30].

Due to lack of Glacier, the production of hydroelectric will be lacked. Once this flow will be stopped the production of electricity will be reduced. Thus, it will effect on several stations. And as a result the economy of the country will be also be affected.

There are many organisms that depend on glacier for its survival, because certain animals need certain range of temperature which is observed at Glaciers. Some bird's species depend on fish species for their survival. So if their aquatic animal will extent then many of bird's species will be affected. Corals require sunlight for photosynthesis, thus as sea level will rise then sunlight will not reach to corals. This will effect on the people who depend on fish as a food for survival in these areas.

It is very necessary to understand the problem of Glacier melting. Because it is accurate time to make some strategy to control or reduce the speed of melting of Glaciers for better future. Otherwise this will create hazardous effect on the ecological system.

6 Conclusion

Importance of Glaciers along with their formation, color, different zone and phenomenon behind melting are discussed in this paper. Discussion on the melting and fast melting of Glaciers with their prominence is presented in the article. This review clearly verifies that Glaciers play an imperative role in sustainability of life on earth. Glaciers sustain our planet's coolness; provide a life to aquatic animals and plants; provide water for irrigation, hydrological electricity and for drinking purposes. Moreover, glaciers are on the urge of danger for its disappearance. It has been believed from retrospective studies that many of these glaciers will melt in coming 300 years. This will be the foremost reason behind the drastic future of planet Earth, if the precautions are not been taken instantaneously. It has been concluded from the studies that excluding one or two glaciers in Indian Himalayas, remaining show negative mass balance. The immediate change in glacier does not always mean that it is increasing or retreating rather needs to be observed for prolonged time to comment on its status. National and International studies clearly indicate that no doubt, scientist are seriously thinking about the behavior of Glacier Health but studies will be required for better understanding and for the formation of some strategies to control the melting of Glaciers.

Neither the glaciers retreat nor a precaution to save them is a simple task. None of the precautions taken could solve this problem through root. This is known that melting of glaciers is essential but their discontinuity to restructure is dangerous. Therefore for better future we need to be prepared with efficient water management and hazard mitigation systems to minimize and control the hazardous effects of glacier melting on the life and ecology on earth. Moreover, further studies are needed to be conducted so that modeling of data can lead to design of some strategies for the better health of the Glaciers.

References

1. Kamb, B., et al. (1985). Glacier surge mechanism: 1982–1983 Surge of Variegated Glacier, Alaska. *Science, 227*(4686), 469–479.
2. Broecker, W. S., & Denton, G. H. (1989). The role of ocean-atmosphere reorganizations in glacial cycles. *Geochim. n Cosmohmica Acfa, 53*, 2465–2501.
3. Kumar, V., Venkataramana, G., & Høgda, K. A. (2011). Glacier surface velocity estimation using SAR interferometry technique applying ascending and descending passes in Himalayas. *International Journal of Applied Earth Observation and Geoinformation, 13*(4), 545–551.
4. Cogley, J. G. (2011). Present and future states of Himalaya and Karakoram glaciers. *Annals of Glaciology, 52*(59), 69–73.
5. Boone, D. J. (1980). The construction of an artificial drilling island in intermediate water depths in the Beaufort sea. In *12th Annual OlC* (pp. 187–197).
6. Piao, S., et al. (2010). The impacts of climate change on water resources and agriculture in China. *Nature, 467*(7311), 43–51.
7. Agarwal, R., Awasthi, A., Singh, N., & Susheel, K. (2012). Epidemiological study on healthy subjects affected by agriculture crop- residue burning episodes and its relation with their pulmonary function tests. *International Journal of Environmental Health Research, 23*(4), 281–295.
8. Fountain, A. G. (1996). Effect of snow and firn hydrology on the physical and chemical characteristics of glacial runoff. *Hydrological Processes, 10*(4), 509–521.
9. IPCC. (2001). IPCC Third Assessment Report: Climate Change 2001. IPCC.
10. Sabine, C. L., et al. (2004). The oceanic sink for anthropogenic CO_2. *Science, 305*(July), 367–372.
11. Caldeira, K., & Wickett, M. E. (2005). Ocean model predictions of chemistry changes from carbon dioxide emissions to the atmosphere and ocean. *Journal of Geophysical Research, 110*, 1–12.
12. Austin, J., Butchart, N., & Shine, K. P. (1992). Possibility of an Arctic ozone hole in a doubled-CO2 climate. *Nature, 360*, 221–225.
13. Muller, F. (1961). Zonation in the accumulation area of the glaciers of Axel Heiberg Island, NWT, Canada. *Journal of Glaciology, 4*(33), 302–311.
14. W. S. B. Paterson. *The physics of glacier: 2002.*
15. Raina, V. K. (2009). *Himalayan glaciers: A state of art review of glacial studies, glacial retreat and climate change.* New Delhi, India: Ministry of Environment & Forests, Government of India.
16. Fairbanks, R. G. (1989). A 17,000-year glacio-eustatic sea level record: Influence of glacial melting rates on the Younger Dryas event and deep-ocean circulation. *Nature, 342*, 637–642.
17. Eisen, O., Harrison, W. D., Raymond, C. F., Echelmeyer, K. A., Bender, G. A., & Gorda, J. L. D. (2005). Variegated Glacier, Alaska, USA: A century of surges. *Journal of Glaciology, 51*(174), 399–406.
18. Galka, M., & Sznel, M. (2013). Late Glacial and Early Holocene development of lakes in northeastern Poland in view of plant macrofossil analyses. *Quaternary International, 292*, 124–135.
19. Hormes, A., Muliler, B. U., Schl-chter, C., & Institut, G. (2015). The Alps with little ice: Evidence for eight Holocene phases of reduced glacier extent in the Central Swiss Alps. *The Holocene, 3*(2001), 255–265.
20. Berthier, E., Arnaud, Y., Rajesh, K., Sarfaraz, A., Wagnon, P., & Chevallier, P. (2009). Remote sensing estimates of glacier mass balances in the Himachal Pradesh (Western Himalaya, India). *Remote Sensing of Environment, 108*(3), 327–338.
21. Bhambri, R., Bolch, T., & Chaujar, R. K. (2012). Frontal recession of Gangotri Glacier, Garhwal Himalayas, from 1965 to 2006, measured through high-resolution remote sensing data Frontal recession of Gangotri Glacier, Garhwal Himalayas, from 1965 to 2006, measured through high-. *Current Science, 102*(3), 489–494.

22. Kulkarni, A., & Bahuguna, I. M. (2002). Glacial retreat in the Baspa basin, Himalaya, monitoredwith satellite stereo data. *Journal of Glaciology, 48*(160), 171–172.
23. Dobhal, D. P., Gergan, J. T., & Thayyen, R. (2008). Mass balance studies of Dokriani Glacier from 1992 to 2000, Garhwal, Himalya, India. *Bulletin of Glaciological Research, 25*, 9–17.
24. Vincent, C., Ramanathan, A., Wagnon, P., Dobhal, D. P., Linda, A., & Berthier, E. (2012). Mass gain of glaciers in Lahaul and Spiti region (North India) during the nineties revealed by in-situ and satellite geodetic measurements. *The Cyrosphere Discussions, 6*, 3733–3755.
25. Hewitt, K. (2005). The Karakoram anomaly? Glacier expansion and the 'elevation effect', Karakoram Himalaya. *Mountain Research and Development, 25*(4), 332–340.
26. Kaab, A. (2005). Combination of SRTM3 and repeat ASTER data for deriving alpine glacier flow velocities in the Bhutan Himalaya. *Remote Sensing of Environment, 94*, 463–474.
27. Bahr, B., Meier, F., & Peckham, S. D. (1997). The physical basis of glacier volume-area scaling. *Journal of Geophysical Research: Solid Earth, 102*, 20355–20362.
28. Scherler, D., Bookhagen, B., & Strecker, M. R. (2011). Spatially variable response of Himalayan glaciers to climate change affected by debris cover. *Nature Geoscience, 4*(1), 1–4.
29. S. S, "IPCC (2007): Climate Change The Physical Science Basis," 2007.
30. Dai, A., & Trenberth, K. E. (2002). Estimates of freshwater discharge from continents: Latitudinal and seasonal variations. *Journal of Hydrometeorology, 3*(6), 660–687.

Impact of Global Warming on Indian Electricity Demand

Madhu Sharma, M. R. Ayyappadas, and D. E. Dhinesh Raj

1 Introduction

India is one of the fastest growing economy in the world where thirst for energy is on the rise. When it comes to climate change, the current climatic variations, increased frequency of natural disasters and constant rise in global mean temperature are the leading global threat to the world in this regard. In India, climate change is categorized into increase in atmospheric greenhouse gas (GHG) concentration levels, rising sea levels, anomalies in precipitation patterns and rise in global mean temperature[1]. Predictions are modeled by World bank that a rise in mean temperature of 0.5°C is likely over all India by the year 2030 and a warming of 2–4°C by the end of this century [2].

The rise in mean temperature along with increasing income for middle class sector of the society will drive energy demand for thermal comfort in future. According to World Economic Forum, India is one of the country with the most cooling degree days (CDD), more than 3,000 per year. At present there are around 290 million households are in India, among them less than around 20 million have air conditioning. Which is domestic ownership of ACs in India today is just 14%, the demand for cooling is expected to push the requirement of room air conditioners to over 100 million by 2050, a 40-fold growth from 2016. Currently ACs responsible for about 40–60% of peak electricity demand in summers in major cities like New Delhi and Mumbai [3]. According to world economic forum, it is estimated that by 2050 energy demand required for thermal comfort would be approximately 600 GW of newly added power generation capacity.

M. Sharma (✉)
University of Petroleum and Energy Studies, Dehradun, India
e-mail: madhusharma@ddn.upes.ac.in

M. R. Ayyappadas · D. E. D. Raj
M-Tech Renewable Energy Engineering, University of Petroleum and Energy Studies, Dehradun, India

N. A. Siddiqui et al. (eds.), *Advances in Behavioral Based Safety*,
https://doi.org/10.1007/978-981-16-8270-4_20

This report provides insights into the role of global warming on the electricity demand of the nation. In this study, relation between mean temperature and electricity demand of five states of India is studied. Along with climatic data, socioeconomic parameters like Per capita income, Per capita electricity consumption and state gross domestic product values of these states are also taken to consideration for the analysis because these are the factors which determine how people react to thermal discomfort. Selected state/regions are Jammu & Kashmir, Kerala, New Delhi, Rajasthan and West Bengal. States are selected for study are from different climatic zones of India. Since Bengaluru is the only city in Karnataka state which belongs to temperate climatic zone particular climatic zone is not considered, instead West Bengal is considered as the fifth state from Hot and Humid climatic zone.

2 Literature Review

A few researches have been done around the world over to inspect the effect of different essential climatic variables, like humidity, wind speed, and solar radiance etc. on power demand. While auxiliary climatic parameters, for example, heating and cooling degree days are additionally considered. On other hand, many socioeconomic and indices such as the local Gross Domestic Product (GDP), energy prices, growth rate and level of production etc. are also used as input parameters to estimate electricity demand most examinations have reasoned that surrounding temperature is the parameter that influences on the variations in electricity demand.

In majority of the studies effect of temperature is considered in terms of Cooling Degree Days 7 Heating Degree Days. The heating degree day measure is the measure used by heating engineers commonly [4]. Cooling degree day (CDD) and Heating degree day (HDD) are measurement records shown to relate demand for energy to cool or heat spaces. The basis to determine whether a particular day is a heating degree day or a cooling degree day is the daily mean temperature. Each day with an average temperature below 18 °C is considered as one heating degree day [5]. The heating degree day is the winter counterpart of the cooling degree day. Heating degree days are calculated over a year by summing up the differences between every day's average daily temperature and the temperature of 18 °C. Degree day is represented for change in the outdoor temperature and excludes those times when heating or cooling systems do not need to work for a day. There are studies carried out in European countries where demand for heating is larger than demand for cooling because of climatic peculiarity of the region. These studies prove that rise in average temperature of those region would reduce the consumption for heating and thereby overall consumption for thermal comfort [6, 7].

Since India's geography is diverse in nature, climate of each region is different in their own way throughout the year. A place is designated to one among the first five climatic zones only when the defined conditions exist there for more than 6 months. There are cases when a region cannot be identified in any of the category, then the place is defined as Composite [8]. Study Carried out by Minke & Bansal reported

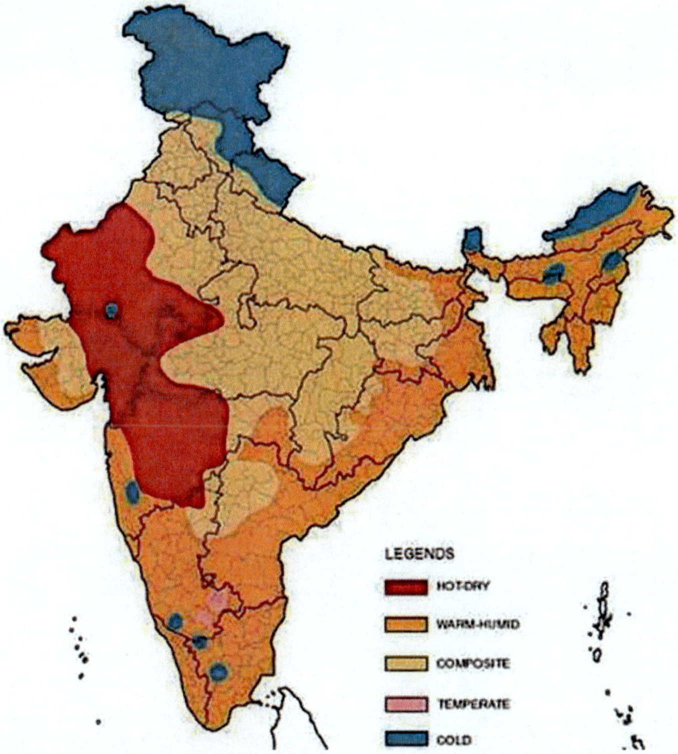

Fig. 1 Climatic zones of India. *Source* User guide Energy conservation building code 2011 [8]

that India can be divided into five climatic zones, which are Temperate, warm and humid, hot and dry, composite and cold (Fig. 1).

Classification and description of climatic zones are explained in the following table with annual range of temperature, precipitation, humidity etc. (Table 1).

In most of the studies, reference projection of climatic scenarios to end of the century is using the Representative concentration pathways (RCP) [9]. The RCPs are the result of an innovative collaboration between integrated assessment modelers, terrestrial ecosystem models, climate models and emission inventory experts. The resulting product creates a data set with sectoral and high spatial resolutions for the period continuing to year 2100. Socioeconomic and emission scenarios are currently implemented in climate research to give reasonable descriptions of how the future scenarios with respect to a series of parameters including, technological change, socioeconomic change, land and energy use, and emissions of GHG and air pollutants [10].

Four RCPs designed from Integrated Assessment Models were selected from the published literature and are used in the Fifth Intergovernmental Panel for Climate Change (IPCC) Assessment as a basis for the climate predictions and projections.

Table 1 Classification of different climatic zones in India (*Source* User guide Energy conservation building code 2011 [8])

Climate zone	Desperation	Mean temperature (C)					Mean relative humidity	Annual precipitation	Sky condition	Places
—		Summer (high)	Summer (low)	Winter (high)	Winter (low)	Diurnal variation	—			—
Hot and dry	High temperature I Low humidity and rainfall I Intense solar radiation I Hot winds during the day and cool wind at night I low underground water table	40–45	20–30	5–25	0–10	15–20	Very low 25–40%	Low < 500 mm/year	Cloudless skies with high solar radiation, causing glare	Rajasthan, Gujarat western Madhya Pradesh, central Maharashtra etc.
Warm and humid	Temperature is moderately high during day and night I very high humidity and rainfall I Calm to very high winds from prevailing wind directions I	30–35	25–30	25–30	20–25	5–8	High 70–90%	High >1200 mm/year	Overcast (cloud cover ranging between 40 and 80%), causing unpleasant glare	Kerala, Tamil Nadu, Costal parts of Orissa and Andra Pradesh etc.

(continued)

Table 1 (continued)

Climate zone	Desperation	Mean temperature (C)					Mean relative humidity	Annual precipitation	Sky condition	Places
—		Summer (high)	Summer (low)	Winter (high)	Winter (low)	Diurnal variation	—			—
Temperate	Moderate temperature | Moderate humidity and rainfall | high winds during summer depending on topography |	30–34	17–24	27–33	16–18	8–13	High 60–85%	High >1000 mm/year	Mainly clear occasionally overcast with dense low clouds in summer	Bangalore, Goa and parts of the Deccan
Cold (sunny/cloudy)	Moderate summer temperature and very low in winter | low humidity in cold/sunny and high humidity in cold/cloudy region	17–24/20–30	4–11/17–21	(−7)–8/4–8	(−14)–0/(−3)–4	25–25/5−15	Low: 10–50%/High: 70–80%	Low: <200 mm/year/Moderate 1000 mm/year	Clear with cloud cover <50%/overcast for most of the year	Jammu & Kashmir, Ladakh, Himachal Pradesh, Uttaranchal, Sikkim, Arunachal Pradesh

(continued)

Table 1 (continued)

Climate zone	Desperation	Mean temperature (C)					Mean relative humidity	Annual precipitation	Sky condition	Places
—		Summer (high)	Summer (low)	Winter (high)	Winter (low)	Diurnal variation	—			—
Composite	This applies when 6 months or more don't fall with in any of the above categories | high temperature in summer and cold in winter | Low humidity in summer and high in monsoons |	32–43	27–32	10–25	4–10	35–22	Variable dry periods = 20–50% wet periods = 50–95%	Variable 500–1300 mm/year, during monsoon reaching 250 mm in the wettest month	Variable overcast and dull in the monsoon	Uttar Pradesh Haryana, Punjab, Bihar, Jharkhand, Chhattisgarh, Madhya Pradesh etc.

Following are the Scenarios modeled after considering numerous factors and climatic models.

RCP2.6: One pathway where radiative forcing maximums at approximately 3 W/m^2 before 2100 and then declines. Reflects a future only achievable by aggressively reducing global emissions through a rapid transition to low carbon energy sources. GHG emission in this pathway is predicted to be very low.

RCP4.5 and **RCP6.0**: Two intermediate pathways in which radiative forcing is stabilized at approximately 4.5 W/m^2 and 6.0 W/m^2 after 2100. Represents intermediate pathway is consistent with a modest slowdown in global economic growth or a shift away from fossil fuels. Medium baseline & high mitigation of GHG emission in this case.

RCP8.5: It is a high pathway for which radiative forcing reaches greater than 8.5 W/m^2 by 2100 and remains to rise for some quantity of time. This represents a world where fossil fuels continue to power the future. High level of Green House Gases emission and air pollution is predicted in this pathway [11].

According to the 5th assessment report by the Intergovernmental Panel on Climate Change (IPCC), the average temperature increase in India is mostly to be in the order of 1 °C to 1.5 °C (RCP 2.6) for the period 2016–2035 (relative to 1986–2005) and in the range of 1.5–3 °C(RCP 4.5) for the period 2046–2065. According to World Bank's projection India's average summer temperature to increase by 5°C by the year 2100 under the RCP8.5 emission scenario, which is the worst scenario projected.

Studies have demonstrated that temperature heterogeneously affects electricity demand contingent upon the geographic area of the nations and the job of electricity in warming and cooling. There are papers which identify nonlinear relationship between outdoor temperature and energy consumption parameters like daily load demand and peak demand [9]. Studies carried out in cold countries indicate that rise in global mean temperature would reduce the energy demand for thermal comfort in long-term future, where in warm and humid countries like India rise in temperature would play an important factor in rise in energy demand for thermal comfort. The mean, aggregate electricity demand in India increases by 11% or more at temperatures level above 30°C from demand at temperatures between 21 and 24 °C [12]. The impact of temperature increment on electricity demand in summers is higher in a more sweltering atmosphere as individuals adjust with the utilization of higher cooling equipment though there is a higher contrary reaction to temperature increment in winters in colder atmospheres as individuals adjust utilizing higher warming equipment. Since electric heating is not the only heating technique in colder region the penetration of heating load in electricity demand is lesser compared to heating load.

The analysis by Centre for Science and Environment (CSE) on electricity demand and its relation with weather parameters during the various stages of lockdown and unlocking in New Delhi in 2020. Every degree rise in temperature index resulted in 187 megawatt increase in electricity demand during COVID-19 lockdown, which was 6% higher than in 2019 [13]. The trend was driven by mostly residential cooling load since other major load was absent during lockdown phases.

There are few papers which consider multi variables to project the future energy demands considering climatic as well as socioeconomic factors which influence the electricity consumption. Urbanization, rising incomes, and the price elasticity of energy services in emerging countries like India will greatly influence the trajectory of energy demand over the coming years [14]. Peak demand indicates infrastructure requirements for generation, transmission and distribution. The state of the load curve is significant for grid the board, particularly as we have a rising entrance of renewable energy into the grid, where grid adjusting turns out to be additionally testing because of progress in demand as well as in stockpile.

Most of the studies show that, in a hotter and a richer future economy, there is more chance for rapid adoption of energy using equipment for thermal comfort. The effects of both the colder and the hotter climates on energy demand are predicted to be more reflected at the higher income sectors. A study by Eshita Gupta proposes that climate change will increase electricity demand by 6.9% with 4% per annum (p.a.) GDP growth and 8.6% with 6% p.a. GDP growth rate in 2030 over the model scenario of no climate change. Also 1% rise of per capita income reflects in about 1 to 0.7% increase in everyday electricity demand in most scenarios. This mirrors the way that the assessed minor impact of a more sizzling atmosphere is more prominent when pay is higher. As pay decides how individuals adjust to environmental change, both a worldwide temperature boost and pay development will affect electricity consumption in summers and winters.

3 Data and Summary Statistics

3.1 Data Sources

Five Indian states from different climatic zones are selected for the study. States considered for study are: Jammu & Kashmir (Cold), Kerala (Warm and Humid), New Delhi (Composite), Rajasthan (Hot–Dry) and West Bengal (Warm and Humid).

Monthly data for five states of India is collected for a period of 10 years (2010–2019). The dependent variable is the monthly electricity demand of the state, measured in Million Units (M.U) as obtained by the operator of the national electricity grid, National load dispatch center (NLDC). State wise consumption reports of every month from 2010 to 2019 were obtained from Power System Operation Corporation Limited (POSOCO).

The explanatory variables are of two categories:

1. Climate/Weather variable.
2. Socioeconomic Characteristics.

First category includes monthly mean temperature (in °C) and monthly mean humidity (%). In case of temperature monthly maximum and minimum temperature for each location is collected from Climatology Lab, an open source online weather data portal. Average of Maximum and minimum temperature was calculated for every month and year too. Data for average humidity of each month is collected from World Weather Online, an open source online weather portal. Using monthly humidity values average of a year is calculated.

Second type of explanatory variable is Socioeconomic indices which are Gross State Domestic Product (GSDP), State wise Per Capita Income (in Rupees) and Per capita Electricity Consumption (in Kwh). These data are collected in yearly basis from 2010 to 2019. Gross state domestic product for a period of 10 years of five states is obtained from report of Central Statistical Organization, New Delhi. Per capita income of Indian states for 10 years is collected from Niti Aayog's reports. Per capita electricity consumption for five states is obtained from Central Electricity Authority's Executive Summary on Power Sector report.

3.2 Summary Statistics

Projection of electricity demand is essential for power sector planning. This projection requires multiple dependent and independent variables among which most of them affect energy demand in a nonlinear way. This non-linearity may be due to geographical differences, seasonal variations and different usage of fuels [15].

Figure 2 represents the scatter plot of mean temperature versus energy consumption of five states. Plots are obtained using Spyder (Python). Graph is plotted using temperature and energy consumption data of past 10 years.

Scatter plot represents the impact of temperature on electricity demand is nonlinear with low values of R^2 in linear regression results. Jammu & Kashmir shows clustered points which indicate high non-linearity. J&K is a cold region throughout the year where mean temperature doesn't go above 18°C. Minimum and maximum average temperature clocked in the decade was 14.34 °C (2012) and 15.50 °C (2018). There seems to be a rise of 0.5°C in mean temperature. During this span of 10 years, Jammu & Kashmir's electrical consumption rises by 27.33%. During the decade (2010–2019) a rise of 0.72 °C in average temperature is observed and a rise of 31.11% electricity consumption also occurred during this period. Kerala is a state which lies in southern most part of India which has a long coastal region throughout the western part of the state. Region lies in the Warm & Humid climatic zone. Humidity of the region is consistently above 70% which increases thermal distress results of Kerala show a near linear result where points are scattered in a linear fashion. New Delhi, national capital of India is a region under composite climatic zone. Region has an average maximum temperature of 40 °C in peak summers and minimum average temperature of 6 °C in winters. Data shows a rise of 0.5°C during the span of last 10 years while electricity consumption of the state had 25% rise in electricity consumption. Rajasthan is one of the states in India which comes under Hot-Dry climatic zone. As name suggests humidity in the region is consistently low through the year. Even though state feels extreme temperature during summer (up to 43 °C), temperature in the region drops to below 10°C during the peak winters.

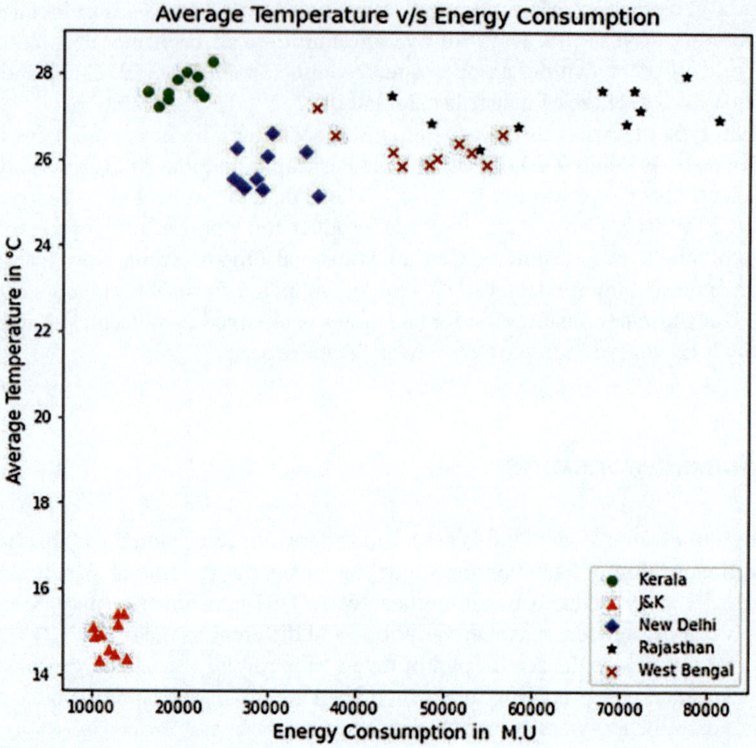

Fig. 2 Average temperature v/s Energy consumption

This brings annual average temperature of the state to nearly 27 °C. During the past 10 years, there was SSA rise of 1 °C in average temperature. State witnessed 46% rise in overall electricity consumption during last decade. West Bengal lies in the east coastal part of India within Warm-Humid climatic zone. As the case of Kerala increased level of humidity aids thermal discomfort. Average temperature of the region does not follow a pattern. It varies in order of 0.5 °C in both direction during last decade. But energy consumption is on steady rise year after year with a rise of 3–4% annually and 38% during 2010–2019.

Per capita income and energy consumption are studied in many papers because their relation is linear in nature. Income of the individual determines how they adapt to climate change or global warming. People who could not afford energy intensive thermal comforting equipment like air conditioners goes for less energy consuming equipment like ceiling fans and those who could opt for air conditioners. This trend is not only applicable to cooling/heating appliances but also for almost every other energy consuming equipment. Region with high per capita income will see exponential growth in energy demand in coming years.

Figure 3 explains the relation between per capita income of five states with their energy consumption for last ten years. Plot is generated in using Spyder (python) software. Growth of energy consumption in region like Rajasthan & West Bengal is faster compared to other three states. One of the reason may be for this is hot, dry and humid climate of those states respectively. Per capita income of Rajasthan and West Bengal rises by 55% and 39% respectively during last 10 years. Energy consumption of New Delhi does not follow the trend of exponential growth in energy consumption because of high rate in per capita income from 2010 onwards. Jammu & Kashmir which comes under cold climatic zone is low income state and per capita income of the region rises 48.37% during last decade. Kerala with hot and humid climatic condition is one of the high income region shows relatively less growth in energy in accordance with their per capita income. This could be due to lack of growth in industrial sector in the region. Per capita income of the state increases by 59.27% during the period of 2010–2019.

Per capita electricity consumption is the ratio of total electricity consumption of the region to the total population of the region. Currently India has a per capita electricity consumption of 1181 Kwh as on 2019 which is way below global rate

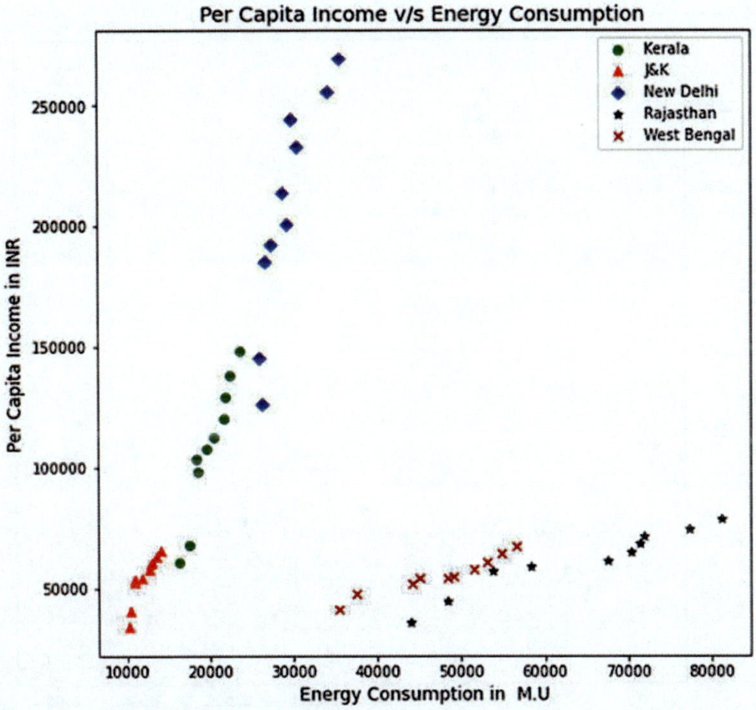

Fig. 3 Per capita income v/s energy consumption

(3130 Kwh). Figure 4 represents the relation of five state's per capita energy consumption to gross electricity consumption. Scatter plot is plotted with the help of data from last 10 years. From the plot, it is evident that relation between these two variables is linear in nature. During the first half of decade, Per. Capita energy consumption of the Kerala rises in 5% every year but due to recent modernization and urbanization in the region a growth of 9% is visible in 2019. This could be due adoption of thermal comforting equipment in recent years. In the case of Kerala Growth in Per capita, electricity consumption was between 1 and 3% during the decade. New Delhi has one of the highest per capita energy consumption rate in the country, which is around 1898 Kwh (As on 2018) and had been 1448 Kwh during beginning of the decade. The growth in Per capita electricity consumption of Rajasthan was linear until 2017, where large rise in the index is seen during last couple of years. West Bengal with low value of Per capita energy consumption during the beginning of the decade is on a steady rise from recent years.

The growth of electricity consumption has strong implications on economy of the country. There are number of studies which researches on relationship between GDP and electricity consumption and each has different findings. This will help to

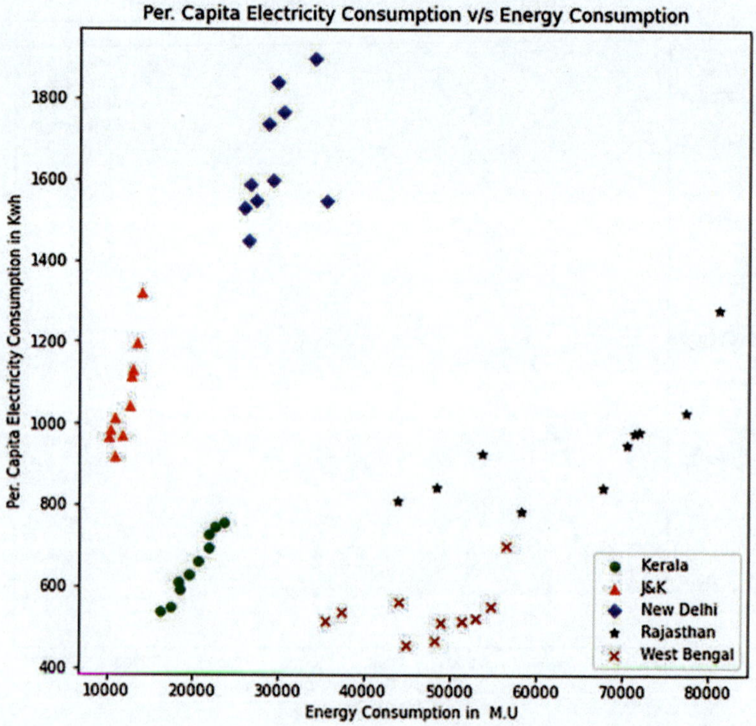

Fig. 4 Per capita electricity consumption v/s energy consumption

predict impact of electricity demand on country's economy and policy makers could formulate energy policies accordingly.

Figure 5 shows the scatter plot of State GDP and Electricity consumption of each of the five states. Graph is plotted using Python and data sets of last ten years. Since two parameters considered have strong relation each other every state gives a linear plot. Among the considered variables, State GDP shows the perfect linear relationship with Electricity consumption in almost all cases. Every state has witnessed trend of rise in GDP during the decade. Among the five states, Delhi has the highest growth in SGDP, a rise of 63.52% during 2010–2019. Rajasthan and West Bengal had a growth of 60.75% and 46.06% over the last decade. Even though Jammu & Kashmir has low GDP there was significant growth during last decade of 57.08%. Kerala also had a similar growth over ten years, which is 58.52%.

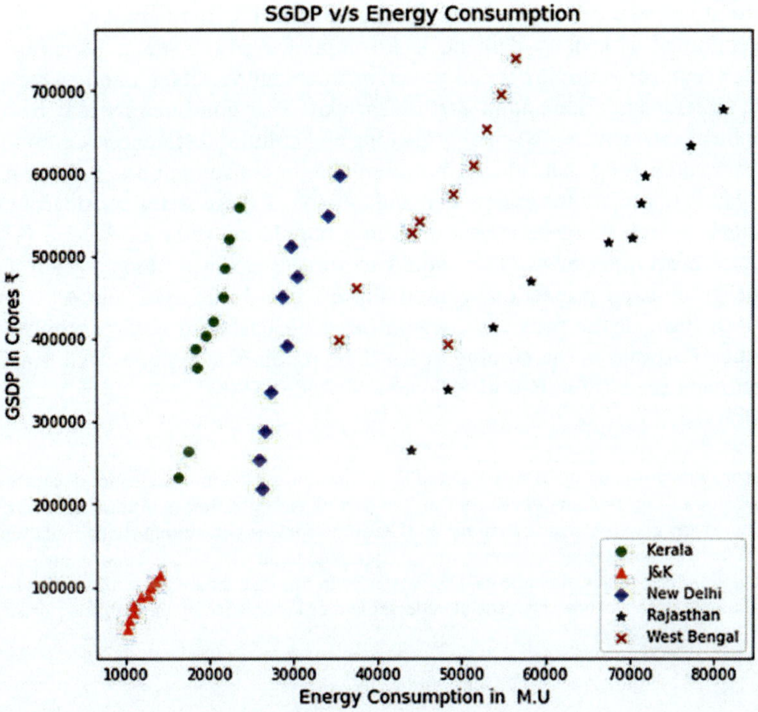

Fig. 5 State GDP v/s electricity consumption

4 Conclusion and Future Work

The statistical evidence from Indian states from different climatic zones in the paper suggests that the climatic sensitivity of electricity demand in a developing country like India is mostly to be sensitive to income growth. From the study parameters influencing energy demand including climatic and socioeconomic are identified. Each climatic zones has their own unique behavior toward tackling temperature distress. Long-term effect of climate on electricity demand which is conditional on each state depends numerous socioeconomic factors considered and also other factors like population, energy tariff, energy policies cultural development in long term which varies from region to region. The energy consumption rate of rich hot states will see a massive growth in future when mean temperature rises approximately by 1–2 °C as projected by World Bank and IPCC. It focuses to the basic need to participate in electricity demand management and lift proficiency in the efficiency of electricity to turn into a low energy consuming society in the future.

Projection of electricity demand is essential for power sector planning. This projection requires multiple dependent and independent variables among which most of them affect energy demand in a nonlinear way. This non-linearity may be due to geographical differences, seasonal variations and cultural differences. From Figs. 4 and 5 it is understood that relation between energy consumption is nonlinear. The future plan is to project the energy demand of each of these states considering these parameters as well as some others including population price of energy, price of thermal comfort equipment. This would require to make a model which should give energy demand projection with conditions like 1 °C, 2 °C and 4 °C by the year 2100. Each of the projection would be again based on both pessimistic and optimistic GDP rate in the coming years. This requires techniques like Multivariable nonlinear regression, Partial Adjustment Model (PAM) or seemingly unrelated regression (SUR) [16].

Disclaimer The presentation of material and details in maps used in this chapter does not imply the expression of any opinion whatsoever on the part of the Publisher or Author concerning the legal status of any country, area or territory or of its authorities, or concerning the delimitation of its borders. The depiction and use of boundaries, geographic names and related data shown on maps and included in lists, tables, documents, and databases in this chapter are not warranted to be error free nor do they necessarily imply official endorsement or acceptance by the Publisher or Author.

References

1. Daniel Burillo (December 10th 2018). Effects of Climate Change in Electric Power Infrastructures, Power System Stability, Kenneth Eloghene Okedu, IntechOpen, https://doi.org/10.5772/intechopen.82146.
2. Mani, M., Bandyopadhyay, S., Chonabayashi, S., Markandya, A., & Mosier, T. (2018). *South Asia's hotspots: The impact of temperature and precipitation changes on living standards.* Herndon, VA: World Bank Publications.

3. World Economic Forum. (2019). INDIA: A report by the World Economic Forum's system initiative on shaping the future of consumption. www.weforum.org
4. Somani, G., & Bhatnagar, M. (2015). *Stringency analysis of building envelope energy conservation measures in 5 climatic zones of India.* New Delhi, India: Environmental Design Solutions.
5. Bhatnagar, M., Mathur, J., & Garg, V. (2019). *Climate zone classification of India using new base temperature* (pp. 4841–4845). Jaipur, India: Malaviya National Institute of Technology.
6. Damm, A., Köberl, J., Prettenthaler, F., Rogler, N., & Töglhofer, C. (2017). Impacts of +2 °C global warming on electricity demand in Europe. *Climate Services, 7*, 12–30. https://doi.org/10.1016/j.cliser.2016.07.001
7. Liu, X., and Sweeney, J. (2007). Impacts of climate change on energy demand in the greater Dublin region, Ireland. In *The 20th Conference on Climate variability and Change* (pp. 1–9). Boston: AMS.
8. Bureau of Energy Efficiency, India (2011). *Energy Conservation Building code User guide.*
9. Larsen, K., Larsen, J., Delgado, M., Herndon, W., and Mohan, S. (2017). Assessing the effect of rising temperatures: The cost of climate change to the US power sector. https://rhg.com/wp-content/uploads/2017/01/RHG_PowerSectorImpactsOfClimateChange_Jan2017-1.pdf
10. Wayne, G. (2013). The beginner's guide to representative concentration pathways, Skeptical cience
11. van Vuuren, D. P., et al. (2011). The representative concentration pathways: An overview. *Climate Change, 109*(1), 5–31. https://doi.org/10.1007/s10584-011-0148-z
12. Harish, S., Singh, N., & Tongia, R. (2019). Impact of temperature on electricity demand: Evidence from Delhi and Indian states. *Energy Policy, 140*(June), 2020. https://doi.org/10.1016/j.enpol.2020.111445
13. Roychowdhury, A., and Somvanshi, A. (2020). Power and the pandemic. *Cent. Sci. Environ.*, 1–32.
14. Waite, M., Cohen, E., Torbey, H., Piccirilli, M., Tian, Y., & Modi, V. (2017). Global trends in urban electricity demands for cooling and heating. *Energy, 127*, 786–802. https://doi.org/10.1016/j.energy.2017.03.095
15. Gupta, E. (2014). *The effect of development on the climate sensitivity of electricity demand in India.* Indian Statistical Institute, Delhi, India: Economic and Planning Unit.
16. Central Electricity Authority. (2019). Long term electricity demand forecasting. http://cea.nic.in/reports/others/planning/pslf/Long_Term_Electricity_Demand_Forecasting_Report.pdf

Design and Fabrication of Solar Powered Agricultural Pesticide Sprayer

Madhu Sharma and Deepali Yadav

1 Introduction

Pesticide spraying forms an integral part of agriculture for crop protection against insects and hence ensuring crop yield. Traditionally, farmers hand sprayed pesticides or used a fuel based sprayer pump system. Former method is tiresome and exhausting due to heavy and bulky size. Additionally, fuel based method affects the environment adversely. Hence, there is need for a new easy to use environment friendly spray pump system. The existing pesticide spray pump systems also suffer with lower flow rate and faster battery discharge (if wireless) [1]. With wired systems, there is always a danger of electric shock to the farmers/users. The proposed spray-pump system design aims to overcome these challenges. Some traditionally used spray systems also suffer with pesticide chemical effect, fertilizer wastage due to aerial spraying, incomplete crop area coverage and not suited for small farms.

Above stated challenges, call for new model design that utilizes renewable energy source and a means to reduce or eliminate muscle fatigue for farmer. In this article, the system design is based on solar power as the fuel source. The absence of engine run by fuel operated sprayer and fossil fuel makes the complete system noise/vibration free and environment friendly. Additionally, the use of RES makes it cost-effective.

2 System Layout

The block diagram for the proposed pesticide sprayer design is shown in Figure 1 below. The major components include a solar panel as the primary component for tapping solar power for subsequent conversion into electrical power, phenomenon

M. Sharma (✉) · D. Yadav
School of Engineering, University of Petroleum and Energy Studies, Dehradun, India
e-mail: madhusharma@ddn.upes.ac.in

N. A. Siddiqui et al. (eds.), *Advances in Behavioral Based Safety*,
https://doi.org/10.1007/978-981-16-8270-4_21

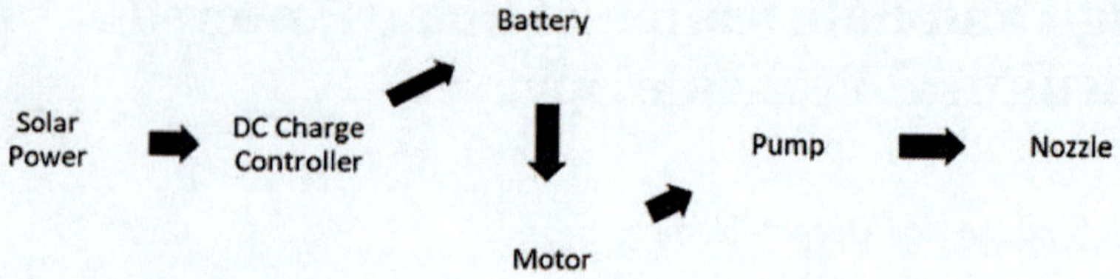

Fig. 1 Block diagram for proposed solar powered sprayer

known as photovoltaic effect [2]. The panel output is stored in a lead acid battery via a dc charge controller. The battery stores power required to run the motor-pump set. The battery-powered motor runs the pump, which in turn lifts the pesticide solution from a tank, further to be fed into the nozzle. The nozzle used furnishes for the spraying pattern on the field. The complete setup is carried on a trolley system for effortless and smooth spraying process across the farm field.

The DC charge controller is essential to regulate the dc current into the battery to prevent over charging or over discharging against over voltage and fast discharging. The controller implements pulse width modulation (PWM) technique, wherein the modulated signals gradually stop charging the battery as the battery voltage drops fall below or rises above the safe limit or maximum level respectively. PWM offers the advantage of very low switching power loss. Hence, when compared to the maximum power point tracking (MPPT) method, power requirement in PWM is much less. In this design, the charge controller is furnished with a USB charging port for up to 5 V charging. The charge controller output is fed into the battery via a socket mounted on the controller unit itself through an electrical network. Subsequently, the battery is then connected to the motor-pump set as dc input to the motor input terminals.

The motor-pump set employs a diaphragm type pump, which gives positive reciprocating displacement [3, 4]. In this type, a pulling back action creates a vacuum within the chamber in front of a diaphragm as shown in Figs. 2 and 3 [3]. This vacuum further forces the discharge valve against its seat. With pressure equivalent to atmospheric pressure, the vacuum pushes the pesticide fluid against the outside of the suction valve, thereby opening it and filling the chamber. With pressure returning to the diaphragm, it is forced in the direction of chamber's front, and hence the increased pressure forces the suction valve to close and discharge valve to be forced

Fig. 2 Air operated single diaphragm pump

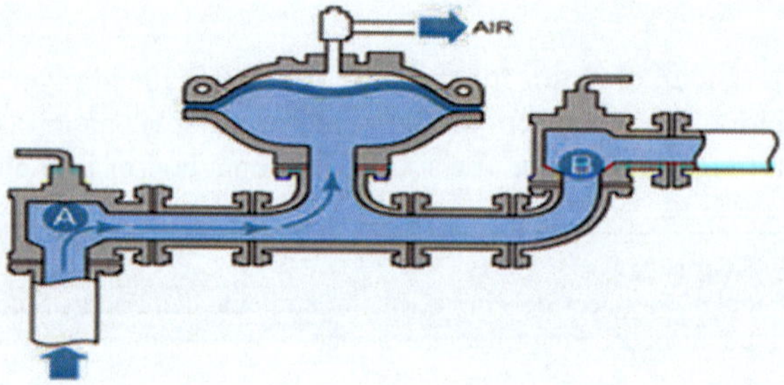

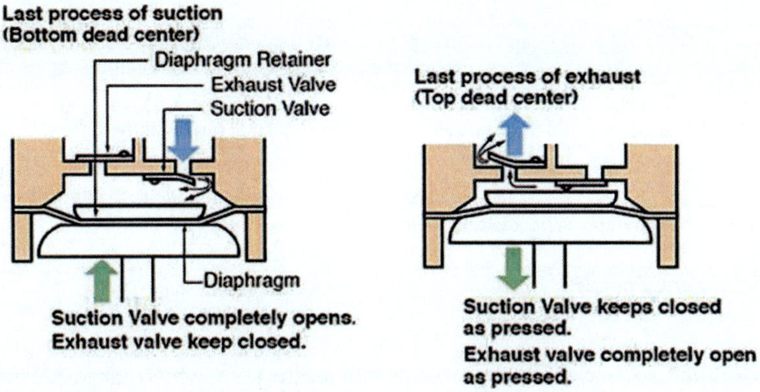

Fig. 3 Working model of a diaphragm pump

open. This action forces the pesticide out of the chamber and the same pumping cycle repeats all over again. With this there is an added velocity to the fluid flow, which helps, in efficient and high discharge from the pump, and hence meeting the fulfilling the intended purpose.

3 Design Description

The system designed aims to keep it cost-effective. An effort to meet the objective with limited resources and compete with contemporary system designs has been made. A suitable modification, meeting these criteria, in the existing conventional spray system modules helps meet the standard quality requirements. The proposed design adopts the principle of solar powered spray pump system. The complete spraying process initiates with fluid flow measurement and then solar power application for spraying the liquid at an optimum rate.

Since this design does not employ meters for flow rate and fluid velocity measurements, the process of reverse engineering is adopted to serve the required purpose. Hence, a major challenge of parameter measurement is overcome with reverse engineered solar powered spray system design. The subsequent sections present the required component dimensions and their inter-relation for achieving the described design.

3.1 Area Coverage During Spraying

Table 1 below presents the dimensional measurements and assumptions made for required area to be covered in spraying.

Table 1 Component details

Parameter	Dimension	Condition/assumptions
Height of delivery pipe containing the nozzle	0.42 m	1. Above ground 2. Held horizontally
Coverage area	–	1. In the form of the sector of a circle 2. Path controlled by operator
Angle at which sector area is covered for spraying	30°	On both sides
Cross section of the nozzle	–	Circular in shape

In addition to above parameters, the portion of the pipe, which carries the nozzle, is held in horizontal and the nozzle in vertical position with respect to the ground. The discharged fluid from the nozzle is in the form fine atom type particles following the trajectory of a projectile motion.

Displacement as given by kinematic equation:

$$S = ut + 0.5at^2$$

Displacement along y-axis direction;

$$H = u_y t + 0.5at^2$$

Hence,

$$t = \sqrt{(\frac{2H}{9.8})}$$

$$V_x = u + at \quad (1)$$

$$V_y = \sqrt{(2 * 9.8H)/\tan\theta}$$

So,

$$V_R = V_y/\sin\theta \quad (2)$$

For horizontal component of velocity,

$$u_x = V_R * \cos\theta \quad (3)$$

$$u_x = \sqrt{(2 * 9.8H)/\tan\theta}$$

$$S = u_x * t \quad (4)$$

Hence, the area covered during spraying operation is given by;

$$A = (170\pi * S * S)/360$$

In addition, by considering the assumption of circular nozzle cross section following equation is obtained,

$$A = 90\% of A \quad (5)$$

$$A = (1.7\pi * H * H)/\tan\theta$$

3.2 *Nozzle Dimensions*

The discharge from nozzle is represented by 'Q' in liters per minute. The diameter of the pipe d_1 is given by:

$$Q = \mathrm{A1v1} = 4\mathrm{Q}/(\pi * d1 * d1)$$

where v1 represents nozzle inlet velocity.

Hence, if v2 represents nozzle outlet velocity, then

$$v_2 = u_x$$

$$Q = A_2 * v_2 \quad (6)$$

$$\mathrm{d2} = \sqrt{}((4\mathrm{Q} * \tan\theta)/(\pi * \sqrt{}(2 * 9.8 * \mathrm{H})))$$

Here the flow rate is assumed to 5 liters per minute, with five nozzles having a diameter of 1centimeteres.

3.3 *Head Developed by the Pump*

Here, the roughness of the pipe is neglected, assuming laminar flow of the pesticide liquid.

For water at 20° C temperature, $\mu = 0.01$ poise, $\rho = 1000\ \mathrm{kg/m^3}$

$$\mathrm{Re} = (\rho * \mathrm{v1} * \mathrm{d1})/\mu$$

If $R_e < 2300$, then it is a laminar fluid flow.

$$\lambda = 64/Re \tag{7}$$

$$v2 = \surd((2hg)/(1 + (4\lambda(d2/d1)2)))$$

If the required head is considered to be more than 20%, then the frictional losses can be accounted for. Hence,

$$(h) = 1.2 * h \tag{8}$$

3.4 *Power of the Motor-Pump Set*

The electrical power input of the motor-pump set is given by following equation no. 9,

$$P = \rho * g * Q * (h) \tag{9}$$

Battery Selection
For this design, a 12 V, 9 Ah lead acid type of battery is employed. The battery rating is given by,

$$P = V * I = 12 * 9VAh = 108Watts$$

3.5 *Selection of the Solar Panel*

The choice of solar panel depends on its weight and also on its ability to efficiently charge the battery as per the power wattage [5, 6]. The output dc current of the solar panel is obtained by maximum power (P) of the solar panel and the rated voltage (V) of the battery. The battery charging time (T) is measured as the ratio of ampere-hour (Ah) rating of the battery to the total current produced by the solar panel. The Table 2 below gives the current, charging time and solar panel weight for varied power ratings of solar panels for a rated voltage of 12V.

The table suggests that, as the solar panel rating increases, the charging time of the battery increases and the corresponding weight of the solar panel increases.

Hence, a solar panel of high power rating is preferred, so as to reduce the charging time of the battery, in turn allowing for fast battery charging [7]. Hence, with solar panel weight and battery charging time as the as primary selection criterion, a 40 Watt solar panel is selected for the sprayer design.

Table 2 Solar panel parameters

S.NO	Solar panel rating (Watts)	I = P/V (Amps)	T = Ah/I (h)	Weight of the solar panel
1	6	6/12 = 0.5	18	0.6
2	8	8/12 = 0.67	13.5	0.8
3	10	10/12 = 0.83	10.8	1.2
4	15	15/12 = 1.25	7.2	1.5
5	20	20/12 = 1.67	5.3	2
6	30	30/12 = 2.5	3.6	3.6
7	40	40/12 = 3.33	2.7	5.0

4 System Fabrication

4.1 Component Specification

System fabrication involves considering design specifications including necessary assumptions as described in previous sections. Table 3 below presents the specifications of the components used for the spray model.

4.2 Testing

The final fabricated model of the solar powered spray pump system is shown in Fig. 4. below.

The final prototype was tested in broad daylight from 10 am to 1 pm in the field of Bidholi area in Dehradun city. The battery charged completely from 10.55 V to 13.56 V. The system proves to be highly efficient in reducing human muscle fatigue considerably. The pesticide fluid discharge is measured by considering fluid of 1-L capacity in a measuring cup.

The prototype successfully demonstrates that system efficiency is directly related to the measure of time taken and corresponding battery voltage for a certain volume of pesticide discharged, for a given crop field area. Factors like crop height and simultaneous battery charging will also affect the system efficiency. From cost-effective viewpoint, motor, battery, solar panel and nozzle give a good life cycle, however, the trolley wheels used to carry the complete system may be the only running/maintenance cost due to wear and tear. Considering the same the system provides the advantage of very low running cost in addition to relieved muscle effort of the farmer as compared to conventional pesticide sprayers [8].

In addition, as a future improvement, the new and latest technology developments can be adopted in the spray pump system design like compact modules with high capacity in terms of motor, battery and solar panel advancements.

Table 3 Component specifications for the components used in the proposed model

S.No	Component	Specification/s
1	Tank	PVC, 15 L, 450 g
2	Solar panel	Solar PV panel 40 W Dimensions: 48 cm × 65 cm Weight: 2 kg Maximum voltage: 17 V Maximum current: 1.18 A Open circuit voltage: 21.47 V Short circuit current: 2.56 A Tolerance: ± 5%
3	Battery	Sealed lead acid type Capacity: 12 V, 7.6 Ah Dimensions: (15 cm × 6.5 cm × 9.2 cm) Weight: 2.5 kg Constant charge with regulation Standby use: 13.5 V–13.8 V Cycle use: 14.5 V–14.9 V Maximum initial current: 2.4 A
4	Charge controller	Capacity: 12 V, 5A Pulse width modulation based
5	Diaphragm pump set	Plastic and Metal material Dimensions: 6.42 in. × 18 in. (or) 16.3 cm × 5 cm Voltage: 12 V Direct current power: 60 W Pressure rating: 0.6 MPA, 100 PSI (5.5 Bar) Flow rate: 5 L per minute
6	Pipe	Diameter: 1.1 cm Length: 200 cm

5 Conclusion

In this article, a solar powered pesticide spray pump system is discussed with complete fabrication and testing details. The design parameters for the fabricated model are presented. The prototype was operated in the field under standard conditions. This system offers the advantage of low cost and is easy to move in the farm field. The testing results confirm that the model is effective in reducing the user fatigue and also improves the quality of pesticide spraying in the field.

Fig. 4 Final fabricated model for the solar powered spray pump system

References

1. Vasantharao, V., Mathapati, S., & Amarapur, B. (2013). Multiple powered supplied fertilizer sprayer. *International Journal of Scientific and Research Publications (IJSRP), 3*(8), 1–5.
2. Rao, G. D. 0000. Solar energy, solar photo-voltaic: electric power generation, *15*, 434–484
3. Best Process Equipment, How the ODS pump works, Dorr-Oliver ODS Air-Operated Single-Diaphragm Pump – Best Process Equipment.
4. Hicks, T. G. (2001). *Pumps operation and maintenance.* New Delhi: Tata McGraw Hill.
5. Joshua, R., Vasu, V., & Vincent, P. (2010). Solar sprayer: An agriculture implement. *International Journal of Sustainable Agriculture, 2*(1), 16–19.
6. Khan, Z. (2014). Unique solar operated spray jet. *IOSR Journal of Mechanical and Civil Engineering*, 43–46.
7. Rai, G. D. (1996). *Solar energy utilization* (pp. 434–479). Delhi: Khanna Publishers.
8. Raut, L., Jaiswal, S., & Mohite, N. (2013). Design, development and fabrication of agricultural pesticides sprayer with weeder. *International Journal of Applied Research & Studies (IJARS), 2*(11), 1–8.

Training Performance Assessment and Analysis of Delivery Riders in a Logistic Company

Juno Reju, Rishi Dewan, and Surendar Varadharajan

1 Introduction

Road Traffic Injuries are the main source of death universally among 15–19 aged youngsters, while for those in the 10–14 years and 20–24 years age sections they are the subsequent driving reason for death [1] WHO 2007. Accident rates in developing countries keep on increasing and are becoming a global reason for concern. This worry is advocated on the grounds that development has the most noteworthy setback rates in numerous nations. In India, the number of motor vehicle production is increasing quickly than the financial and citizenship development [2]. This increase in motor vehicle production and its use on roads have resulted in unfavorable factors like road accidents which further results in injury or death. 4,49,002 road accident cases were noted by different States and Union Territories in 2019 [MORTH].

Logistics industry today is apparently significantly special. Wonderful, regarded, non-extensive and unique definitive capacities in the Logistics marks radiates an impression of being a critical technique to isolate the strategic limits and finally to achieve viable high ground. Delivery riders are accounted for to acknowledge high danger taking, ordinarily more than the overall rider population. Among other reasons, such conduct is likely connected with the frequently casual character of the business. High danger taking is an all-around perceived wonder in different sorts of casual work, described by short agreement span, less employer stability, unpredictable working hours, different types of independent work and furthermore

J. Reju
Health Safety and Environment with Specialization in Disaster Management UPES Dehradun, Bidholi, India

R. Dewan
Department of Physics, UPES Dehradun, Bidholi, India
e-mail: rishi@ddn.upes.ac.in

S. Varadharajan (✉)
Department of Health Safety Environment & Civil Engineering, UPES Dehradun, Bidholi, India

N. A. Siddiqui et al. (eds.), *Advances in Behavioral Based Safety*,
https://doi.org/10.1007/978-981-16-8270-4_22

false/casual work courses of action. The dangers associated to delivery workers/riders on roads are due to improper use of helmet, over speeding, harsh accelerations, harsh cornering etc. Two potential reasons for self-protection negligence are (i) individual characteristics identified with low schooling and financial status of the specific population and (ii) business related elements connected to pressure for quick conveyances as well as pay increment. Mostly the workers have to adapt to a few orders of various clients simultaneously. In view of the time requirement, for each request and likely consequences for late delivery, it is important for them to design an optimum conveyance course and limit the time spent on riding during work. These conditions all convert into an expansion in their psychological responsibility and ensuing pressure [3]. Studies have shown more than 10 h of working by the delivery workers and in some cases, they work on their Week-off's. This immense working pressure leads to neglecting the safety aspects and results in the at-risk behaviors during driving. Delivery riders are impermanent utilized, ineffectively paid, and frequently paid 'by the work', e.g., paid continuously or the measure of conveyance products. This will in general actuate extraordinary work pace for extended periods of time, without breaks yet in addition higher work pressure, work weakness, and perilous driving practices, e.g., running a red light or a stop sign. Moreover, conveyance riders much of the time drive without wearing a head protector for efficient and comfort or utilize uncertified caps only for consistence purposes. The previously mentioned discoveries propose that more than simple obliviousness, the consolidated pressing factor applied both by managers for quick conveyances and by workers for higher benefits perhaps pushes conveyance riders to dangerous conduct. Simultaneously, it is incredibly hard to indicate hard administrative measures for reducing such joined pressing factor (e.g., speed limit, most extreme number of conveyances each hour, and so on) This is both unrealistic due to high variety in movement distances and request tops among the stores and unsounded on the grounds that such hard limitations don't address the perplexing interrelationship basic the previously mentioned risk-taking rehearses.

Internet of Things (IoT) can be used for making different smart solutions to prevent road accidents. This is done by shrewd observing, revealing, and control of various spaces of our day by day lives [4]. IoT executions are on the ascent because of the headways in web speed with higher transfer speed and more modest telecom hardware [5]. IOT in Mobile applications has given some astonishing results in capturing the driver behavior and the at-risk behaviors. Mobile applications with IOT sensing can be further more used to Road Safety as it can help in avoiding accidents through early warning systems, detect weather event etc. [6].

The objective of this research paper is to analyze the driving behavior of delivery riders through a IOT based Mobile application that is installed in the participants mobile phones. This research will be conducted for one month followed by a training session. The participants will be monitored after the training and the data will be collected to analyze the behavioral changes both pre and post training.

2 Literature Review

Road accidents have been increasing in the past few years as the technology and automobiles industries are growing. Studies show that there is an increase in road accidents in India and therefore they suggest to focus on the real issues behind the accidents and the measures to prevent them [7]. Several statistics presented by government also show the increase in accident rate in India [2]. Delivery riders are an integral part of logistics industry and without them the business will be left out. Growth of logistics industry has increased the number of job opportunities in this field with a greater number of young riders coming for the delivery rider jobs. Several surveys have been conducted on the delivery riders in order to measure or monitor their behavior while riding their bikes. Survey on delivery riders from China suggested that attitude, traffic environment, and workload affected the behavior of riders [8]. Similarly, age group also plays an important role in driver's change in behavior. Younger riders show more at-risk behaviors as compared to older riders [9]. Fatigue and average working hours also play an important role in change of behavior [3]. Internet of Things has emerged as a part of rising technologies which includes things that are connected with sensors and different technologies which connect them to Internet for data. IOT is being used for gathering and monitoring the data related to road safety and is been useful for managing traffic accidents as well as monitoring driving behaviors [4, 10]. Driver's health condition and drowsiness can also be monitored through IOT which also gives us smart solutions for the same [6, 11].

Othe factors like overspeeding also act a dangerous form of at-risk behavior shown by the driver's which leads to several misjudgments [1, 12]. Phone usage has now become a trend for the coming generation of riders, most of the riders use their phone while driving, causing distraction [13, 14].

3 Methodology

The proposed research was completed in four parts (1) Study of various types of At-Risk behavior during driving. (2) To measure the Road Safety parameter performance of Delivery riders for one month. (3) To conduct/participate in a training program along with the delivery riders chosen for the research work. (4) To evaluate the effects of training on the delivery riders based on their performance.

3.1 *Study of Various At-Risk Behavior*

Over Speeding

Speeding is the most common cause of road crash injuries and death [15]. Studies have shown that 80% of road accidents are a result of over-speeding/human fault, as the given speed limits on roads are not followed by the individuals [12]. Traffic accidents causing fatalities will go up to 4–5% if the average speed is increased to 1%. Over speeding results in improper judgment of traffic movement/information, visual angle is changed, thinking time is reduced which eventually affects the braking time [16]. Higher the speed of vehicle higher will be the fatalness rate [17]. To reduce or monitor these over speeding conditions several solutions are been made, which use Overspeeding detectors for accurate detection. These high-tech solutions can be used as smart solutions for reducing the frequent accidents on roads.

Phone Usage

Humans have limited attention to the things they perform and tend to multitask during driving as well. However, the invention of new technologies has made the communication easier while driving. Studies show that, usage of cellphone weakens the driving performance [14]. Phone usage distracts the driver which leads to visual distraction, driver tends to remove their hands from their wheels/handle and difficult to process the vehicle information [13]. Studies in Finland have shown that the dangerous situations that arose from phone usage rose from 44–50% [18]. Driving speed automatically tends to increase if the phone conversation time increases [19]. Young adults and students are more disinclined to follow the road safety rules whereas males are more frequent users of phone than females while driving [20].

Use of Drugs

Liquor contributes profoundly to the inclusion of youthful drivers in mishaps. Albeit the age class of youthful drivers shifts fairly in the writing. With Blood Alcohol Concentration crossing the limit of 0.08 mg%, the younger drivers are prone to accidents than the older ones [21]. An Australian study shows that age group of 16–35 years old males have higher fatal accidents caused due to alcohol than females. Another reason for such drink and drive incidents may be due to peer pressure or benefits that are been awaited by the drivers [22]. It was assessed that the individuals who occupied with successive beverage driving practices had paces of dynamic auto collisions that were 2.5 occasions higher than the individuals who didn't drive under the influence [23].

Harsh Acceleration, Harsh Braking, and Harsh Cornering

When more force is applied to the vehicle than the usual during driving then it's an aggressive behavior. Such conditions arise due to at-risk behavior shown by the drivers. A driving conduct is forceful in the event that it is purposeful, prone to build the danger of impact furthermore, is persuaded by anxiety, inconvenience, aggression, or potentially an endeavor to save time. Such a conduct incorporates extreme speed increases and decelerations [24].

Drowsy/Sleeping While Driving

Sleepy driving likewise is more normal among drivers who now and again, sometimes, or never wear safety belts while driving or riding in a vehicle contrasted and the individuals who consistently or quite often wear safety belts. Nodding off while driving is obviously perilous, yet drowsiness additionally debilitates the capacity to drive securely regardless of whether drivers do not nod off [25]. More than one of every four drivers confesses to having driven when they were 'lethargic to the point that [they] struggled keeping [their] eyes open'. Work and rest plans were both firmly connected with association in a rest related accident. Contrasted with drivers in non-rest crashes, drivers in rest crashes were almost twice as liable to work at more than one work and their essential occupation was a lot bound to include non-standard hours. Working the night shift expanded the chances of a rest related (versus non-rest related) crash by almost multiple times. Time spent snoozing each night was likewise a solid danger factor: the less the hours dozed, the more noteworthy the chances for association in a rest related accident [10].

Road Safety Trial

Cell phones are outfitted with an assortment of sensors, for example, movement sensors (for example accelerometer and gyrator), position sensors (for example magnetometer), Global Navigation Satellite System (GNSS) collectors, natural sensors (gauges, photometers, and thermometers), receiver, cameras, and so forth Albeit, hypothetically every sensor may add information to the driver's conduct and conditions during driving. These sensors therefore help in detecting the driver's behavior. The study was conducted by selecting 5 participants from an organization. All five participants are from Delhi and work in the Delivery Centers. The participants were asked to install an application which will be used to detect the data while they are on trips. Once the data is detected it is then analyzed through the website which displays their data of at-risk behavior. Five different delivery drivers were monitored for one month for the data. All had Android mobile phones including OPPO, REDMI, and VIVO. Participants have daily trips given to each one of them for shipment delivery. The mobile phones can be placed in any directions or in any place. While installation of the mobile application they have to give permissions including location and sensor permission which is necessary for detecting the trips.

Once the permissions are given, trips will be captured through their mobile phones. Daily captured trips will be shown the next day in their personal mobiles and in the website. Table 1 shows the captured data through the mobile application.

Therefore, the data that was captured was made available by 493 trips that were done between January 2021 and February 2021. Figure 1 shows the first implementation stage where the delivery riders are monitored. It is noticed from the experiment that phone usage and over-speeding are two main at-risk behaviors shown by the delivery riders, followed by harsh cornering and harsh braking. Figure 2 shows a statistical view of at-risk behaviors on the weekdays and weekends. It is noticed that

Table 1 At-risk driving behavior

At-risk behavior	Description
Phone usage	Percentage of number of times a user uses his mobile phone
Overspeeding	If speed exceeds the prescribed speed by road authorities, it is captured dynamically
Harsh acceleration	Sudden increase in speed by applying more force than normal
Harsh braking	Sudden decrease of speed by applying more force than normal
Harsh cornering	Vehicle enters a curve/corner at a greater speed than safe turning in given time interval
Distance traveled	Number of kilometers a vehicle is said to travel

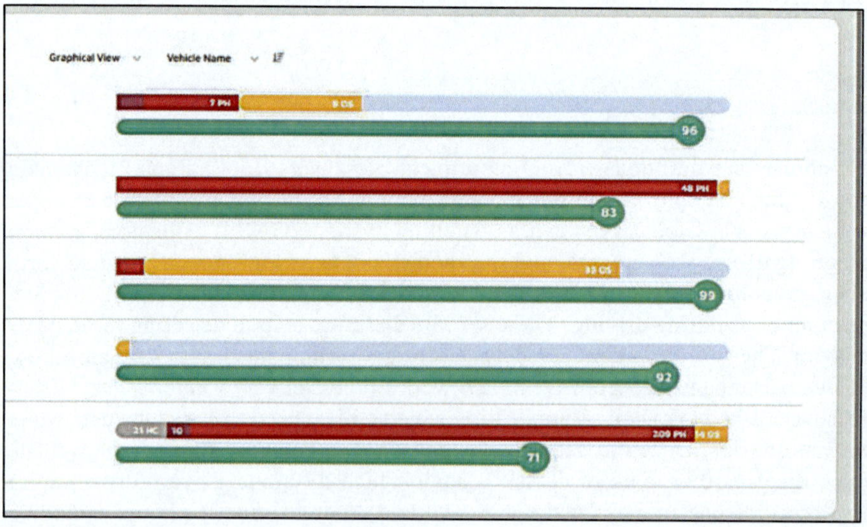

Fig. 1 Driver score based on the driving performance

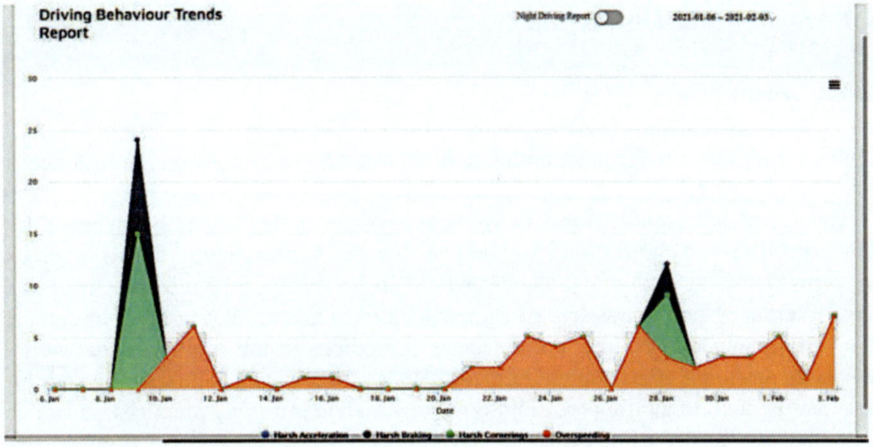

Fig. 2 Driver behavior trends report

the at-risk behavior tends to increase toward the end of the week including overspeeding and phone usage. The server also allows us to see the inactive participants. This is helpful to check if all the participants are active during the work time. Figure 3 shows three active and two inactive participants.

3.2 Road Safety Training

All five participants attended the training/counselling session conducted on road safety. The meeting was conducted after the one month trial run along with the delivery riders. Due to the COVID-19 pandemic situation around the world, the training session was held on virtual platform. The training session was theoretical and included information related to road safety. Participants were given training on defensive driving, safety rules, and information related slow driving. During the session the driving scores of each participant were shown individually. Training was given by professionals in the field of Environment, Health and Safety, and Mobile application development. The agenda for training was to interact with the employees and make them understand the importance of the app and educate them about it.

3.3 *Result and Discussion*

Safety Scores

Tables 2 and 3 shows the pre-training and post-training driving score of employees 5 and employee 2.

The scores represent that there have been changes in the driving behavior of the delivery riders and the training has helped them in understanding the road safety concerns. Prior to training, members would in general accept mishaps to others were inferable from an absence of care or lack of concern. During the training, all five participants believed that accidents are unpredictable and would become worse due to the unsafe actions shown by the drivers during driving. The training session was instrumental in guiding the employees toward better driving practices. In Delhi, distracted driving is the most at-risk behavior observed, and upon interaction with the employees, it was indeed found that they were distracted over the phone mostly on calls or WhatsApp and work-chat. With proper counseling the employees understood and also have been working toward making sure they reduce phone usage while traveling. The next thing that was focused during training was the Harsh braking, Overspeeding, Harsh Cornering, and Phone Usage. Harsh braking was occurring due to overtaking, lane changing, traffic avoidance, and un-mindful driving practices. During the training session participant were willing to accept the at-risk behavior shown by them pre-training. They also interacted by giving examples of situations that led them to these at-risk behaviors. Three out of five participants have accepted that the at-risk behavior shown by them was because of pressure to get the work completed.

Table 2 Pre-training score

Employee	Phone usage	Harsh acceleration	Harsh braking	Harsh cornering	Over speeding	Score
Employee 5	209	0	10	21	14	71
Employee 2	48	0	0	0	1	83

Table 3 Post-training score

Employee	Phone usage	Harsh acceleration	Harsh braking	Harsh cornering	Over speeding	Score
Employee 5	37	0	0	0	1	91
Employee 2	0	0	0	0	12	94

3.4 Behavior

The safety score suggests there was improvement in the driving behavior of the participants after training. Studies have shown that risky driving behavior can be a result of low self-control, as it can also cause distraction which is another factor leading to risky driving. There are several factors that lead to behavioral changes in driver [26]. Participants are given with conditions like working hours which are long, insufficient rest, and large number of orders that have to be delivered by them. Participants have to work more than 8.5 h per day with one day off in a week. It was noticed that the participants were asked to give more than 30–35 orders for delivery.

Another aspect to consider in change of behavior is time pressure. Time pressure not necessarily be from the organizational side but also from the customer side. There are instances when the delivery riders have to rush to pick the returning orders as per the time favored by the customer. Such routines create a pressure on the delivery riders for completing the work on time. These conditions lead to risky driving on roads. These time pressure and intensive workloads give rise to another factor called the fatigue. Fatigue is common in drivers working for hours, same as in the case of delivery rider who work more than 8.5 h a day. The trip routes of the participants have shown that common risky driving behaviors were shown during the afternoon time post or pre-lunch. Driving for a longer period itself is not only a reason for fatigue but also carrying the delivery shipment bag along with driving makes it more tiring. The bag contains shipments allotted to delivery riders to finish the delivery on the given day as this reflects in their performance.

3.5 Motivation

Inspiration includes the excitement, bearing, and ingenuity of conduct. Utilizing this definition, wellbeing inspiration can be characterized as the excitement, bearing, and industriousness of conduct that decreases the probability of word related injury or disease. There are two viewpoints the first identifies with a person's assumption that a given outcome will happen because of their conduct, while the subsequent angle identifies with a person's convictions about the probability or likelihood of that result happening in the event that they act in a specific way.

As per the first aspect, training on road safety has made the participants aware that their own behavior is responsible for any accidents caused during driving. Before the training was conducted the participants were less concerned about the consequences that may occur due to risky behavior. As per the second aspect mishaps are rare and numerous illness/injuries have long dormancy time period. In most cases, delivery riders tend to ignore the safety aspects/rules as they don't have any personal experience with mishaps. Therefore, delivery riders may be overconfident about their skills of driving and believe that such accidents or incidents will never happen to them and such motivational strategies will not help them. Studies have shown that drivers tend to believe on their driving skills more and tend to ignore the rules and safety practices based on the belief.

Before the training delivery riders were unaware about the mobile app and on further explanation also they tend to continue the at-risk behavior. But, after the training was given participants themselves have mentioned that it has helped them in understanding the road safety aspects and have also explained the importance of safe driving. Personal/ intrinsic motivation comes by personal choice which allows the person to act in a safe manner while driving but apart from this external motivation can also be provided by giving some incentives or rewards to the best performer.

3.6 Age Group

Participants selected for the research work were between 20 and 45 age group and junior and senior high school as education background. Studies have shown that younger drivers are more prone to risky driving than older drivers. The research work also clearly shows that the employee 3 with 40 age had good safety score even before the training. Youthful respondents tended to encounter more significant levels of time pressure thought about with more seasoned ones, and time pressing factor could likewise be set off by heavier responsibilities concerning orders conveyed per day. Respondents with less working experience detailed more successive commitment in hazardous riding practices in examination with experienced ones, and those who deliver more shipments per day were more inclined to perform hazard taking practices. Additionally, riders' discernment of time pressure was decidedly connected with their at-risk practices. Training session noticed that the younger participants were more accepting their at-risk behavior and scores have shown that they have improved after the training session.

4 Conclusion

This study provides us an understanding of at-risk behaviors shown by the delivery riders during shipment delivery. For this a one month trial was conducted with five participants using a mobile application. The mobile application was installed in each participants cellphones, using which their driving behaviors were monitored. The at-risk behaviors noted during the trial were overspeeding, harsh braking, harsh acceleration, phone usage, and harsh cornering. A training session was conducted after the trial which included all participants and proffesionals. The outcomes of the training have suggested that there is a positive effect on the participants as their scores have improved. It is noticed that on each trip drivers behave differently that leads them to unsafe and safe behavior. Phone usage is one of the major unsafe acts shown by the delivery riders as they get calls from the customers, which leads to unsafe behavior. Participants have showed interest in improving the behaviors. Behavior changes in delivery riders are also due to fatigue, time pressure, and shipment load in their bags. Age group can play an important role in driving behavior as younger drivers are more prone to rash driving than the older drivers. Motivating the delivery riders can improve their driving conditions and their performance. The understanding of risky driving and its consequences solely depends on how the delivery riders perceive these conditions. However motivation can create a positive impact on the drivers, which clearly depends on how an organization provides them health and safety services and also reward recogonitions.

5 Limitations and Future work

First limitation is that the research was conducted on a small group of participants hence its effectiveness on a large group is unknown. Second limitation is that the installation process in mobile phones is a challenging task as due to covid all work is done through calls which makes it difficult to check if participants have given permission to all the settings. The incoming data may get halted in case of network issues. If the mobile phone sensors are not activated then the data will not be captured during the trips. This will lead to low safety scores.

Third limitation is that the research is limited to some certain aspects, for future work this can be extended to emergency alerts in case of any critical situations on roads. Incident reporting can also be added to the mobile application for better usage.

References

1. Mondal, P. (2011). A silent tsunami on Indian Road: A comprehensive analysis of epidemiological aspects of road traffic accidents. *British Journal of Medicine and Medical Research, 1*(1), 14–23. https://doi.org/10.9734/bjmmr/2011/106
2. Ruikar, M. (2013). National statistics of road traffic accidents in India. *Journal of Orthopedics, Traumatology and Rehabilitation, 6*(1), 1. https://doi.org/10.4103/0975-7341.118718
3. Zheng, Y., Ma, Y., Guo, L., Cheng, J., & Zhang, Y. (2019). Crash involvement and risky riding behaviors among delivery riders in China: The role of working conditions. *Transportation Research Record, 2673*(4), 1011–1022. https://doi.org/10.1177/0361198119841028
4. Taha, A. E. M. (2018). An IoT architecture for assessing road safety in smart cities. *Wireless Communications and Mobile Computing, 2018*. https://doi.org/10.1155/2018/8214989
5. Vishal, D., Afaque, H. S., Bhardawaj, H., & Ramesh, T. K. (2018). IoT-driven road safety system. *International conference on electrical, electronics, communication computer technologies and optimization techniques, ICEECCOT 2017, 2018-January*, 862–866. https://doi.org/10.1109/ICEECCOT.2017.8284624.
6. Jabbar, R., Shinoy, M., Kharbeche, M., Al-Khalifa, K., Krichen, M., & Barkaoui, K. (2020). Urban traffic monitoring and modeling system: An IoT solution for enhancing road safety.
7. Mohanl, D. (2009). Road accidents in India. *IATSS Research, 33*(1), 75–79. https://doi.org/10.1016/s0386-1112(14)60239-9
8. Shen, X., Zhang, F., Lv, H., Wei, S., & Sun, Z. (2020). The application and extension of the theory of planned behavior to an analysis of delivery riders' red-light running behavior in

China. *Accident Analysis and Prevention, 144*(August 2019), 105640. https://doi.org/10.1016/j.aap.2020.105640.
9. Papakostopoulos, V., & Nathanael, D. (2020). The complex interrelationship of work-related factors underlying risky driving behavior of food delivery riders in Athens, Greece. *Safety and Health at Work, 0–6*. https://doi.org/10.1016/j.shaw.2020.10.006
10. Stutts, J. C., Wilkins, J. W., & Vaughn, B. V. (1999). *Why do people have drowsy driving crashes: Input from those who just did* (pp. 1–81).
11. Iqbal, A., & Rana, M. E. (2019). Adoption of IOT in automobiles for driver's safety: Key considerations and major challenges. *International Journal of Scientific and Technology Research, 8*(9), 1378–1384.
12. Goel, G., & Sachdeva, S. N. (2016). Analysis of road accidents on NH-1 between RD 98 km to 148 km. *Perspectives in Science, 8*, 392–394. https://doi.org/10.1016/j.pisc.2016.04.086
13. Shaaban, K., Gaweesh, S., & Ahmed, M. M. (2018). Characteristics and mitigation strategies for cell phone use while driving among young drivers in Qatar. *Journal of Transport and Health, 8*(February), 6–14. https://doi.org/10.1016/j.jth.2018.02.001
14. Strayer, D. L., Drews, F. A., & Crouch, D. J. (2006). A comparison of the cell phone driver and the drunk driver. *Human Factors, 48*(2), 381–391. https://doi.org/10.1518/001872006777724471
15. Newnam, S., Lewis, I., & Warmerdam, A. (2014). Modifying behavior to reduce over-speeding in work-related drivers: An objective approach. *Accident Analysis and Prevention, 64*, 23–29. https://doi.org/10.1016/j.aap.2013.10.032
16. Cheng, W., Han, C., & Xi, J. (2010). Mechanism analysis of over-speeding impact on the road safety. *2010 International Conference on Intelligent Computation Technology and Automation, ICICTA 2010, 2*, 652–656. https://doi.org/10.1109/ICICTA.2010.804.
17. Jurewicz, C., Sobhani, A., Woolley, J., Dutschke, J., & Corben, B. (2016). Exploration of vehicle impact speed—Injury severity relationships for application in safer road design. *Transportation Research Procedia, 14*, 4247–4256. https://doi.org/10.1016/j.trpro.2016.05.396
18. Lamble, D., Rajalin, S., & Summala, H. (2002). Mobile phone use while driving: Public opinions on restrictions. *Transportation, 29*(3), 223–236. https://doi.org/10.1023/A:1015698129964
19. Rosenbloom, T. (2006). Driving performance while using cell phones: An observational study. *Journal of Safety Research, 37*(2), 207–212. https://doi.org/10.1016/j.jsr.2005.11.007
20. Isa, K. A. M., Masuri, M. G., Abd Aziz, N. A., Isa, N. N. M., Hazali, N., Tahir, M. P. M., Noor, S. H., Danis, A., & Fansuri, H. (2012). Mobile phone usage behavior while driving among educated young adults in the Urban University. *Procedia—Social and Behavioral Sciences, 36*(June 2011), 414–420. https://doi.org/10.1016/j.sbspro.2012.03.045.
21. Marcil, I., Bergeron, J., & Audet, T. (2001). Motivational factors underlying the intention to drink and drive in young male drivers. *Journal of Safety Research, 32*(4), 363–376. https://doi.org/10.1016/S0022-4375(01)00062-7
22. Greening, L., & Stoppelbein, L. (2000). Young drivers' health attitudes and intentions to drink and drive. *Journal of Adolescent Health, 27*(2), 94–101. https://doi.org/10.1016/S1054-139X(99)00114-7
23. Horwood, L. J., & Fergusson, D. M. (2000). Drink driving and traffic accidents in young people. *Accident Analysis and Prevention, 32*(6), 805–814. https://doi.org/10.1016/S0001-4575(00)00005-1
24. Mantouka, E. G., Barmpounakis, E. N., & Vlahogianni, E. I. (2019). Identifying driving safety profiles from smartphone data using unsupervised learning. *Safety Science, 119*(November 2018), 84–90. https://doi.org/10.1016/j.ssci.2019.01.025.

25. Wheaton, A. G., Shults, R. A., Chapman, D. P., Ford, E. S., Croft, J. B., & Division of Population Health, National Center for Chronic Disease Prevention and Health Promotion. (2014). Drowsy driving and risk behaviors—10 States and Puerto Rico, 2011–2012. *MMWR. Morbidity and Mortality Weekly Report*, *63*(26), 557–562. http://www.ncbi.nlm.nih.gov/pubmed/24990488%0A, http://www.pubmedcentral.nih.gov/articlerender.fcgi?artid=PMC4584902.
26. Ferreira, A. I., Martínez, L. F., & Guisande, M. A. (2009). Risky behavior, personality traits and road accidents among university students. *European Journal of Education and Psychology Eur. j. Educ. Psychol. E-ISSN*, *2*(2), 79–98. www.ejep.es.

Usage of Behavioral-Based Safety Approach for Improving Worker Performances in Construction Sector: A Review

Bikarama Prasad Yadav, Nikhil Verma, Akshay Kant Mishra, N. A. Siddiqui, and Devendra Gill

1 Introduction

Construction projects are active, evolving and accommodate immanent risks [1]. The construction industry enrolls 7% of the world's personnel, yet it is in charge of for 30–40% of causalities [2]. Regardless of technological events and the usage of robust safety management systems, the construction industry's chronic level of causalities, genuine injury and weakness appear to be impervious to change. The International Labour Organization approximates that in few countries, 30% of construction workers experience the ill effects of back pain or musculoskeletal disorders. Furthermore, the event rate for non-fatal occupational wounds and illness in construction is 30% excessive than the average industry. These circumstances require expansion of the limitations of safety research to amalgamate interdisciplinary views and pivot on worker safety behavior, since risky worker behavior hand out above 80% of accidents [3].

The behavioral practices that workers perform in their daily occupation can have an immediate and prompt impact on health and safety. In the progress of safety theories (or accident causation versions) over the last ten years, understanding and overseeing unsafe behavior has continually been a prime feature of research. In spite of the reality that reasons of accidents are quite extended up to superintendence and top management level in further accident causation models (e.g., Swiss cheese model), unsafe behavior endured as mainstream research topics in safety science. Guided by these models, critical research attention has been set on distinguishing determinants of risky behavior and studying how to lessen unsafe behavior [4].

Behavior-Based Safety management is an approach which aims to recognize and alter critical unsafe behavior through a mix of perception, feedbacks, trainings and

B. P. Yadav (✉) · N. Verma · A. K. Mishra · N. A. Siddiqui
University of Petroleum & Energy Studies, Dehradun, India

D. Gill
Delhi Metro Rail Corporation, Delhi, New Delhi, India

N. A. Siddiqui et al. (eds.), *Advances in Behavioral Based Safety*,
https://doi.org/10.1007/978-981-16-8270-4_23

objective setting. In the construction industry, scrutiny of BBS can be productively used to raise safety at two building sites. Results proposed that BBS was efficient to lessen accident rate and severity. It was concluded that critical feedback and objective setting affected positively on performance toward safety at two construction areas in Finland [5]. Outcomes appeared that the objective setting, and response tools developed critical improvements in almost all safety behavior classifications, except PPE. Recently, it was devised and used BBS initiatives on construction platforms in Hong Kong which suggested that BBS is highly efficient measure to slow down unsafe behaviors with proper application by the committed management [6].

Earlier paper has discussed upon BBSM and proposal for the designed for the Supervisory-based intervention cycle and the tracking system for the BBS and their challenges [7]. Although work was done on the feedback basis where communication accidents were discussed [8]. Casual loop diagrams were made but were not that effective as it deals with punishment and incentives [9]. Some of the research works has been done on characteristics but were applicable for accident rate reduction [10]. Questionnaire based upon the SEM-based equations were made to analyze the performance but was no longer applicable for improving the workers BBS [11]. Safety Indexes were also trying to improve the system behavior rather than individual [12]. Some works were also done on impact of safety stressors on safety performance and how communication in the organization also plays its role [13]. PsyCap with its connections between authentic leadership, safety climate and outcomes in safety critical organizations (SCOs) were also discussed [14].

Safety leadership is the quality to enhance the intentions of the management for the workers safety. Psychological capital of the workers is a typical direct factor to influence the behavior. It helps the individuals for personal development toward safety. Psychological contract is a new concept in the study of safe behavioral psychology which implies that that employee with employer is involved under reciprocating behavioral aspects together for enhancing safety. Work stress is also a factor that promotes unsafe behavior indirectly on sites due to interaction between the human being and environment. Few safeties-related stressors include role ambiguity, role conflict and interpersonal safety conflicts which are the barriers for employees in their performance and can lead to accident-causing injury. Another important factor to influence behavior is safety compliance [15]. Compliance in construction industries is crucial and relates to the enforcement of standards meant to protect the worker by adhering to the requirements of the standard. It can act as the activator that encourages safe behavior of workers, it can act a trigger for motivating the workers either in positive or negative way to influence worker's behavior toward one' own safety. Lastly, effective safety training and motivation are also chief factors because they provide knowledge and awareness for the workers to act accordingly toward the typical construction activities which pose enormous risk to their safety. Motivation of worker is dependent on two aspects like positive reinforcement where rewards, job securities, etc. are given for their performance and other negative reinforcement where workers may be punished by salary cut, negatively criticized, etc. Although BBS methods are consistent for triggering the safe behaviors, above mentioned factors can be optimized with BBS for enhancing the safety throughout the organization [16].

This paper presents systemized review on the crucial factors which are important toward playing a role in the usage of BBS programs. Objective lies in the scanning of expository factors with the latest work criteria in construction industries to scrutinize the behavior-based safety from a conceptual perspective. These factors are implemented on a broad scale nowadays for affecting the behavior of workers, includes performance evaluation of BBS like approach to target behavior, time to time interaction, feedback, involvement of top management, proper safety management system, etc. Organization Behavior has key roles such as positive attitude of management, commitment, applicability of continuous improvement, safety committee. Worker participation actually involves the workers to face potential hazards, risks and their assessment simultaneously [7].

2 Factors Affecting the BBS at Construction Work

2.1 Performance Evaluation of BBS

Various methods are considered for evaluating the safety behavior at the site. This includes quantitative or qualitative analysis of the performance. Some of the approaches include setting the target behavior and analyzing it by the safety supervisor. Also monthly reviews and comparison were done to see the trend [17]. Time to time interaction and feedback can help in analyzing the performance of the construction industry. Also, once the process is started, further pathway for improvements can be decided [18]. Previous studies were proven that benchmarking is an important criterion to decide [18]. Involvement of top management has a key role in the safety performance. Providing proper SMS (safety management system), engagement of employees and workers, continuous improvement are the key players for evaluating the safety performance. Developing the guidelines by the top officials and successful implementation at the shop floor can be more effective [19].

It has been seen that methods including proper goal settings and feedback system helped the organization in reducing the unsafe behavior and increase in the performance [17, 20]. Quantitative methods include designing the SEM-based equations for analyzing the performance where various parameters like questionnaire, relation between the safety climate and personal characteristics were identified. Based on the logical equations, results were obtained and compared [11]. Some of the construction sites include rewards and incentive programs for the workers to encourage them for improving the safety culture. Certain goals were given for the workers for analyzing the performance [21]. In the era of advancement analysis part has become more accurate using the computer modeling which includes analysis the capture image (2D) and to identify the unsafe behavior, having deep learning on it. Also there were proposed system for predict, learn and analyze the behavior [22]. Some companies include identifying the critical behaviors. Then these are analyzed and are recorded. Once the baseline is set, then intervention is done to change the behavior. The duration, rate of target are recorded and based upon that the performance is evaluated [23]. The ABC model of BBS is also one of the efficient method for BBS. Several aspects were considered during performance evaluation which includes Organization commitment, Training, leadership, motivation, etc. were considered and subsequent weightage was given for each parameter on a scale of 10 [24]. The above methods can be used for analyzing the implementation of BBS and to identify the gaps to better improvement [24].

2.2 *Organization Behavior*

Organization roles are important in the implementation of the BBS at the site. Positive attitude, Commitment of top management and continual improvement are the key roles for BBS [6]. The previous researchers used it as a key factor and quantified it as a part of their performance parameter. Under this, various factors like the Union involvement, Worker management relation, Site contractor behavior, safety committee, Talk by management [25]. The safety leaders of the organization can influence their role more in this. Higher the position involved in it, more effective implementation the BBS is. This can be achieved by setting the goals and linking under the organizational and individual factors. Also organizational leaders have the ability to change the goals as per the hindrance observed before doing the influencing and motivation [26].

Studies have proven that the support from the organization toward the employee safety and quality of communication has helped the organization to reduce the accidents [10]. The organizations are responsible to maintain a good quality relation between the employee and employer. Under the research, it was found that almost 88 percent organizations agreed for management commitment toward safety needs to be driven from top level [24]. Various failures of organization includes poor designing, decision making which lead to errors and violations [27]. Accidents like Chernobyl tell about the working culture and management practice. It has been analyzed that there is a relation between the individual and organization factors. Some results also shown that some of the accidents occurred due to high level influence of the organization. Safety climate as an organization factor plays important role as it influences the individual factor [11]. Some researchers also find the organizational commitment is divided into parts viz. affective, normative and continual. Successful implementation of the BBS also changes the assumption in the organization views. The same was imprinted on the loop diagrams for analyzing the same. It can be seen that the safety behavior is a complex function and places BBS in a wide organizational context [17].

2.3 Workers Participation

During working on the construction sites, workers use to face the potential risks for the safety. Studies have proven that safety programs and processes have improved the number of accidents [12]. In addition, assessment methods need to be modified to attain better result. In the era of advancement, the VR method (virtual reality) can help the organization in training their people. Active participation and method of teaching need to be adhered. The participation should also include the seriousness of the occupational health and safety [18]. Another part of the participation includes proper feedback to the top officials, which denotes safety attitudes and behaviors [28]. Through participation of workers, the organization will be able to understand workers knowledge or skills toward safe behavior. Communication level is necessary, as it will lead to how a worker will respond in case of emergency. The mental process model helps in improving the feedback mechanisms [29] Safety Motivation, team work and participation in decision making will help promoting the participation of worker [19]. The implementation can be best achieved by involving the workers during the goal setting process. The goal proposed by the workers needs to be attained, discussed and should be agreed [29]. Periodic review of the safety culture using the workers feedback can help organization in improving the overall BBS of the site. They will develop the interest until their concerns are listened and corrective actions are implemented [11].

2.4 Safety Leadership

Leadership is important in the construction industry and without it, it can deflect from its path. The behavioral factors are measured using the leadership. Effective leadership includes role model, visibility, provide leadership, supportive, improve employees in goal setting, participate decision making, responsibility [21]. Leadership is that quality which helps to change your dreams into actions. Management actions play major role in shaping the behavior of the organization. It totally depends on the quality, consistency with which leadership is done for safety. In the researches, it has been analyzed that effective leadership has changed the workers behavior [24]. Effective leadership has emphasized the temporary organization to success. Leadership should be important for those countries where the construction is done majorly. Though it has been seen that effect of leadership is quite low when comparing to other industries. There are various conceptions related to the leadership that it is up to the supervisory level. But it can be seen that leadership is at all the levels i.e., from top to bottom. Project management teams are key role partners in this as they are representatives [30]. Leadership should be designed as a subsystem where leaders, followers discuss with the workers to achieve the goals. Various types of leaderships are involved which include transformational and transactional. Transactional includes rewards and monitoring, whereas transformational includes motivation, influencing, safety controlling [31]. Leadership is one of the important factors which can lead the subordinates to their goals. For analyzing the safety leadership, questionnaire can be developed for identification. From the previous research, safety culture and the safety behavior are interlinked. The owner and contract safety leaderships can only lead to effective implementation of the safety culture in the organization [26].

2.5 Psychological Capital Aspect of Safety

Psychological capital (PsyCap) is a constructive capacity that advances during one's growth and development. This function can be evaluated, created and utilized for improving the performance [32]. It is known to have an affect directly on employee's behavior. Research shows that PsyCap not only helps workers complete the work within their role, but also inspires them to perform organizational citizen behavior [33]. Many researchers recognized four sub-proportions of PsyCap: self-efficacy, hope, optimism and resilience [34]. Self-efficacy mentions to a faith that individuals have the scope to use cognitive and persuading resources to attain goals [5]. This conviction is not only obtained from individual's social credence, but is also accomplished from their learning, escapade, positive feedback and psychological arousal. Optimistic workers usually have the readiness to prosper. For the cause of safety objectives, they proactively take interest in safety gatherings and consent with the operation's guidelines and regulations. Optimistic workers are inclined to dissect the reasons of hindrances and amass positive emotions. In view of exhaustive investigations and exact judgments, they settle on reasonable decisions to meet necessary safety requirements [14]. At last, the idea of resilience alludes to the individual propensity to continue and recoil and even beyond when assailed by issues and bad luck. This tendency to never yield and consistently look to defeat problems and obstacles is immensely an important asset for targeted accomplishing outcomes as opposed to giving up or falling back when opposing difficult issues. From this concise review of the factors of PsyCap, it is a feature to note that factors of PsyCap are independent in their own place, but that they may interrelate and go in the same way, e.g., hope/optimism and hope/resilience [35]. The elements of PsyCap will coordinate comfortably all together and impact upon human working through control and coordination of various attention and memory operations into higher request psychological functions.

2.6 Psychological Contract of Safety

Psychological contracts have been currently applied to the grasp of employment connections and are accepted to be chief determinants of employee's viewpoints and behavior. Employees form presumptions about the employment correlations that guide them to believe that certain measures will be retaliated; this encompasses their psychological contract [36]. Psychological contract theory is relied upon social exchange theory and the standard of reciprocity. A psychological contract is fundamentally a recognized trade relationship between the two parties in an employment relation, specifically the employee and the employer [37]. Proposal for the psychological contract as a symbol of investigating this relationship that safety was already based on reciprocity requiring a duty of protection on the portion of the employer and a reciprocal responsibility to endorse safety standards on the part of the employee [38]. Attention on employees is commonly in relation with the safety perspective and behavior, comparatively employee safety responsibilities and obligations [36]. Distinguished social exchange between employers and employees is added in the safety framework with respect to psychological contract of safety [37]. The assumptions about workplace safety that employees get from both societal and hierarchical impacts will not really constitute a psychological contract. Seen obligations of safety will possibly become psychological commitment when the individual accepts that anticipated employee safety obligations and anticipated employer safety obligations are unforeseeable upon one another [38].

2.7 Safety Stressors in Behavior-Based Safety

Work stressors make appearance from the interconnections between an individual and the workplace. Safety-related work stressors prevent employee's safety performance and trigger accidents bringing about injury [39]. Safety stressors are reviewed and arranged into three forms: safety role ambiguity, safety role conflict and interpersonal safety conflict [40]. Safety role ambiguity refers to cases where obtainable information and assets regarding safety roles are unclear or insufficient. Safety role disputes reflect the presence of unpredictability between safety performance supposition and assessment basis. Finally, social safety disputes arise with safety issue dissents between organization members, such as safety operations [8].

As anticipated, construction workers encounter safety barriers and safety unreliability was less likely to captivate in safety involvement. Stressors make the guideline and execution of objective-oriented activities more unmanageable [40]. Thus, when confronted with circumstances demanding more exertion or cognitive assets, construction workers are probably focusing their limited energy on required safety behaviors, for example, safely using machines and instruments. Therefore, workers are probably going to have less energy to engage in extra, wilful safety behaviors such as encouraging others to partake in safety initiatives. Past research on stressors (e.g., interpersonal conflict) and intentional, organizational behaviors has additionally shown that workers facing stressors are less inclined to engage in behaviors that exceed all expectations [41].

On other hand, there could be an essential contrast between safety obstacles and safety vulnerability. This is a comparable difference of opinion proposing that few stressors, particularly role ambiguity, are more damaging than additional stressors. Safety hurdles might be effortless to overcome. If a construction worker is facing hindrances with a co-worker, he or she might be avoiding that person. Conversely, safety unreliability might lie down further outside of a construction worker's control.

2.8 *Safety Compliance for Behavior-Based Safety*

Safety compliance is differing from adequate to substandard where obeying with safety requirements is observed as good safety compliance and not obeying with safety requirements is observed as poor safety compliance. Compliance with safety requirements will help the work can be done both effectively and safely [24]. Consequently, much could be upgraded with sufficient room for improvement with regards to behavioral safety compliance by the way of strong endeavor to observe with safety prerequisites from both employers and employees.

Top administration of the organization should lead industriously the employees toward achieving organization-safety objectives by depicting that organization is sincere about safety by ensuring compliance to safety requirements, everyone in the corporation needs to be firm for their safety and health liability [24].

Effective communication is another consideration for necessary thought for promoting safety compliance at workplace. Leaders demonstrate their intuitions and worth by reciprocating and communicating through standards which lead them to keep one going at different levels for achieving safety objectives [42]. Communication is attributed in different ways: (a) through visible behavior, employees can understand the seriousness of compliance and employer can talk about the significance of safety and health. It influences employees directly to acknowledge what employer demands on priority basis and will effectively embrace their personal behavior accordingly. (b) Written policies of health and safety codes statements, statements with reference to health and safety roles and responsibilities, performance standards and findings from risk assessments also form a decisive part of communication [19].

2.9 *Effective Safety Training and Safety Motivation*

Safety training is highly effective to coach employees on prospective of hazards involved in jobs, risks, accidents and how to put a stop to accidents. Hence, training and education initiatives play a notable part in strengthening of safety in construction and prime to expand safety cognizance and changed behavior of employees [43]. Inadequate safety training of the employees is one of the major root bases of accidents on construction sites as they lack knowledge, education and skills to admit probable hazards at site [44].

Top management engagement such as connection with employees, talk on safety and guidance on safety proceedings related to refine safety motivation and will uplift employee's safety behavior [37]. Two types of motivation; (a) positive reinforcement which provides employees results such as monetary rewards, bonuses and job promotion, (b) other one is negative reinforcement where employers may denounce, punished and intimidate the employees to motivate them to carry-out their jobs in the safe mode. Nevertheless, reinforcement on positive motivation is further encouraging by many safety professional to maintain improve employees' good safety behavior [45]. Safety improvement also will only be achieved if incentives schemes are carried out to motivate employees to change their behaviors [24]. The organization that creates and maintains good quality employer and employee relationships will benefit from higher levels of the employee motivation, commitment and job satisfaction, which in turn impacted positively on the intention to stay and employee performance [46].

3 Conclusion

For additional safety improvement, the construction industry needs to provide more consideration to eradicate unsafe acts and conditions which are more dominant reasons of accidents and will add noteworthy value for enhancing safety on construction sites. This paper has concluded the need of safety behavior to be incorporated from the top management to the frontline workers to help the organization continuously in improving safety performance in the construction industry. Above factors such as organizational behavior, workers participation, work stressors, psychological contract for safety contribute to workplace safety management particularly through proactive safety behaviors and have demonstrated their importance for improving the behavioral change of workers in construction industry. Interdependency of factors is crucial to understand the roles of employers and employees toward enhancing the safety culture and safety climate of the company throughout. Taken together, these factors propose a preferable cognizance of the relationships between factors and their implementation in behavior-based safety approaches. Factors affecting safety behavior must be considered to enhance safety by engaging employees, communication and top management commitment toward the long-term incident-free workplace. Implementation of above factors can cultivate safety-specific trust to promote employee and employers to engage in proactive safety behaviors which reduces the risks in construction industry.

References

1. Newaz, M. T., Jefferies, M., Davis, P., & Pillay, M. (2016). Using the psychological contract to measure safety outcomes on construction sites. In *Proceedings of the 32nd Annual ARCOM Conference ARCOM 2016*, no. September, pp. 487–496.
2. Sunindijo, R. Y., & Zou, P. X. W. (2012). How project manager's skills may influence the development of safety climate in construction projects. *International Journal of Project Management, 4*(3), 286–301. https://doi.org/10.1504/IJPOM.2012.048220
3. Shi, Y., Du, J., Ahn, C. R., & Ragan, E. (2019). Impact assessment of reinforced learning methods on construction workers' fall risk behavior using virtual reality. *Automation in Construction, 104*(April), 197–214. https://doi.org/10.1016/j.autcon.2019.04.015
4. Guo, B. H. W., Yiu, T. W., & González, V. A. (2016). Predicting safety behavior in the construction industry: Development and test of an integrative model. *Safety Science, 84*, 1–11. https://doi.org/10.1016/j.ssci.2015.11.020
5. Laitinen, H. (1996). The effects of feedback and goal setting on safety performance at two construction sites, *24*(1), 61–73.
6. Choudhry, R. M. (2014). Behavior-based safety on construction sites: A case study. *Accident Analysis and Prevention, 70*, 14–23. https://doi.org/10.1016/j.aap.2014.03.007
7. Zhang, M., & Fang, D. (2013). A continuous behavior-based safety strategy for persistent safety improvement in construction industry. *Automation in Construction, 34*, 101–107.
8. Shin, M., Lee, H. S., Park, M., Moon, M., & Han, S. (2014). A system dynamics approach for modeling construction workers' safety attitudes and behaviors. *Accident Analysis & Prevention, 68*, 95–105.
9. Guo, B. H., Goh, Y. M., & Wong, K. L. X. (2018). A system dynamics view of a behavior-based safety program in the construction industry. *Safety science, 104*, 202–215.
10. Mullan, B., Smith, L., Sainsbury, K., Allom, V., Paterson, H., & Lopez, A. L. (2015). Active behaviour change safety interventions in the construction industry: A systematic review. *Safety science, 79*, 139–148.
11. Seo, H. C., Lee, Y. S., Kim, J. J., & Jee, N. Y. (2015). Analyzing safety behaviors of temporary construction workers using structural equation modeling. *Safety Science, 77*, 160–168.
12. Chen, D., & Tian, H. (2012). Behavior based safety for accidents prevention and positive study in China construction project. *Procedia Engineering, 43*, 528–534.
13. Sampson, J. M., DeArmond, S., & Chen, P. Y. (2014). Role of safety stressors and social support on safety performance. *Safety Science, 64*, 137–145.
14. Eid, J., Mearns, K., Larsson, G., Laberg, J. C., & Johnsen, B. H. (2012). Leadership, psychological capital, and safety research: Conceptual issues and future research questions. *Safety science, 50*(1), 55–61.
15. Wang, D., Wang, X., & Xia, N. (2018). How safety-related stress affects workers' safety behavior: The moderating role of psychological capital. *Safety Science, 103*(Oct 2017), 247–259. https://doi.org/10.1016/j.ssci.2017.11.020.
16. Osborne, S. (2017). Effective employee engagement in the workplace, *16*(1), 50–67. https://doi.org/10.5590/IJAMT.2017.16.1.04.
17. Guo, B. H. W., Miang, Y., Le, K., & Wong, X. (2018). A system dynamics view of a behavior-based safety program in the construction industry. *Safety Science, 104*(Sept 2017), 202–215. https://doi.org/10.1016/j.ssci.2018.01.014.
18. Zhang, M., Shi, R., & Yang, Z. (2020). A critical review of vision-based occupational health and safety monitoring of construction site workers. *Safety Science, 126*(July 2019), 104658. https://doi.org/10.1016/j.ssci.2020.104658.
19. Skeepers, N. C., & Mbohwa, C. (2015). A study on the leadership behaviour, safety leadership and safety performance in the construction industry in South Africa. *Procedia Manufuring, 4*(Iess), 10–16. https://doi.org/10.1016/j.promfg.2015.11.008.
20. Patel, D. A., & Jha, K. N. (2016). Evaluation of construction projects based on the safe work behavior of co-employees through a neural network model. *Safety Science, 89*, 240–248. https://doi.org/10.1016/j.ssci.2016.06.020

21. Ismail, F., Ahmad, N., Afida, N., & Janipha, I. (2012). Assessing the behavioural factors' of safety culture for the Malaysian construction companies, *36*(June 2011), 573–582. https://doi.org/10.1016/j.sbspro.2012.03.063.
22. Fang, W., Love, P. E. D., Luo, H., & Ding, L. (2020). Advanced engineering informatics computer vision for behaviour-based safety in construction: A review and future directions. *Adv. Eng. Informatics*, vol. 43, no. November 2019, p. 100980, 2020, doi: https://doi.org/10.1016/j.aei.2019.100980.
23. Depasquale, J. P., & Geller, E. S. (1999). Critical success factors for behavior-based safety: A study of twenty industry-wide applications, *30*(4), 237–249.
24. Zin, S. M., & Ismail, F. (2012). Employers' behavioural safety compliance factors toward occupational, safety and health improvement in the construction industry. *Procedia, Social and Behavioral Sciences*, *36*(June 2011), 742–751. https://doi.org/10.1016/j.sbspro.2012.03.081.
25. Sawacha, E., Naoum, S., & Fong, D. (1999). Factors affecting safety performance on construction sites, *17*(5).
26. Wu, C., Wang, F., Zou, P. X. W., & Fang, D. (2016). ScienceDirect How safety leadership works among owners, contractors and subcontractors in construction projects. *JPMA, 34*(5), 789–805. https://doi.org/10.1016/j.ijproman.2016.02.013
27. Haslam, R. A. et al. (2005). Contributing factors in construction accidents. *Appied Ergonomics, 36*(4), 401–415. https://doi.org/10.1016/j.apergo.2004.12.002.
28. Shuang, D., Heng, L., Skitmore, M., & Qin, Y. (2019). An experimental study of intrusion behaviors on construction sites: The role of age and gender. *Safety Science, 115*(March), 425–434. https://doi.org/10.1016/j.ssci.2019.02.035
29. Shin, M., Lee, H. S., Park, M., Moon, M., & Han, S. (2014). A system dynamics approach for modeling construction workers' safety attitudes and behaviors. *Accident Analysis and Prevention, 68*, 95–105. https://doi.org/10.1016/j.aap.2013.09.019
30. Fang, D., Huang, Y., Guo, H., & Lim, H. W. (2020). LCB approach for construction safety. *Safety Science, 128*(March), 104761. https://doi.org/10.1016/j.ssci.2020.104761.
31. Zhang, P., Li, N., Jiang, Z., Fang, D., & Anumba, C. J. (2019). An agent-based modeling approach for understanding the effect of worker-management interactions on construction workers' safety-related behaviors. *Automation in Construction, 97*(April 2018), 29–43. https://doi.org/10.1016/j.autcon.2018.10.015.
32. Ergeneli, A. (2013). The role of psychological capital and trust in individual performance and job satisfaction relationship: A test of multiple mediation model, *99*, 173–179. https://doi.org/10.1016/j.sbspro.2013.10.483.
33. Luthans, F., Youssef, C. M., & Avolio, B. J. *Psychological capital: Developing the human competitive edge.*
34. Luthans, F., & Youssef, C. M. (2004). Human , social , and now positive psychological capital management: Investing in people for competitive advantage. https://doi.org/10.1016/j.orgdyn.2004.01.003.
35. He, C., Jia, G., Mccabe, B., Chen, Y., & Sun, J. (2019). Impact of psychological capital on construction worker safety behavior: Communication competence as a mediator. *Journal of Safety Research, 71*, 231–241. https://doi.org/10.1016/j.jsr.2019.09.007
36. Walker, A., & Hutton, D. M. (2006). The application of the psychological contract to workplace safety, *37*, 433–441. https://doi.org/10.1016/j.jsr.2006.06.001.
37. Newaz, M. T., Davis, P., Jefferies, M., & Pillay, M. (2019). Using a psychological contract of safety to predict safety climate on construction sites. *Journal of Safety Research, 68*, 9–19. https://doi.org/10.1016/j.jsr.2018.10.012
38. Tanvi, M., Davis, P., Je, M., & Pillay, M. (2019). The psychological contract: A missing link between safety climate and safety behaviour on construction sites, *112*(March 2018), 9–17. https://doi.org/10.1016/j.ssci.2018.10.002.
39. Wang, D., Wang, X., Gri, M. A., & Wang, Z. (2020). Safety stressors, safety-specific trust, and safety citizenship behavior: A contingency perspective, *142*(Sept 2019). https://doi.org/10.1016/j.aap.2020.105572.
40. Action as the core of work psychology. A German Approach.pdf.

41. Zellars, K. L., Tepper, B. J., & Duffy, M. K. (2014). Abusive supervision and subordinates' organizational citizenship behavior. https://doi.org/10.1037//0021-9010.87.6.1068.
42. Al-alawi, A. I., Al-marzooqi, N. Y., & Mohammed, Y. F. (2015). Organizational culture and knowledge sharing: Critical success factors, (April 2007). https://doi.org/10.1108/13673270710738898.
43. Wong, J. Repetition in conversation: A look at 'first and second sayings', *33*(4), 407–424.
44. Construction sites safety roles.
45. Li, H., Lu, M., Hsu, S., Gray, M., & Huang, T. (2015). Proactive behavior-based safety management for construction safety improvement. *Safety Science, 75*, 107–117. https://doi.org/10.1016/j.ssci.2015.01.013
46. Gravina, N. E., King, A., Austin, J., & States, U. (2019). Training leaders to apply behavioral concepts to improve safety. *Safety Science, 112*(Feb 2018), 66–70. https://doi.org/10.1016/j.ssci.2018.10.013.

Consequence Modelling and Risk Assessment for Hazardous Substance Release in Fertilizer Plant Using Aloha

Bikarama Prasad Yadav, Nikhil Verma, Umang Kumar Yadav, Akshay Kant Mishra, and Anand K. Jayan

1 Introduction

Technological advancement in regards to the process industries and plants has been in place from a large amount of time and hence it has raised the importance of technical process safety, which prominently approached as a teaching in 1960s and 1970s and is following continuous development since then [1]. Process industries like ammonia synthesis and production are also combined with many critical activities providing exceptional progress in various sectors as well as involvement of many dangerous and toxic chemical during manufacturing processes while making use of considerable amount of hazardous substances for ammonia synthesis. The only chemical discovery in the manufacturing of ammonia and ammonium nitrates is the Haber Bosch method discovered in 1909 [1]. The formation of ammonia involves both high-rise pressure and low temperature to achieve the equilibrium for the reaction [2]. Synthetic Ammonia (NH_3) or Ammonia is synthesized from natural gas which in turn is reduced to hydrogen and carbon. Following this the hydrogen is subsequently cleansed and made to react with nitrogen to form ammonia. Ammonia (approx. 75%) is implemented as fertilizer either directly or indirectly after processing it into urea, Ammonium Nitrate and Diammonium Phosphates [3]. Ammonia is almost transported in liquid state, for which it is required to be compressed, refrigerated or semi-refrigerated. Its blending for storage is performed under different states within the storage tanks like pressurized, refrigerated or semi-refrigerated states [2]. In such industries, prospective grave aftermath can peduncle from fire, explosion and toxic hazards from unexpected releases and hazards corresponding with failures in high-rise pressure equipment's [4]. Incidents which involve the danger of loss of containment factors can create major consequences having huge potential for harming people and environment [5]. Review of various papers based on quantitative and qualitative analysis justified that identification and determination of scenarios

B. P. Yadav (✉) · N. Verma · U. K. Yadav · A. K. Mishra · A. K. Jayan
University of Petroleum & Energy Studies, Bidholi, Dehradun, India

N. A. Siddiqui et al. (eds.), *Advances in Behavioral Based Safety*,
https://doi.org/10.1007/978-981-16-8270-4_24

and their respective consequences are found to be less in number. In addition, data on the impact of consequences due to toxic release during ammonia production is yet to be determined. Suggestions for implementation of various measures will be provided through the conducted study for the workers and nearby population. Consequence analysis will be helpful in minimizing the risk in emergency scenarios while similarly implementing in ERCP [6]. The proposed study will perform entire sequence of consequence modelling involving the potential accidents arising out of fire, explosion and toxic release of hazardous chemicals from plants, during ammonia production [7]. The purpose of modelling extends for determining dispersion release of toxic chemicals and estimating the radius of dispersion due to fire and explosion [8]. The accident scenarios for consequence analysis are modelled using (Aerial Location of Hazardous Atmosphere) aka ALOHA software [9]. Software can compute the toxicity, heat radiation and shock wave originated by gas dispersion, fire and explosion after chemical discharge with various parameters. It is a conveniently used tool for emergency groundwork, tutelage, response and academic research in most developed and developing countries. ALOHA's software directory incorporates the physical parameters of almost 1000 types of dangerous substances, having accurate and timeliness extent of calculations.

2 Review of Literature

Previous researches showed about accidents during transferring waste thinner from the paint shop and its simulation is done with the help of ALOHA to analyse the consequences arising due to the process and take preventive actions. The results showed that the administrative units possess high danger and certain remedial actions were incorporated [10]. ALOHA modelling was considered for the fertilizer industry in and risk was being estimated, though mitigation plans for such risks were still a question [11]. Even method of chromatography was introduced to determine the traces which caused the main nuisance for odour secretion in the environment [12]. Also the same was considered for the LPG storage tanks where risk assessment was carried out for various scenarios of BLEVE, VCE and jet fire [13]. ALOHA modelling was considered petrochemical industry only for the leakage as this was the major risk factor, whereas other risk factors were not included [14]. In certain cases, it has been seen that chemical exposure index values were considered to calculate the hazard area for probable toxic release [15]. In countries like Brazil, ALOHA is considered to for creating the ERPG guidelines in oil and gas industry [16]. Some studies which include chlorine gas leak from the petrochemical industry to conduct the QRA studies and to analyse the effects on people and environment [17]. Earlier case studies were also being considered to understand the need to MARPLOT to assess the intensity and extent of damage in case of an accident [8]. It has being observed that on analysing the risks in fertilizer industry, QRA methods were used which include HAZOP and FTA but the extent of damage needs to be focused more [18]. ETA and FTA for analysing uncertain handling formulations along with a case study were also taken into consideration. The first assumption is related to likelihood values of input events, and the second assumption is regarding interdependence among the events (for ETA) or basic events (for FTA) [19]. Based on the QRA done, severity of hazards will be calculated and the additional insights are developed to minimize the risk exposure [4]. FTA analysis was also considered same for the root cause analysis [20]. In industries, people go for numerical-based assessment to categorize the risk into low, medium and high and based on that they act on the particular risk [21]. FTA analysis for failure in the Emergency Shut Down system in LNG terminal. This will help identify the root cause of the ESD failure so as pre-action can be decided to avoid accidents [22]. Risk assessment technique like TORAP was developed keeping in mind of the real-life situation and in a cost-effective manner which can be introduced in a petrochemical industry [23]. Some researchers tried to compare between ALOHA and PHAST though they have their certain pros and cons in terms of data requirement [9]. It has been observed that domino effect and the implementation of fuzzy inference system and dynamic Bayesian network were considered for scenario for fire and explosion but toxic release was not considered [7]. Safety practises as well as the recommendations should also be taken care while transporting and storage of ammonia [24]. Assessments through ALOHA has helped the industries to assess the collateral damage in case of catastrophic event and to improve their approach towards the safety [17]. In some cases, ALOHA was done for designing the evacuation route by determining the worst-case scenario [25]. ALOHA software has been considered for rapid and accurate predetermination of hazardous area in case of accidental release and initiate direction of evacuation in different areas by the in case of emergency caused during accidents due hazardous chemicals release [26]. The result helped in the deductions consequences and risk assessment, which will help in reducing the potential impact in future [27]. Certain books describe about technical process safety, the new challenges which the industry face regarding process safety. BBS with respect to process safety is also discussed [1].

The software's maturity can be evaluated by its decision-making function which provides dependable technical information in response to input parameters. It's justification includes: Firstly, simulating the accident scenarios, e.g., Fire, Explosion and Toxic Release; Furthermore, determining the approximate scope of toxic chemical discharge; Thirdly, prediction of transitions in agglomerations at key points [26]. A adaptable mapping approach is MARPLOT (Mapping, Application for Response, Planning and Local Operational Tasks) which is collaborative with ALOHA results can be consolidated with MARPLOT stringently, bestow amalgamated plotting and modelling abilities.

The learning also encompasses the Risk Assessment using dissimilar tools which regulate basic causes of top event's occurrence and Event Tree Analysis (ETA) a logical modelling approach for both success and failures that prospects responses by means of single initiating event and imprecise the probability in addition to grant of divergent roots para mounting to the undesired event. Finally Individual and Societal Risk calculations are executed to scale the number of practicable mortalities that are likely to occur and pretentious population for a given vulnerability for a particular scenario. Recommendations based on the above suggested study is also provided for onsite, offsite training, in case of extremities and operation safety management activities.

3 Methodology

Risk assessment can be defined as a sophisticated way of recognizing, examining and assessing of hazards which can occur in a fertilizer industry and the prime aim of this paper is to put a stop to all the hazards which possess in the industry. Choosing a particular risk assessment technique is difficult. To attain this objective, we have to gather all the relevant data, understand the information portrayed and with the resources we have in our hand appropriate techniques are decided. In this paper, a combination of HAZOP and ETA techniques are used. During the initial phase of our project, we gathered all relevant data regarding the basic process what is happening in a fertilizer industry with the help of manuals, organization reports, books, internet and research paper. The basic process of ammonia and urea production is understood with the help of the above data. Also, all the MSDS of the chemicals used in the industry are studied. But in the project more focusing is given to ammonia production than urea as the risks associated with the urea plant are comparatively low. After understanding the basic concepts, the next step is to identify certain scenarios which may go wrong and posse's danger to the human as well as to the environment. The hazardous gases considered throughout the process are H_2S, NH_3, CO_2, CO, NOx Sox. The probit functions are calculated for thermal radiation, explosion and toxic release of these hazardous gases with the help of CCPS and HSE guidelines and taking LC (Lethal Concentration) value as 50. Probit functions help in estimating the fatality count. Individual and societal risk is calculated while referring to the CCPS guidelines and population density data as provided by the mentor; it is used to estimate the affected population in case of emergency. After the calculation of probit function, IR and SR certain top events are identified with the help of P&ID, process flow diagram, etc. which can cause deviation and subsequently lead to a disaster (Fig. 1).

3.1 Hazard Identification and Risk Assessment

Hazard Identification and Risk Assessment (HIRA) is a term which involves recognizing all the hazards, assessing its risk at the facility all over the period to rectify certain changes that cause risks to the employees, public or environment. HIRA is a key factor in every industry because it helps to control/ manage risks and when the hazards are pointed out risks are evaluated and found out whether they are at toler-

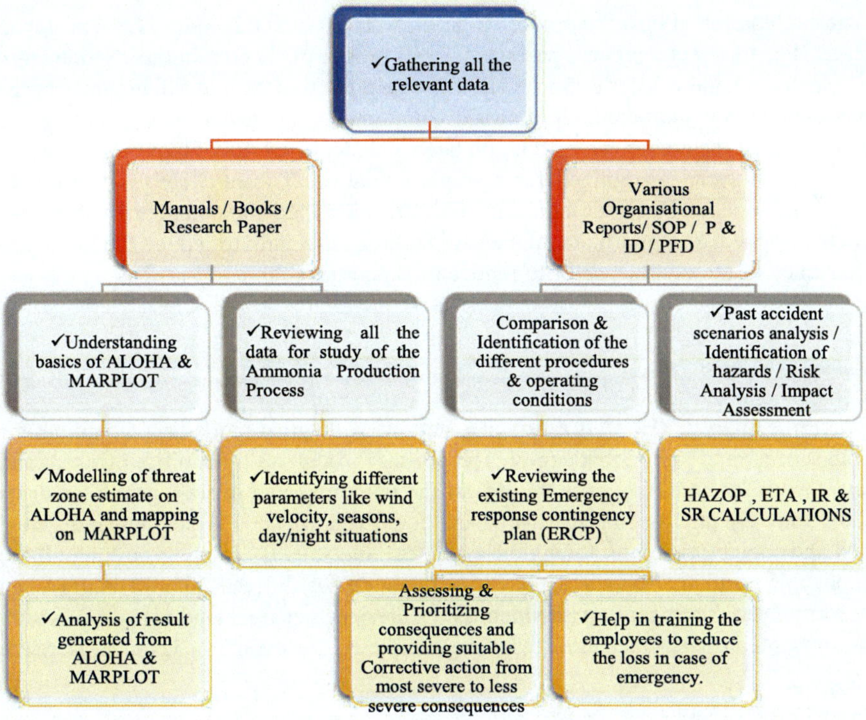

Fig. 1 Methodology flowchart

able level or not. The understanding of the risks associated with the facility forms the basic foundation of process safety. Some of the techniques used in HIRA are mentioned below.

Hazop

HAZOP is a quantitative risk assessment approach which explains hazards, its key words, consequences, sources and actions to be done to avoid an expected dangerous situation. It is constructed taken into account of all the deviations which can happen in the process. For all the process happening in the industry different set of HAZOPs are conducted i.e., for desulphurization, methanation, shift conversion, etc., a separate set of HAZOPs are constructed with the help of process flow diagrams and P&ID's [26]. The HAZOP is constructed by dividing the system into various sections—Desulphurization, shift conversion, ammonia synthesis, etc. Once the sections are identified certain nodes (a node is defined as particular location in the process where we calculate the divergence). Examples are separators, heat exchangers, etc. Once the nodes are identified the design intent of the plant is described. The design intent is an explanation of how the process is predictable to act at the node. The next step in HAZOP is to select a process parameter (e.g., pressure). Then a suitable guide word is selected. Guide word is a short word to create the fascination of a divergence of the process which is intendant. Guide word + Parameter → Deviation. Once the guide words are set the reason why deviation occurs is identified [28]. Numerous numbers of causes may be recognized for a single deviation. The next step is to identify the consequence. Once the consequences are identified the safeguard techniques are recognized. Safeguard techniques are facilities that help to reduce the incidence frequency of the divergence or to reduce its outcomes (Tables 1 and 2).

Event Tree Analysis

Event tree analysis (ETA) is a forward, bottom up, logical modelling technique for both success and failure that explores responses through a single initiating event and lays a path for assessing probabilities of the outcomes and overall system analysis [28]. Since the process industries are hazardous ones, we need to calculate individual risk and societal risk with the aid of event tree analysis. By looking through all the applicable accidental events that have been pointed out with the aid of PHA, ETA can be formulated. The next step is to identify the barriers that are intended to contact with the unwanted event. Once these steps are completed, the event tree is made. Here the

Table 1 Likely emergencies

Locations of the site	Likely emergency 1	Likely emergency 2	Likely emergency 3
NG preheater	Jet fire in pipeline/flash fire	Vapor gas release	Vapor cloud explosion
Desulphurization section	Jet fire in pipeline/flash fire	Vapor gas release	Vapor cloud explosion
Primary and secondary reformer	Jet fire in pipeline/flash fire	Vapor gas release	Vapor cloud explosion
LT and HT shift convertor	Jet fire in pipeline/flash fire	Vapor gas release	Vapor cloud explosion
CO_2 removal section	Jet fire in pipeline/flash fire	Vapor gas release	Vapor cloud explosion
Methanation	Jet fire in pipeline/flash fire	Vapor gas release	Vapor cloud explosion
Drier/purifier	Jet fire in pipeline/flash fire	Vapor gas release	Vapor cloud explosion
Syn gas compressor	Flash fire	Vapor gas release	Vapor cloud explosion
Syn gas converter	Toxic gas release	Flash fire	Vapor cloud explosion
Refrigeration system	Toxic gas release	Flash fire	Vapor cloud explosion
Storage area/dispensing area	Toxic gas release	Flash fire	Vapor cloud explosion

Table 2 HAZOP guide words

Guide word	Meaning	Example
NO	In the absence of	No flow of carbon di oxide
LESS	Reduction in the quantity	Less flow
MORE	Rapid increment in the quantity	More flow
OTHER	Limited or complete replacement	Other compounds were mixed
BEFORE	In a sequential flow	Before the process
REVERSE	Exactly opposite	Reverse flow
LATE	With respect to the time of clock	Late flow

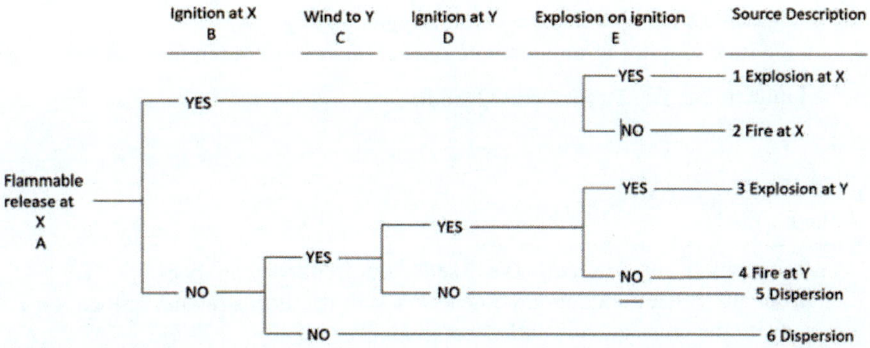

Fig. 2 Sample ETA [28]

barrier techniques include gas leakage detection system and isolation, direct ignition, fire protection, delayed ignition. Supplementary events must also be mentioned in the ETA in an orderly manner. Various after-effects can be recognized from the ETA once the situation has happened. An accidental event may be caused by system-equipment failure, Human error Process upset. Frequencies of outcomes are calculated with the help of CCPS and the data provided by the company (Fig. 2).

Estimation of Individual Risk

Individual risk is defined as a risk to a person in the vicinity of a hazard. This includes the nature of the injury to the individual, the likelihood of the injury occurring and the time period over which the injury might occur. Individual risk can be estimated for the most exposed individual, for groups of individuals at particular places or for an average individual in an effect zone. For a given incident or set of incidents, these individual risk measures have different values. The individual risk is a frequency of fatality, usually chances per million per year.

Individual risk calculation at a particular geographical location near a plant assumes that the contributions of all incident outcome cases are additive. Thus, the total individual risk at each point is equal to the sum of the individual risks, at that point, of all incident outcome cases associated with the industry.

$$IR_{x,y} = \sum_{(i=1)}^{n} IR_{x,y}, I \tag{1}$$

Where,

$IR_{x,y}$ the total individual risk of fatality at geographical location x, y (chances of fatality per year, or yr^{-1})

$IR_{x,y,I}$ the individual risk of fatality at geographical location x, y from incident outcome case *"i"* (chances of fatality per year, yr^{-1})

n the total number of incident outcome cases considered in the analysis.

The inputs to the Eq. 1 are obtained from,

$$IR_{x,y,i} = F_i{}^{*} P_{f,i} \tag{2}$$

Where,

F_i frequency of incident outcome case i, from frequency analysis (yr^{-1})

$P_{f,i}$ probability that incident outcome case i, will result in a fatality at location x, y, from the consequence and effect models.

The inputs to the Eq. 2 are obtained from,

$$F_i = F_I{}^{*} P_{o,i}{}^{*} P_{OC,i} \tag{3}$$

Where,

F_I frequency of incident I, which has incident outcome case i, as one of its incident outcome cases (yr^{-1})

$P_{o,i}$ probability that the incident outcome, having i, as one of its incident outcome cases, occurs, given that incident I has occurred

$P_{OC,i}$ probability that incident outcome case i occurs given the occurrence of the precursor incident I and the incident outcome corresponding to the outcome case "i".

UK-HSE has in the publications "Reducing Risk and Protecting People" and "Guidance on ALARP decisions in control of major accident hazards (COMAH)" enunciated the tolerability criteria for individual risk. The guidance on QRA also can be taken from Bureau of Indian Standards Hazard Identification and Risk Analysis **(IS 15656:2006)**.

Fig. 3 ALARP triangle

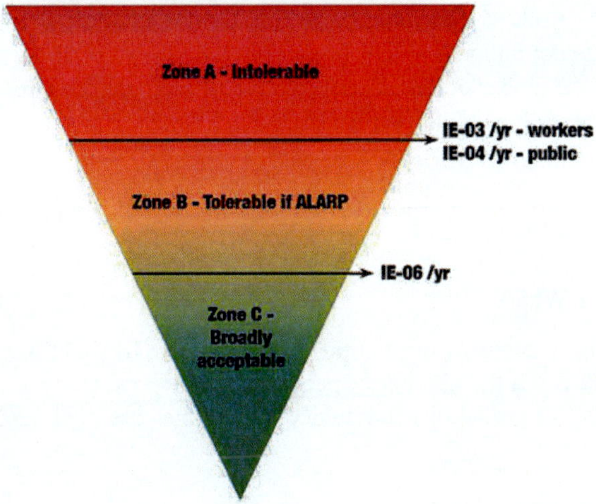

- An individual risk of death of **one in a million (1 x 10^{-6}) per annum** for both workers and the public communicate to a very low level of risk and should be used as a guideline for the boundary between the risk acceptable and ALARP regions.
- An individual risk of death of **one in a thousand (1 x 10^{-3}) per annum** should alone constitute the dividing line between what could be just tolerable for any substantial category of workers for any large part of a working life, and what is unacceptable or intolerable.
- For public members, imposed with a risk "in the wider interest of society" this limit is surrogated to be an order of magnitude lower, at **1 in ten thousand (1 x 10^{-4}) per annum**.

The upper limit of **tolerable risk** to public, **1 x 10^{-4} per year** is within the scope of risk due to transport accidents. The upper limit of **acceptable risk, 1 x 10^{-6} per year**, is within the scope of risk due to natural hazard such as lightning (Fig. 3).

Estimation of Societal Risk

Societal risk is a measure of risk to a group of people. It is most often expressed in terms of the frequency distribution of multiple casualty events (F-N curve.) However, societal risk can also be expressed in terms similar to individual risk. For example, the likelihood of 10 fatalities at a specific location x, y is a type of societal risk measure. The calculation of societal risk requires the same frequency and consequence information as individual risk. Additionally, societal risk estimates require a definition of the population at risk around the facility. This definition can include the population type, the likelihood of people being present, or mitigation factors. Individual and societal risk is the different presentation of the same underlying combinations of incident frequency and consequences.

Both of these measures may be of importance in assessing the benefits of risk reduction measures or in judging the acceptability of a facility in absolute terms.

Number of people affected by all incident outcome cases can be estimated using the following equation.

$$N_i = \sum_{(i=1)}^{n} P_{x,y} * P_{f,i} \tag{4}$$

Where

N_i number of fatalities resulting from incident outcome case I,
$P_{x,y}$ is the number of people at locations x, y
$P_{f,i}$ probability that incident outcome case i will result in a fatality at location x, y

4 Hazards in a Fertilizer Industry

The various hazards which can occur in an ammonia plant are mentioned below.

4.1 Toxic Gas Dispersion

The release of the liquid ammonia coming in contact of air will vaporize rapidly and can have a toxic release. The leakages include from the pressurize pipeline or the area where the ammonia is being manufactured. Also, worst case includes the release of ammonia from the storage tank (10,000 MT) each of two tanks and release of the ammonia from the transferring line to the railway gantry. The toxic scattering is one of the furthermost catastrophic events in all cases as it can travel with the wind, which can be inhaled by multiple people nearby. A small leak (25 mm) can impact for long distance and affects can be seen within as well as outside the plant.

4.2 Flash Fire

It occurs when a vapour cloud of flammable material burns. In fertilizer industry, it occurs for ammonia and natural gas. The cloud is ignited on the edge and burns towards release point. The duration of the fire is for seconds. The overpressure generated by the combustion is not considered significant in terms of damage potential to the person or equipment. Major hazard includes the impingement. The burnt zone which is clarified as that particular region of the vapour which wraps out to half of the

LFL. In situations where the accumulation can be above UFL, turbulent-generated combustion merges with air and the outcome leads to a flash fire.

4.3 Jet Fire

Jet flames are considered as high-pressure liberation of the gas from limited openings. A fireball is a circular fire arising when an unexpected liberation of pressurized liquid or gas that is straight away ignited. The best-known cause of the fireball is the BLEVE. Fireball duration is $5 - 20$ s.

4.4 Vapor Cloud Explosion

When a huge amount of material is released in the atmosphere, mixes with the air to produce the sufficient mass in flammable extent and is ignited, the result is a vapor cloud explosion. Without adequate air mixing up, a diffusion-controlled fireball maybe resulted without important overpressure emerging. The momentum of flame propagation must put on a spurt as vapour cloud flares. Without the acceleration only flash fire will occur. With large release of the huge amount of flammable gas or vapour is accidentally released into the atmosphere it may form a vapour cloud and if its ignition is delayed (5–10 min) could result in a vapor cloud explosion. The damaged consequences are mainly due to overpressure that is formulated from the rapid extension of inflammable products. The overpressure is the key reason of damages to people, equipment and facilities.

4.5 Weather Class Category

Important characteristics in performing the consequence modelling are the weather parameter and its stability. As India is having three major weather viz. summer, rainy and winter. The stability of the weather will decide the existing turbulence and to resist the vertical motion. The tendency will directly affect the ability of the atmosphere to disperse the pollutant emitted into the facilities. Turbulence induced by the buoyancy force atmosphere is closely related to the vertical temperature gradient. Temperature normally decreases with the increase in height. The environmental lapse rate is explained for the relation of temperature in relation to height. It will vary from time to time and place to place. The atmosphere is observed to be stable, neutral or unstable according to the ELR is less, equal or more than the DALR value (Table 3).

Pasquill stability parameter based on the Pasquill Gifford categorization expresses the steadiness of atmosphere, degree of convective turbulence. Also, they described

Table 3 Weather class

Stability class	Definition	Stability class	Definition
A	Very unstable	D	Neutral
B	Unstable	E	Slightly stable
C	Slightly unstable	F	Stable

six different classes from A to F. Wind speeds, intensity of solar radiation and night-time sky cover have been identified as the prime factors defining the stability categories. When the atmosphere is unbalanced and wind speed is high, chances of gas dispersion is more. When the condition of atmosphere is steady and wind current is low, dispersion of the gas is less pollutant, accumulation in the air is high.

5 Top Events for Ammonia Plant

All the scenarios are identified which can lead to a hazard with the help of risk assessment techniques such as HAZOP and ETA. Based upon these most probable occurring events, remedial actions can be taken accordingly. Below are the most probable top events which can go wrong in an ammonia plant.

1. Leak of Natural Gas from Inlet Pipeline of Natural Gas Feed and Desulphurization Section.
2. Leak of Hydrogen from Reforming Section Pipeline Outlet.
3. Leak of Carbon Monoxide from Shift Converter Pipeline Outlet.
4. Leak of Methane from Methanator Pipeline Inlet.
5. Leak of Ammonia from Piping of Ammonia Separator Outlet.
6. Leak from Top of Ammonia Storage Tank (PRV Failure).
7. Leak from Ammonia Transfer / Return Pipeline During Transportation.

6 ALOHA Modelling

After assessing various processes involved in ammonia production and storage of ammonia with the help of different QRA tools (HAZOP, ETA, FTA), the listed scenarios / top event are taken further to perform consequence modelling with the help of ALOHA and MARPLOT. During the modelling, leakage is considered within the vicinity of plant. Modelling is performed mainly for winter seasons, with day and night conditions for particular scenario of dispersion. Winter season is taken into account here because temperature inversion occurs in this weather condition. During winters, the boundary layer is very thin as the cooler air close by the surface of the earth is denser. The cool air is confined under the warm air and forms a lid which

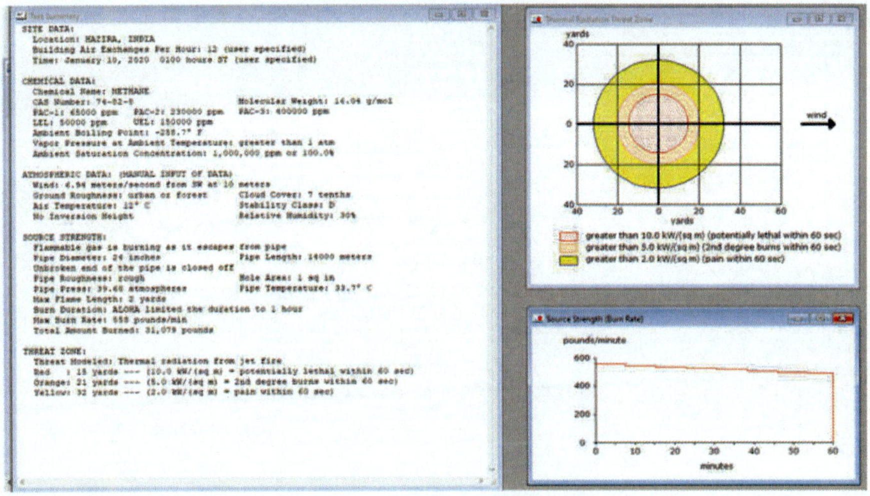

Fig. 4 ALOHA modelling for top event 1 in winter season (Night)

is known as winter inversion. So as a result, pollutants exposed into the atmosphere lack space to get dispersed into the atmosphere. Weather conditions are taken into account with the aid of Surat meteorological data and wind rose diagram of Surat municipality. Specific properties of gases like CAS no. and boiling point are referred from Safety Data Sheets. ALOHA limits the burning duration to 1 h for jet fire and vapor cloud explosion similarly release duration is limited to 1 h for toxic release. The results of the same can be referred below (Figs. 4, 5, 6, 7, 8, 9 and 10).

7 Result

Quantitative Risk Assessment and consequence modelling using ALOHA and MARPLOT are used to simulate the effects of toxic chemicals utilized in the ammonia plant, storage and transportation facilities which have specific hazardous characteristics for human beings and environment. In addition, the above methodologies revealed out the fact that major critical events occur due to loss of containment (LOCs) factors which can lead into catastrophic events such as fire, explosion and toxic release. Such critical top events are identified and a calculation for respective individual and societal risks are also determined. Below represent the top event calculations for individual risk and societal risk and their comparison with the ALARP triangle.

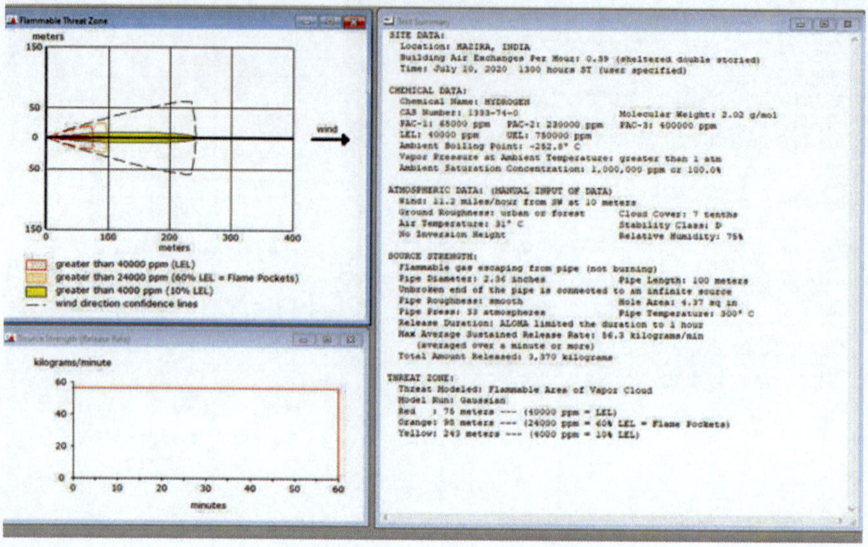

Fig. 5 ALOHA modelling for top event 2 in rainy season (Morning)

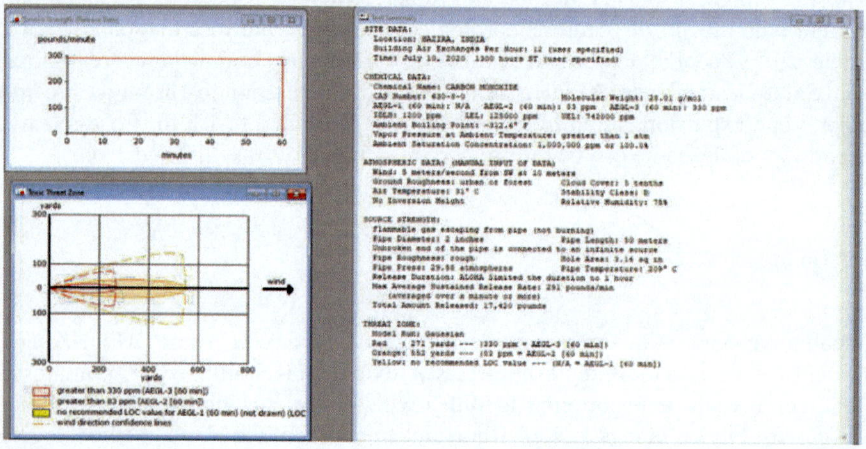

Fig. 6 ALOHA modelling for top event 3 in rainy season (Morning)

7.1 Top Event 1

1. Maximum individual risk to employees and total population due to leak of natural gas from pipeline in desulphurization section are **3.91 * 10**$^{-06}$ **per year** and **1.95 * 10**$^{-07}$ **per year**.
2. Number of fatalities resulting from incident outcome case in societal risk is **47**.

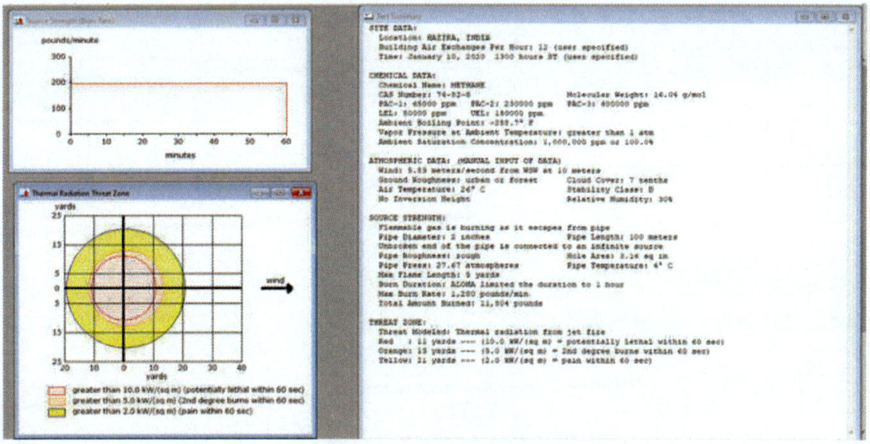

Fig. 7 ALOHA modelling for top event 4 in winter season (Morning)

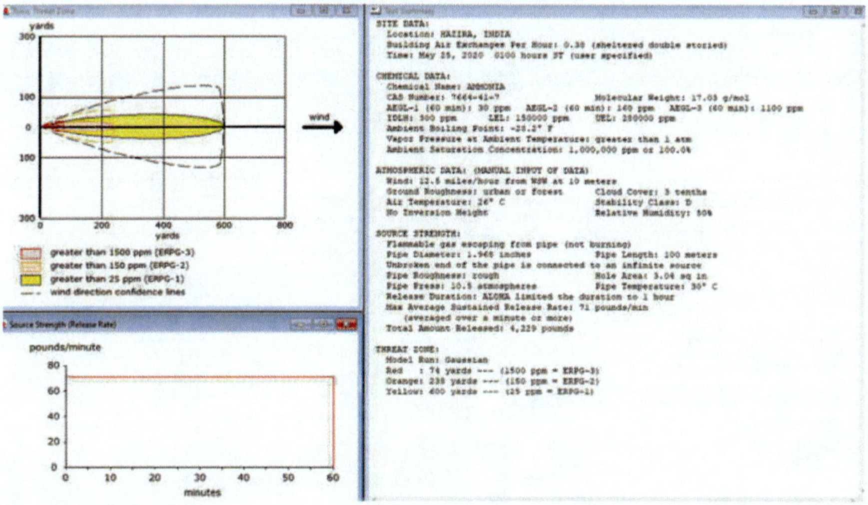

Fig. 8 ALOHA modelling for top event 5 in summer season (Night)

3. Radiation levels of Jet Fire and Flash Fire reach up to **14 m** and **50 m**. Overpressure levels in Vapor Cloud Explosion reach up to **125 m**, respectively.
4. Results of consequence analysis indicate that significantly fire radiation and explosion overpressure effects for current credible leak scenario are **within tolerable region if ALARP**.

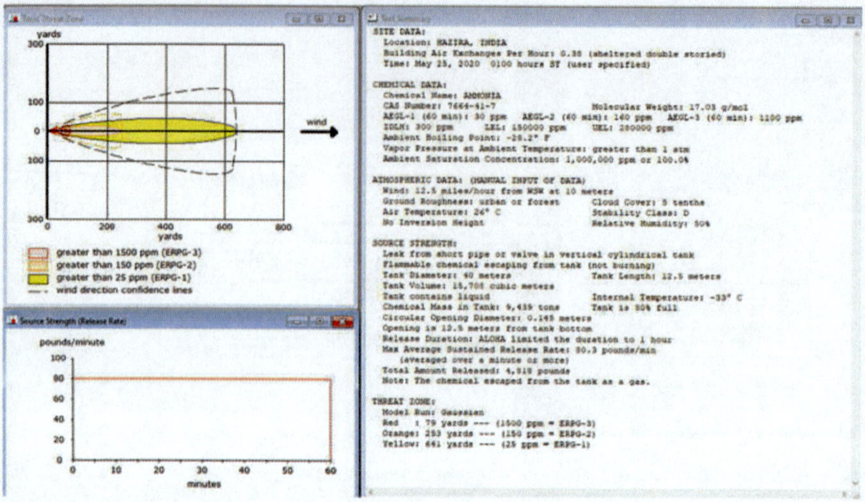

Fig. 9 ALOHA modelling for top event 6 in summer season (Night)

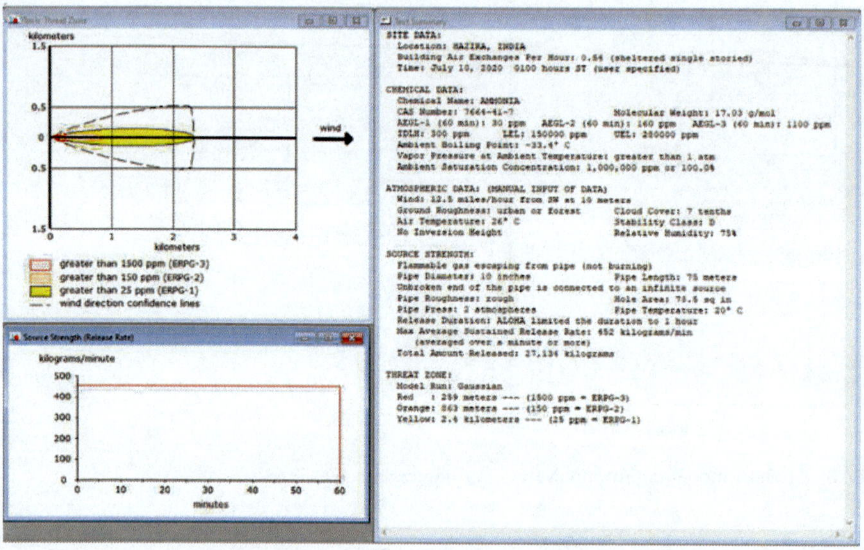

Fig. 10 ALOHA modelling for top event 7 in rainy season (Night)

7.2 *Top Event 2*

1. Maximum individual risk to employees and total population due to leak of hydrogen from reforming section/ shift converter section are **3.84 * 10^{-06} per year** and **3.57 * 10^{-07} per year.**
2. Number of fatalities resulting from incident outcome case in societal risk is **75.**
3. Radiation levels of Jet Fire and Flash Fire reach up to **30 m** and **92 m.** Overpressure levels in Vapor Cloud Explosion reach up to **227 m,** respectively.
4. Results of consequence analysis indicate that significantly fire radiation and explosion overpressure effects for current credible leak scenario are **within the tolerable region.**

7.3 *Top Event 3*

1. Maximum individual risk to employees and total population due to leak of carbon monoxide from shift converter section are **7.72 * 10^{-06} per year** and **9.81 * 10^{-07} per year.**
2. Number of fatalities resulting from incident outcome case in societal risk is **143.**
3. Radiation levels of Jet Fire and Flash Fire reach up to **10 m** and **114 m.** Overpressure levels in Vapor Cloud Explosion reach up to **114 m.** Lethal concentrations for Gas Dispersion reach up to **512 m,** respectively.
4. Results of consequence analysis indicate that significantly fire radiation, explosion overpressure effects and toxic concentration for gas dispersion for current credible leak scenario are **within tolerable region if ALARP.**

7.4 *Top Event 4*

1. Maximum individual risk to employees and total population leak of methane from methanator/ compressor outlet are **2.45 * 10^{-06} per year** and **1.99 * 10^{-07} per year.**
2. Number of fatalities resulting from incident outcome case in societal risk is **62.**
3. Radiation levels of Jet Fire and Flash Fire reach up to **13** and **42 m.** Overpressure levels in Vapor Cloud Explosion reach up to **104 m.**
4. Results of consequence analysis indicate that significantly fire radiation and explosion overpressure effects for current credible leak scenario are **within the tolerable region.**

7.5 *Top Event 5*

1. Maximum individual risk to employees and total population due to leak of ammonia from piping of ammonia separator outlet are **1.47 * 10^{-5} per year** and **2.07 * 10^{-06} per year.**
2. Number of fatalities resulting from incident outcome case in societal risk is **158.**
3. Radiation levels of Jet Fire and Flash Fire reach up to **10 m** and **210 m**. Overpressure levels in Vapor Cloud Explosion reach up to **210 m**. Lethal concentrations for Gas Dispersion reach up to **540 m,** respectively.
4. Results of consequence analysis indicate that significantly fire radiation, explosion overpressure effects for and toxic concentration for gas dispersion for current credible leak scenario are **within tolerable region if ALARP.**

7.6 *Top Event 6*

1. Maximum individual risk to employees and total population due to leak from top of storage tank (PRV failure) are **1.08 * 10^{-04} per year** and **4.06 * 10^{-05} per year.**
2. Number of fatalities resulting from incident outcome case in societal risk is **276.**
3. Radiation levels of Jet Fire and Flash Fire reach up to **10** and **70 m**. Overpressure levels in Vapor Cloud Explosion reach up to **70 m**. Lethal concentrations for Gas Dispersion reach up to **574 m.**
4. Results of consequence analysis indicate that significantly fire radiation, explosion overpressure effects and toxic concentration for gas dispersion for credible leak scenario are **within tolerable region if ALARP.**

7.7 *Top Event 7*

1. Maximum individual risk to employees and total population due to leak from ammonia transfer/ return pipeline during transportation is **2.45 * 10^{-04} per year** and **1.07 * 10^{-04} per year.**
2. Number of fatalities resulting from incident outcome case in societal risk is **325.**
3. Radiation levels of Jet Fire and Flash Fire reach up to **40** and **300 m**. Overpressure levels in Vapor Cloud Explosion reach up to **200 m**. Lethal concentrations for Gas Dispersion reach up to **2400 m.**
4. Results of consequence analysis indicate that significantly fire radiation, explosion overpressure effects and toxic concentration for gas dispersion for credible leak scenario are **within the tolerable region if ALARP** (Table 4).

Table 4 IR and SR calculations result

S.no	Event	Average IR (employee)	Average IR (total population)	Societal risk
1	Leak of natural gas from inlet pipeline of natural gas feed and desulphurization section	3.91E−06	1.95E−07	Region A = 6.40E−06 Region B = 5.45E−06 Region C = 2.18E−06
2	Leak of hydrogen from reforming section pipeline outlet	3.84E−06	3.57E−07	Region A = 8.01E−07 Region B = 6.82E−07 Region C = 2.73E−07
3	Leak of carbon monoxide from shift converter pipeline outlet	7.72E−06	9.81E−07	Region A = 1.58E−06 Region B = 6.91E−07 Region C = 5.89E−07 Region D = 1.47E−06 Region E = 8.84E−07
4	Leak of methane from methanator pipeline inlet	2.45E−06	1.99E−07	Region A = 5.15E−07 Region B = 4.39E−07 Region C = 1.75E−07
5	Leak of ammonia from piping of ammonia separator outlet	1.47E−05	2.07E−06	Region A = 2.94E−06 Region B = 1.29E−06 Region C = 1.10E−06 Region D = 2.75E−06 Region E = 1.65E−06
6	Leak from top of ammonia storage tank (PRV failure)	1.08E−04	4.06E−05	Region A = 2.19E−05 Region B = 1.86E−05 Region C = 1.72E−05 Region D = 2.05E−05 Region E = 3.28E−06

(continued)

Table 4 (continued)

S.no	Event	Average IR (employee)	Average IR (total population)	Societal risk
7	Leak from ammonia transfer/return pipeline during transportation	2.45E−04	1.07E−04	Region A = 4.77E−05 Region B = 4.06E−05 Region C = 3.75E−05 Region D = 4.46E−05 Region E = 7.15E−06

8 Discussion

Ammonia plants conform to the rules and regulations of Factories Act 1948; Indian Standard 15,394 "Fire Safety in Petroleum Refineries and Fertilizer Plants". Industry's HSE management system adheres to ISO 9001, ISO 14001 and ISO 45001 systems. In line of the study for Fire, Explosion and Toxic Release consequence assessment and modelling during ammonia synthesis and transportation extensive methodology is developed in which relevant data from manuals, books, research papers, company SOPs, P&IDs / PFDs are derived for the particular project. Detailed study of the processes involved in ammonia plant and its storage facilities are studied. Different gases properties involved in the production process of ammonia like Natural Gas, Hydrogen, Carbon Monoxide, Hydrogen Sulphide, etc. are examined. Qualitative and Quantitative Risk Assessment tools are chosen for scenario identification gaps in ammonia plant, storage and transportation. HAZOP analysis and ETA identify the risks and their possible causes and consequences. Using these analysis techniques, seven critical top events were identified which may lead to serious or catastrophic failures. Probit calculations determined the probability of fatalities that can arise due to thermal radiation in case of fire, overpressure levels in case of VCEs and lethal concentrations in cases of toxic gas dispersion. For performing the consequence modelling, the probability of fatalities was taken 100%. Discharge calculations are performed for different top events for which possible leakage scenarios are considered. ALOHA engages a variety of models to estimate the rate at which a chemical is released from confinement and enters the atmosphere. ALOHA Modelling for hazards is performed for which required inputs like basic scenario inputs (such as date, time, and location), chemicals from ALOHA library, atmospheric information such as wind speed and direction, air temperature are taken. Source information such as release amount, tank and pipeline dimensions are entered. The area where there is a possibility of exposure to toxic vapours, a flammable atmosphere, overpressure from vapor cloud explosion or thermal radiation from a fire are represented graphically as threat zones. A threat zone indicates the area within which the ground-level exposure exceeds the user-specified level of concern at some time after the beginning of a release. Based on the threat zones, ERPG levels 1, 2, 3 are set. Affected areas that can be dangerous to human exposures at different levels are detected. These affected areas are divided into three levels, namely red zone, orange zone and yellow zone. Red zone's (ERPG Level 3) concentration level of toxic gas and exposure to which may cause life-threatening health effects or even death. Likewise, orange zone (ERPG Level 2) can cause long-lasting adverse health effects and in yellow zone (ERPG Level 1) average individual may feel notable discomfort, irritation but reversible upon cessation of exposure. MARPLOT software is used to draw objects on map or import from other sources including both satellite and street view maps with global coverage. It provides population estimation, elevations and weather conditions of ammonia plant. A set of numerical measures of risk is derived from CCPS guidelines for Chemical Process Quantitative Risk Analysis such as risk to an individual vs. risk to society existing control measures and ERCP plans are equipped with a full-fledged emergency handling capacity. The industry comprises of emergency response team, fire control centre, emergency facilities/gadgets, emergency communication systems, state of the art automatic and manual fire and toxic gas alarm and gas detection systems are installed. New recommendations were provided while for offsite and onsite training, recommendations were also provided with respect to process safety management and in case of toxic release, fire and explosion. Individual risk is estimated for most exposed individual, for group of individuals at particular places average individual risk is estimated. Risk contours are developed for every top event that shows geographical distribution of individual risk. Contour lines connect points of equal risks superimposed over a local map. Measurement of risk for a group of people societal risk is estimated which is most commonly expressed in terms of frequency distribution of multiple causality events (the F-N curve). F-N curve represents the cumulative probability or frequency of events "F" causing "N" or more fatalities, injuries or exposures. After assessing all the top events, individual risk and societal risk has been calculated with the help of CCPS guidelines. Below are the results for the same.

8.1 Individual Risk and Societal Risk

See Figs. 11, 12, 13, 14, 15, 16 and 17.

8.2 Event Tree Analysis

See Figs. 18, 19, 20, 21, 22, 23 and 24.

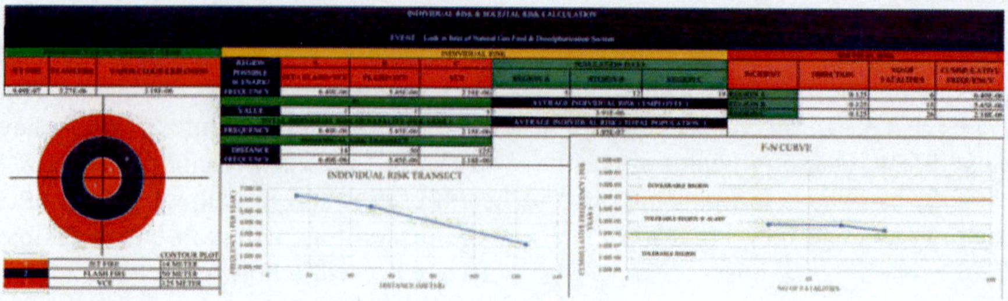

Fig. 11 Individual risk and societal risk calculation for top event 1

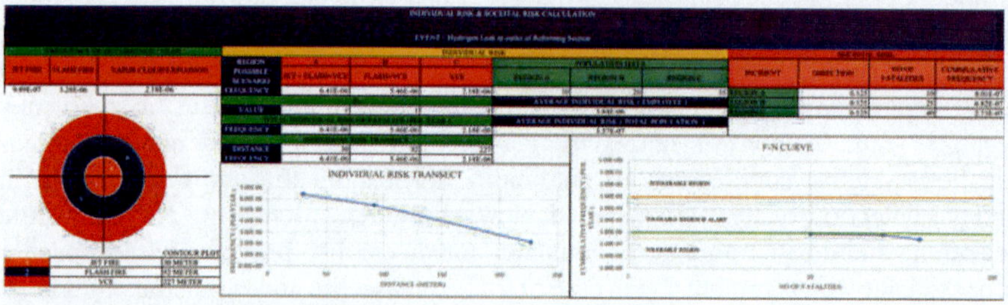

Fig. 12 Individual risk and societal risk calculation for top event 2

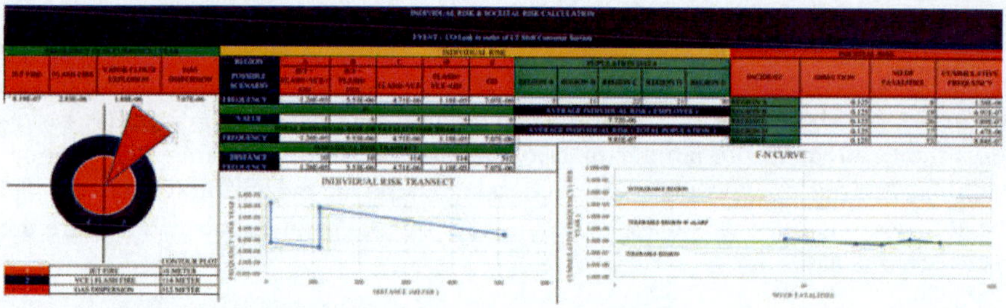

Fig. 13 Individual risk and societal risk calculation for top event 3

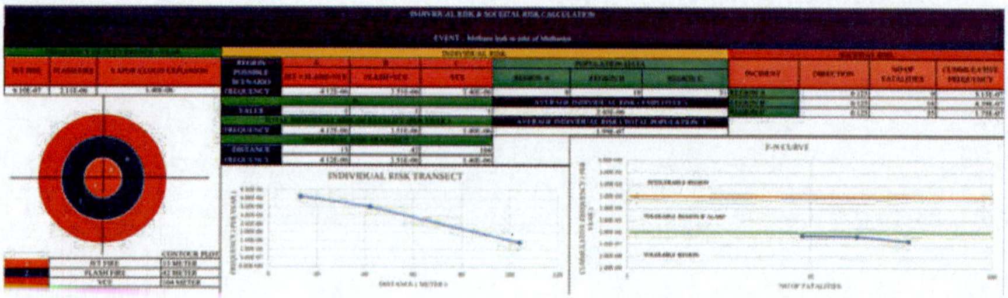

Fig. 14 Individual risk and societal risk calculation for top event 4

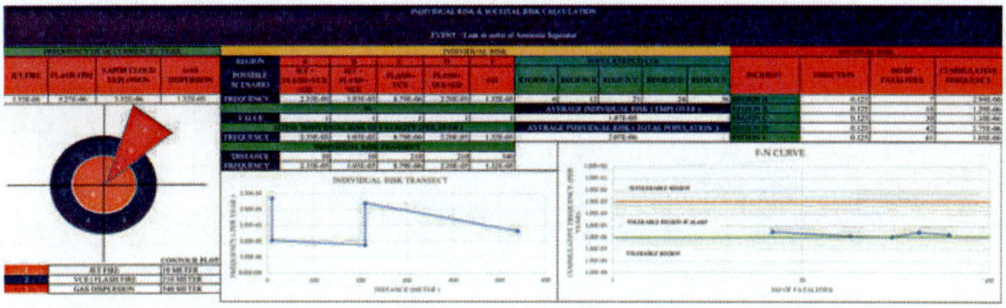

Fig. 15 Individual risk and societal risk calculation for top event 5

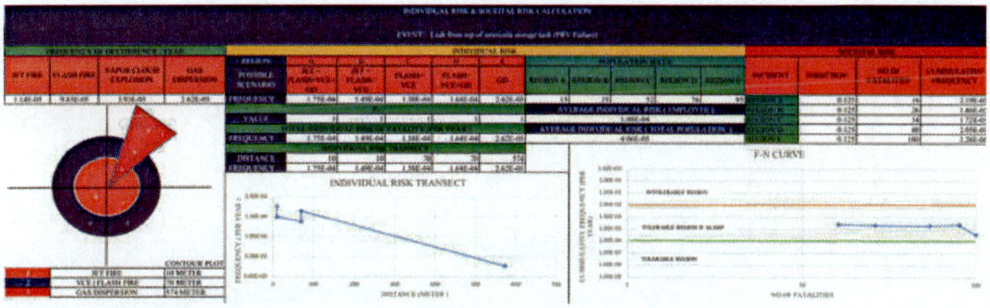

Fig. 16 Individual risk and societal risk calculation for top event 6

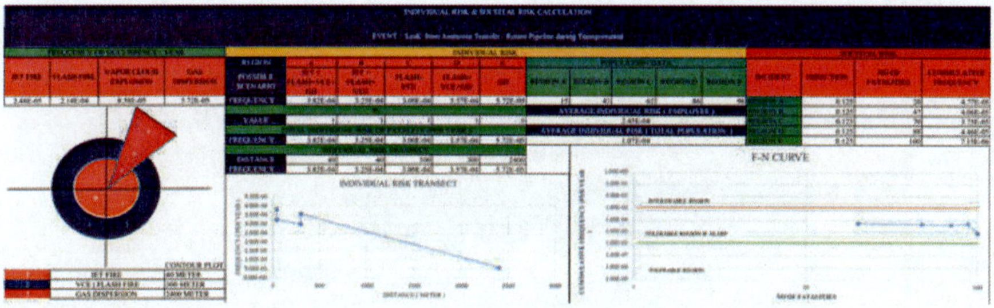

Fig. 17 Individual risk and societal risk calculation for top event 7

9 Conclusion

This work has included the consequence assessment of fire, explosion or toxic release scenarios arsing due to loss of containment during Ammonia production, storage and transportation. Based in the detailed study and analysis of the chemical properties and other parameters associated with ammonia production, various accident scenarios are developed along with the application of both the quantitative risk assessment and qualitative risk assessment tools in order to identify the sequences of events that can lead to accidents or serious fatalities and thus prioritizing the preventive

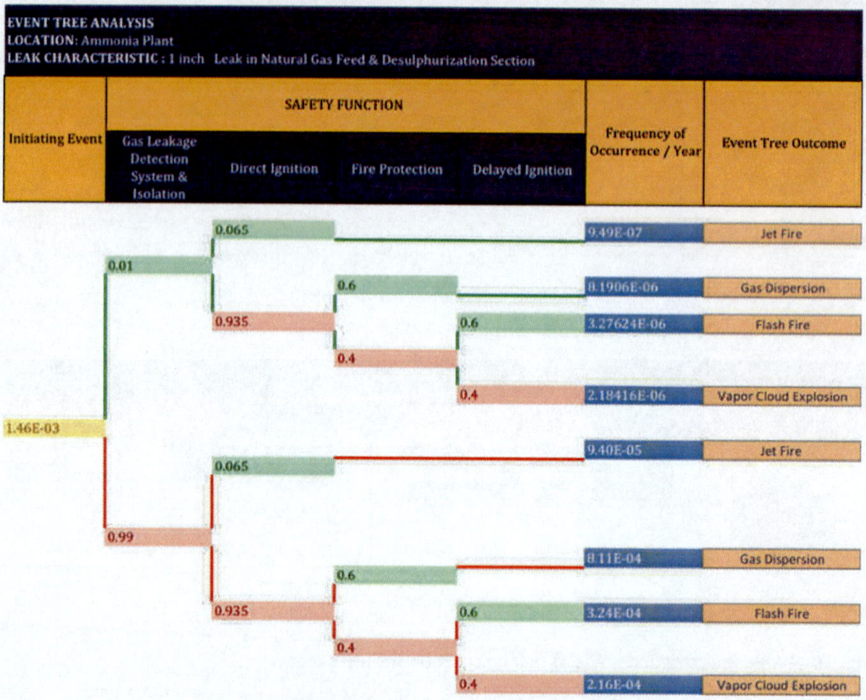

Fig. 18 Event tree analysis for top event 1

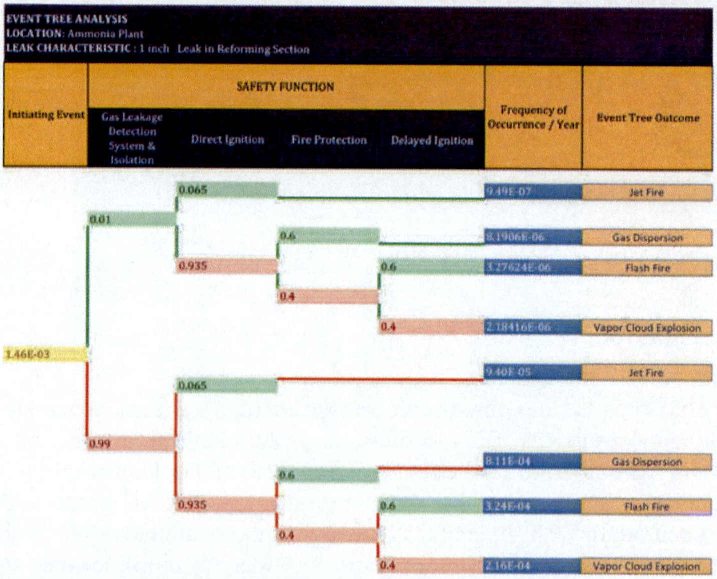

Fig. 19 Event tree analysis for top event 2

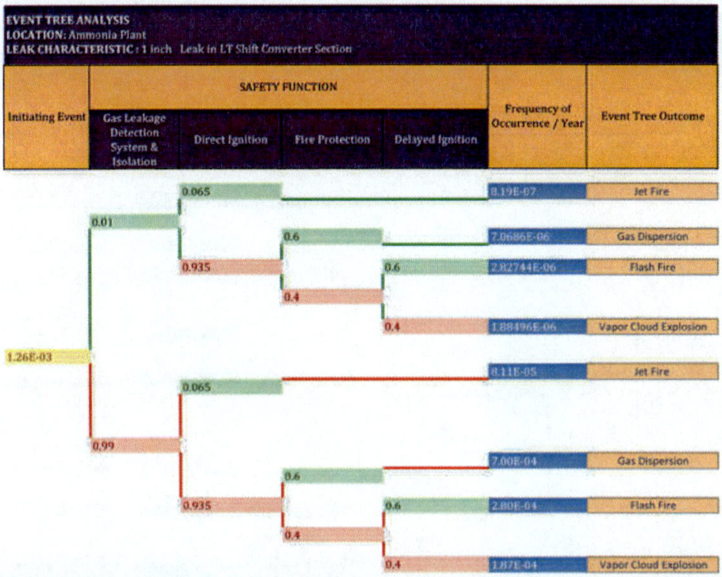

Fig. 20 Event tree analysis for top event 3

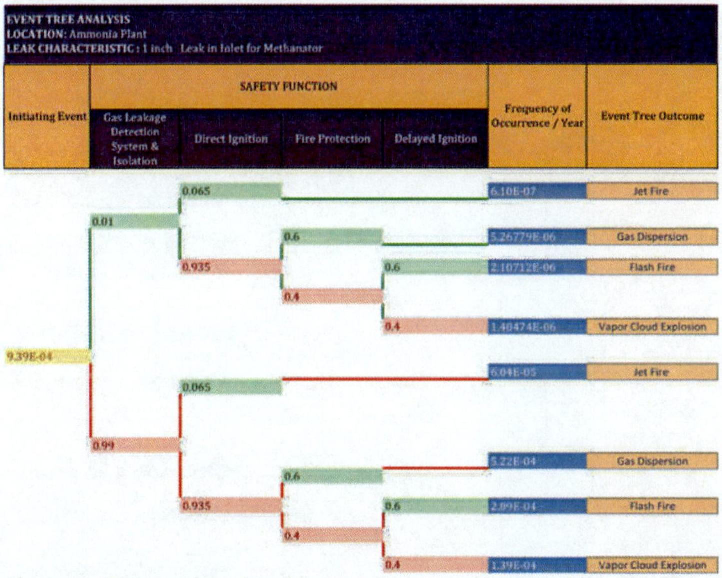

Fig. 21 Event tree analysis for top event 4

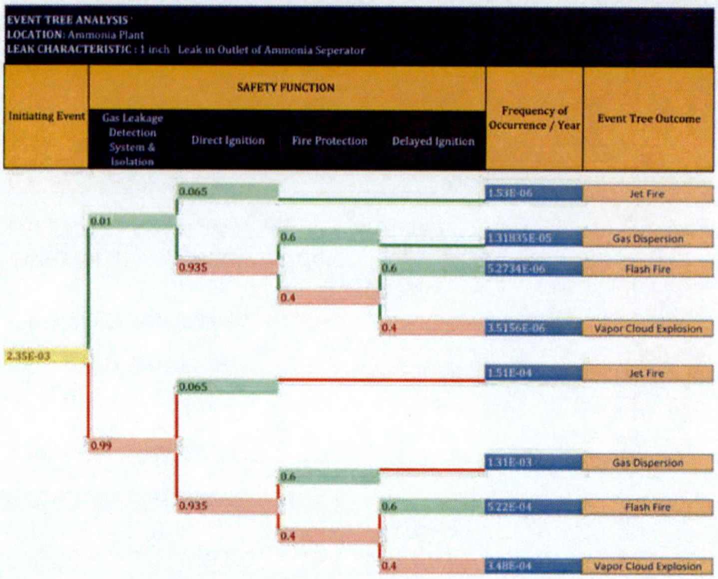

Fig. 22 Event tree analysis for top event 5

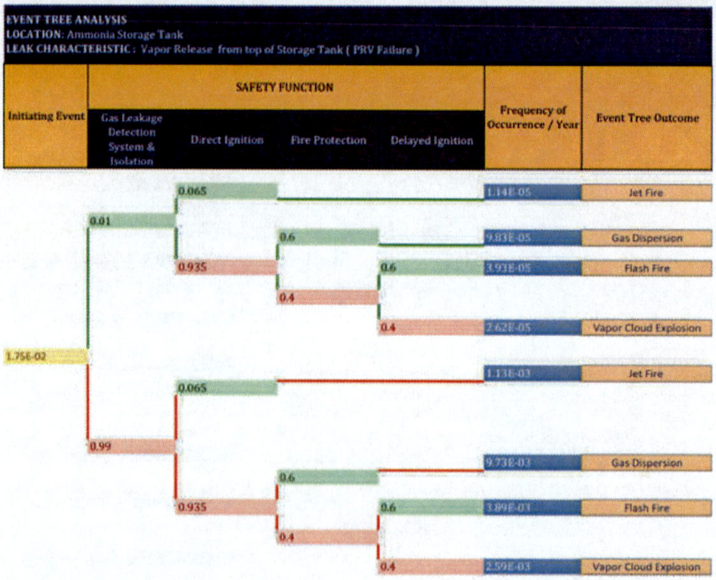

Fig. 23 Event tree analysis for top event 6

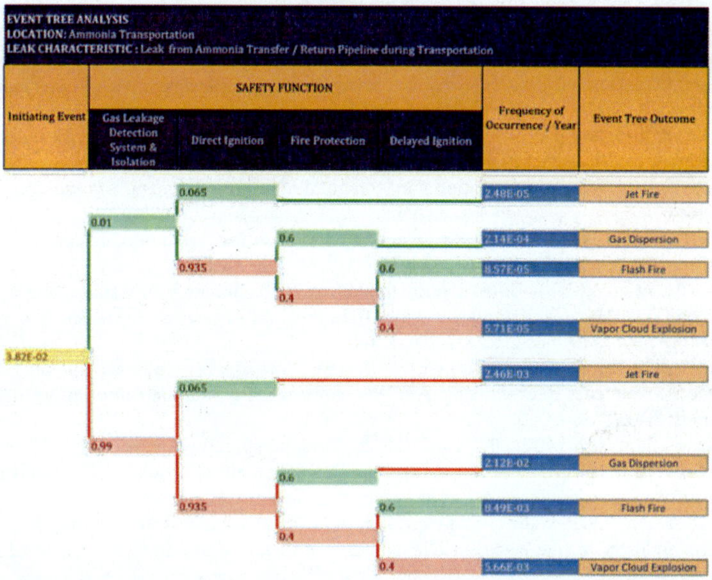

Fig. 24 Event tree analysis for top event 7

and corrective measures in order to minimize the probability of failure. Ammonia toxic vapour when leaked in the atmosphere possess the major risk and is potentially lethal to the staff and people residing in nearby vicinity at concentration > 1100 ppm. The same has been modelled on ALOHA and the results generated by ALOHA and MARPLOT are predictive in nature which were Assessed and certain Corrective action was recommended, which can be added as a part of PSM and Emergency Response and Contingency Plan as it will be effective in training the people to reduce the risk of the personnel as well as property damage in case of emergency. The organization must ensure that the emergency response measures outlined in the ERCP are adhered to along with establishment of audit programmes for verification of handling of emergency procedures. As per recent statistics in India, there have been numerous accidents occurred recently, thus there should be a scope for continual improvement with enhanced process safety management and ERCP in order to reduce the possibility of occurrence of incidents and reduce the associated risks for people located both onsite and offsite.

References

1. Eames, P., & Brightling, J. R. (2010). *Process safety in the fertiliser industry, A new focus.* International Fertiliser Society.
2. Dybkjaer, I. (1995). Ammonia production processes. *Ammonia* (pp. 199–327). Springer.

3. Rafiqul, I., et al. (2005). Energy efficiency improvements in ammonia production—Perspectives and uncertainties. *Energy, 30*(13), 2487–2504.
4. Shah, J., & Shaffer, D.M. (2011). *Risk-based approach for evaluating safety events in fertilizer production.* In *Safety in ammonia plants & related facilities symposium proceedings.*
5. Pittiglio, P., Bragatto, P., & Delle Site, C. (2014). Updated failure rates and risk management in process industries. *Energy Procedia, 45*, 1364–1371.
6. Hignett, T. P. (1985). Transportation and storage of ammonia. *Fertilizer Manual* (pp. 73–82). Springer.
7. Dadashzadeh, M., et al. (2013). An integrated approach for fire and explosion consequence modelling. *Fire Safety Journal, 61*, 324–337.
8. Ramli, A., et al. (2018). Consequence modelling for estimating the toxic material dispersion using ALOHA: Case studies at two different chemical plants. *Multidisciplinary Digital Publishing Institute Proceedings,* 2(20), 1268.
9. Bhattacharya, R., & Ganesh Kumar, V. (2015). Consequence analysis for simulation of hazardous chemicals release using ALOHA software. *International Journal of ChemTech Research, 8*(4), 2038–2046.
10. Allahyari Mehrabani, M., & Ghiyasi, S. (2018). Modeling the consequences of accidents during the process of transferring waste thinner from the paint shop of an automotive industry using aloha software. *International Archives of Health Sciences, 5*(3), 55–61.
11. Danasa, A. et al. (2019). The ammonia release hazard and risk assessment: A case study of urea fertilizer industry in Indonesia. In *IOP conference series: Earth and environmental science.* IOP Publishing.
12. Hoffmann, J., et al. (2009). Emission of the odour substances from fertilizer industry. *Polish Journal of Chemical Technology, 11*(1), 12–15.
13. Anandhan, M. et al. *Quantitative risk assessment in LPG storage area for different fire scenarios.*
14. Abbaslou, H., & Karimi, A. (2019). Modeling of ammonia emission in the petrochemical industry. *Jundishapur Journal of Health Sciences* (In Press).
15. Ema, S.S., Roy, A., & Sowgath, M. T. *Comprehensive hazard identification and safety evaluation for Shahjalal Fertilizer Industry Limited.*
16. da Silva Rodrigues, A.J. et al. (2017). Risk reliability analysis, resulting from explosions in petrochemical industries: A case study using ALOHA software. In *2017 12th Iberian conference on information systems and technologies (CISTI).* IEEE.
17. Ilić, P., Ilić, S., & Bjelić, L. S. (2018). Hazard modelling of accidental release chlorine gas using modern tool-ALOHA software. *Quality of Life, 16*(1–2).
18. Fuentes-Bargues, J. L., et al. (2017). Risk analysis of a fuel storage terminal using HAZOP and FTA. *International Journal of Environmental Research and Public Health, 14*(7), 705.
19. Ferdous, R., et al. (2011). Fault and event tree analyses for process systems risk analysis: Uncertainty handling formulations. *Risk Analysis: An International Journal, 31*(1), 86–107.
20. Liang, S.F.W. (2003). Fault tree analysis on the fire and explosion of pipeline gas. *Industrial Safety and Dust Control, 1.*
21. Guan, Y., et al. (2016). Fault tree analysis of fire and explosion accidents for dual fuel (diesel/natural gas) ship engine rooms. *Journal of marine science and application, 15*(3), 331–335.
22. Cheng, S.-R., et al. (2009). Fault-tree analysis for liquefied natural gas terminal emergency shutdown system. *Expert Systems with Applications, 36*(9), 11918–11924.
23. Khan, F. I., Iqbal, A., & Abbasi, S. (2001). Rapid risk assessment of a fertilizer industry using recently developed computer-automated tool TORAP. *Journal of Loss Prevention in the Process Industries, 14*(5), 413–427.
24. Aneziris, O., et al. (2014). Integrated risk assessment for LNG terminals. *Journal of Loss Prevention in the Process Industries, 28*, 23–35.
25. Law, W., & Gimbun, J. (2020). Modeling the effect of hypothetical chlorine leakage from Malay-Sino Chemical Industries using ALOHA software and development of an emergency evacuation route around Teluk Kalong industrial area. *MS&E, 736*(7), 072015.

26. Jiang, X. J. et al. (2014). Application of ALOHA in emergency response for hazardous chemicals accidents. In *Applied mechanics and materials*. Trans Tech Publ.
27. Tseng, J., Su, T., & Kuo, C. (2012). Consequence evaluation of toxic chemical releases by ALOHA. *Procedia Engineering, 45*, 384–389.
28. Freeman, R. A. (1990). CCPS guidelines for chemical process quantitative risk analysis. *Plant/Operations Progress, 9*(4), 231–235.

A Review on Effect of Covid-19 on Air Pollution in India

Rishi Dewan, Shannon Mark Dsouza, Rakesh Sati, and P. A. Arun

1 Introduction

Covid-19 is one of the major disasters which hit the world in 2019 resulting in lakhs of fatalities as well as economic loss all over the globe. Soon all the countries started to take action to stop this pandemic [1]. When the Covid-19 pandemic arrived in India, the government has decided to implement the lockdown in order to stop the spread of this deadly virus and somehow this step became a major improvement factor in terms of the environment. All the domestic and international flights were put on halt. Vehicular transport was also stopped except the emergency services.

2. The Northern part of India majorly experiences the poor quality of air and atmospheric pollution mainly due to the emission from vehicular movement, industries, coal-based power plants, crop leftover burning, and brick kilns. For example, New Delhi suffers poor air quality and which is higher than the Beijing air quality. India has the second-largest population but accounts for only 6% of total primary energy usage. In the past few years, poor air quality remained a big issue [2]. According to the WHO, 37 cities from India made their place in a global list of 100 world cities with the highest PM10. The poor air quality has persistently increased over the last few decades associated with growing anthropogenic activities. As per the National Ambient Air Quality Standards (NAAQS), 77% of India's population is exposed to air pollution levels above the safe limit (PM 2.5 above 40 g/m^3 [3]. Due to this high level of PM 2.5 and PM 10 exposure, there were approximately 12.5% of the total death according to the state of the Indian environment report (SOE) 2019, 8.5 out of every 10,000 children in India die before they turn five due to this poor air quality. Due to this pandemic, there has been a sudden rise in plastic waste as well as a rise in methane concentration has been noted by many agencies. On March 21, 2020, AQI was moderate in Delhi which gradually improved from higher to lower category.

R. Dewan · S. M. Dsouza · R. Sati · P. A. Arun (✉)
School of Engineering, UPES, Dehradun 248007, India
e-mail: arun.pa@ddn.upes.ac.in

N. A. Siddiqui et al. (eds.), *Advances in Behavioral Based Safety*,
https://doi.org/10.1007/978-981-16-8270-4_25

AQI Class (Range)	Health Impact	PM_{10} 24 hrs (µg/m^3)	$PM_{2.5}$ 24 hrs (µg/m^3)	SO_2 24 hrs (µg/m^3)	NO_2 24hrs (µg/m^3)	O_3 8hrs (µg/m^3)	CO 8 hrs (mg/m^3)	NH3 24 hrs (µg/m^3)
		Concentration Range						
Good (0–50)	Minimal Impact	0–50	0–30	0–40	0–40	0–50	0–1	0–200
Satisfactory (51–100)	Minor breathing discomfort to sensitive people	51–100	31–60	41–80	41–80	51–100	1.1 - 2	201–400
Moderately polluted (101–200)	Breathing discomfort to the people with lung,	101–250	61–90	81–380	81–180	101–168	2.1–10	401–800
Poor (201–300)	Breathing discomfort to people on prolonged exposure	251–350	91–120	381–800	181–280	169–208	10–17	801–1200
Very poor (301–400)	Respiratory illness to the people on prolonged exposure	351–430	121–250	801–1600	281–400	209–748*	17–34	1200–1800
Severe (401–500)	Respiratory illness to the people on prolonged exposure	>430	>250	>1600	>400	>748	>34	>1800

Fig. 1 Effect of varying concentration of pollutants on human health

Overall, up to 44% reduction in PM10 level was observed in Delhi during March 22–23, 2020 compared to the previous day. The PM2.5 level was reduced by 8% on the curfew day but declined to 34% the next day owing to negligible activities related to combustion on March 22–23, 2020, in and around the city. PM10 and PM2.5 levels dropped as low as 67 µg/m^3 and 34 µg/m^3.

The reduction in the number of vehicles on the road resulted in up to 51% reduction in the NOx level and 32% reduction in CO level during March 22–23, 2020, as compared to the previous day. Also due to the reduction in industrial activities as well as residential activities, there was a sharp decline in the NOx and CO level in Delhi city and thus increasing the air quality. In this study, a detailed discussion is presented on the effects of Covid-19 on air quality factors. Also, there is a detailed comparison of different pollutants and their effects on the environment and on human health is concluded in this study (Fig. 1).

2 Methodology

The method involves the comparison of the data which was collected last year and raw data of this year to get the idea of different pollutants levels. This comparison also includes the graph and pictures that represent the various concentration of different pollutants. The various effects of air pollutants on human health have shown in Fig. 2. The main idea is to know the effect of lockdown on the environment and description of various pollutants harmful to the environment.

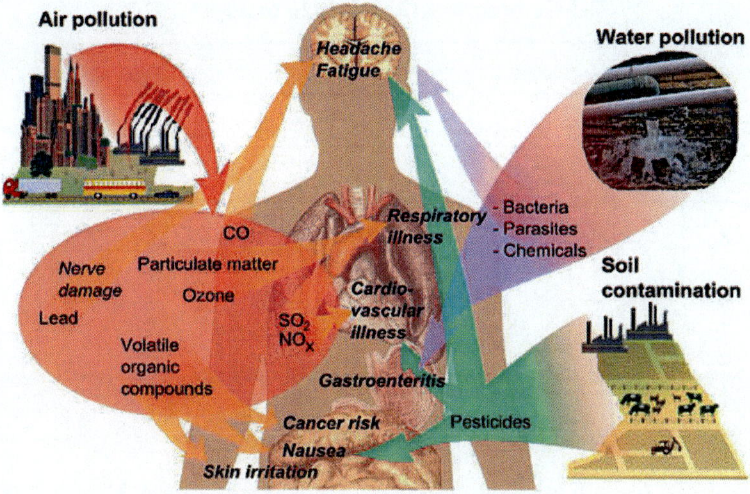

Fig. 2 The bi-products of air pollution and the effects on human health

3 Review of Air Pollution During Lockdown

3.1 Carbon Dioxide and Monoxide Emission

India's CO_2 emission fell by 1% in the financial year ending March 2020, which is the first time since 1982. This analysis was based on the 2019–20 fossil fuel generation and data of consumption of coal and oil. The CO_2 emission fell by an approx. 15% in the month of March. On March 21, the concentration of CO was five times higher than the concentration on March 22, 2020. This sudden downfall highlighted the *absence of vehicular emission.* The major source of emission of carbon monoxide is due to burning of the carbonaceous fuel and the emission from the IC engine and in 2019 (4) carbon monoxide amply exceeded the WHO guidelines (100 mg m^{-3} for 15 min and 10 mg m^{-3} for 8 h) as 15 min average value reached up to 400 mg m^{-3} [4]. In March 2020, there was an average drop of 10% in the CO emission as recently measured the CO level in the month of April was close to 11.60 mg m^{-3} which was way better than the value recorded in the year 2019 and before the lockdown also (Fig. 3).

3.2 Nitrogen Oxide Emission (NO_2)

Nitrogen oxide is an important air pollutant because it contributes to photochemical smog which can have significant effects on human health. (3) Annually 4.6 million people die worldwide due to the problems associated with NO_2 which can cause

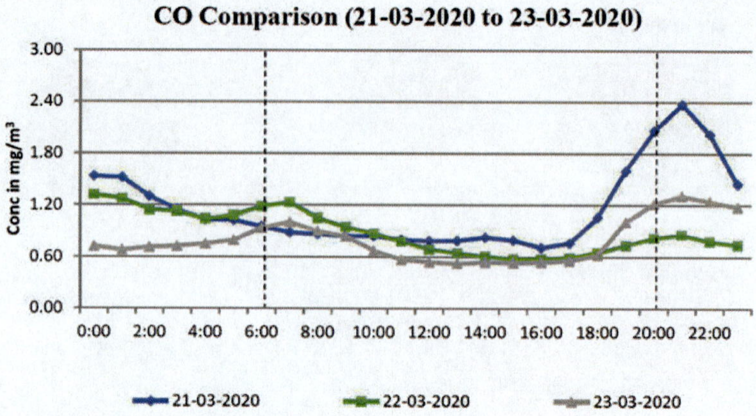

Fig. 3 CO concentrations during lockdown days [5]

cellular inflammation, bronchial hyperresponsiveness, and respiratory problems [3]. Nitrogen oxide went from 52 per cubic meter to 15 per cubic meter, a 71% fall in the concentration of NO_x. Nitrogen oxide is generally emitted in the atmosphere as a result of human activities such as motor vehicles, power plants, and industrial facilities. Due to the lockdown, all these activities were put on halt and result in a significant reduction in the NO_x concentration in the atmosphere. In the New Delhi region before and after lockdown, levels of NO_2 were completely different. Before the lockdown, No_2 was ranged from 4 to 158 $\mu g/m^3$ while after the lockdown it was ranged from 9 to 112 $\mu g/m^3$ (Fig. 4).

The average nitrogen oxide concentration was significantly reduced across major cities in India as Mumbai and Delhi saw drops of around 40–50% as compared to the last year i.e., 2019 (Fig. 5).

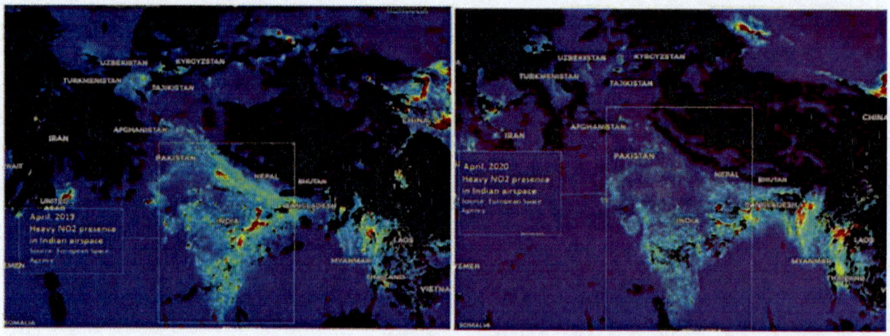

Fig. 4 NO_2 concentrations in 2019 and 2020

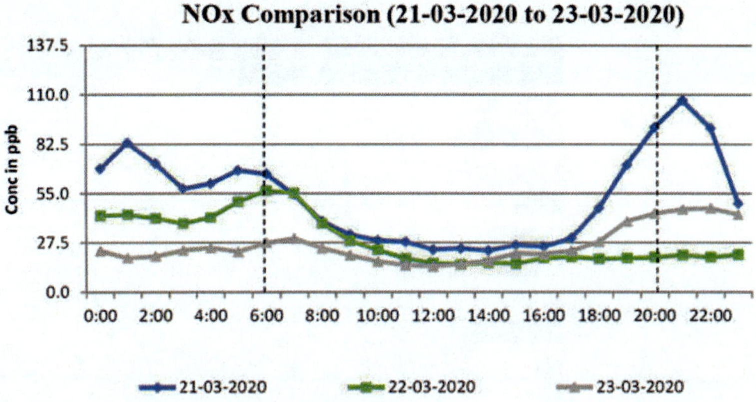

Fig. 5 NO_x concentrations during lockdown days [5]

3.3 *Sulfur Dioxide Emission (SO_2)*

A major contribution of sulfur dioxide in the air is due to industrial activities such as thermal power generation, mineral ores, industrial activities that burn fossil fuels. Major issues caused by SO_2 to human health include cough, shortness of breath, bronchitis, fatigue, etc. and it is also the main reason behind the acid rain due to the formation of sulphuric acid. According to the Greenpeace analysis, India is the largest emitter of SO_2 in the world and most of this emission in India is because of the coal-burning activities. The SO_2 concentration reduced across India by almost 40% when the country was shut down for more than two months. However, the level of SO_2 was not changed much before and after the lockdown.

3.4 *Particulate Matter (PM_{10} and $PM_{2.5}$)*

Airborne particulate matter includes a complex mixture of organic and inorganic substances. Particulate air pollution is a mixture of the solid, liquid, or solid and liquid particles suspended in the air. These suspended particles vary in size, composition, and origin. Particulate matter exposure has various problems including premature death in people with heart or lung disease, it can also cause non-fatal heart attack and irregular heartbeat.

With very few vehicles running on the road and only essential activities functioning due to lockdown has made a significant reduction in the PM 2.5 and PM 10 level in the Northern part of India which generally experiences very poor air quality. Central Pollution Control Board has reported 46% reduction in PM2.5 and more than 50% depletion in PM10 concentration. PM2.5 refers to the particulate matter having a diameter less than 2.5 which can enter into the lungs and in the bloodstream

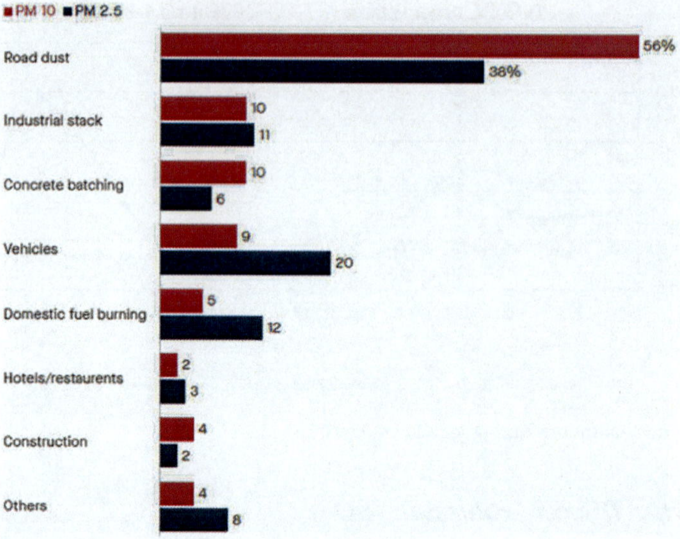

Fig. 6 Sources of PM 2.5 pollution in Delhi

too. Twenty-four-hour average of PM2.5 and PM 10 has been dropped in lockdown period as low as.

24 μg/m^3 and 39 μg/m^3. WHO guideline level for particulate matter 2.5 is 10 μg/m^3 for annual mean and 25 μg/m^3 24-h mean and for PM10 is 20 μg/m^3 annual mean and 50 μg/m^3 for 24- hour mean. Last year (2019) in the month of October, the PM2.5 level in Delhi was 289 μg/m^3 which was very high as compared to the WHO guideline (Fig. 6).

Figure 7 shows the PM2.5 level of Delhi which is the average of 37 readings stations in Delhi [6].

Given below is the graph of the average readings of PM2.5 level of the following cities and it is also showing the readings level after the lockdown in Mumbai and after implementation of the Janta Curfew (Fig. 8).

Similarly, the PM10 levels (an average of PM10 readings taken at a total of 37 stations) were also plotted on the graph showing the variations in the levels before and after the Janta Curfew. As 21 March is one day before the lockdown and 23 March post-lockdown (Fig. 9).

3.5 *Ozone*

Ozone gas can be good as well as bad depending upon its existence as it is found both in the upper atmosphere as well as at ground level. (3) Upper atmospheric ozone is called stratospheric ozone which shields us from the harmful ultraviolet

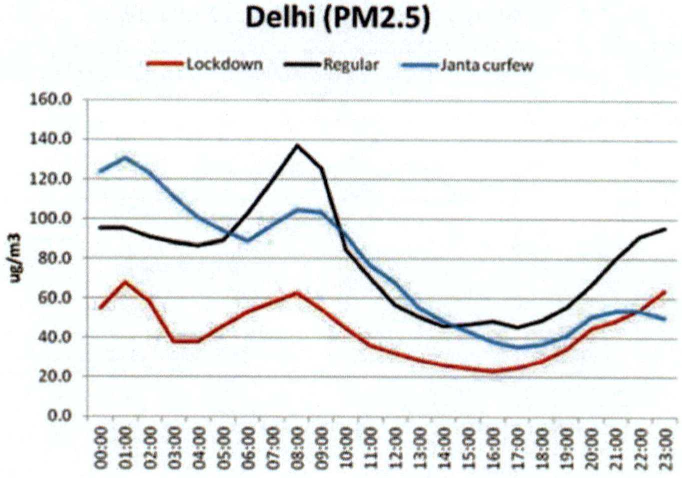

Fig. 7 Average PM2.5 levels in Delhi

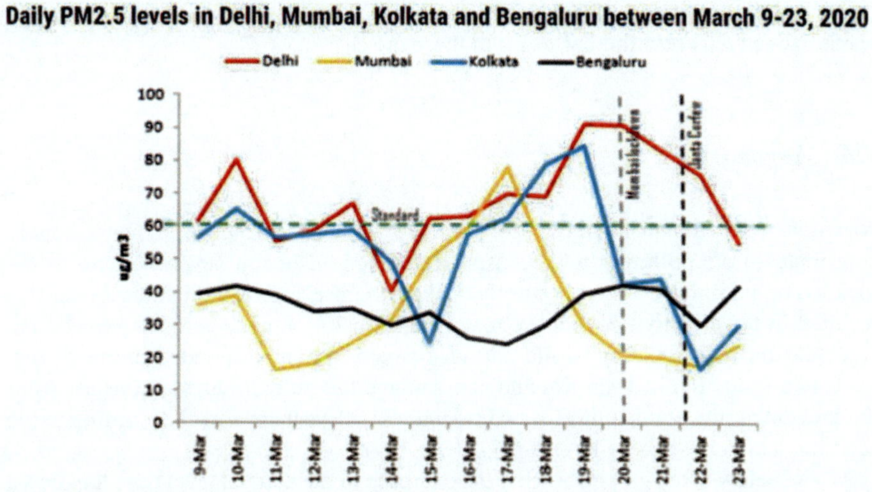

Fig. 8 Average PM2.5 levels in various Indian cities

rays by forming a protective layer [3]. This layer is partially destroyed by human interferences. On the other hand, ground-level ozone is very harmful and it affects both humans and the environment. Ozone at ground level can be created by the reaction between the volatile organic compounds and oxides of nitrogen but not emitted directly into the atmosphere. As per the CSE (Centre for Science and Environment), there has been a significant increase in the level of the ozone gas as it is formed by the photochemical reaction between the oxides of nitrogen and VOCs (Volatile organic

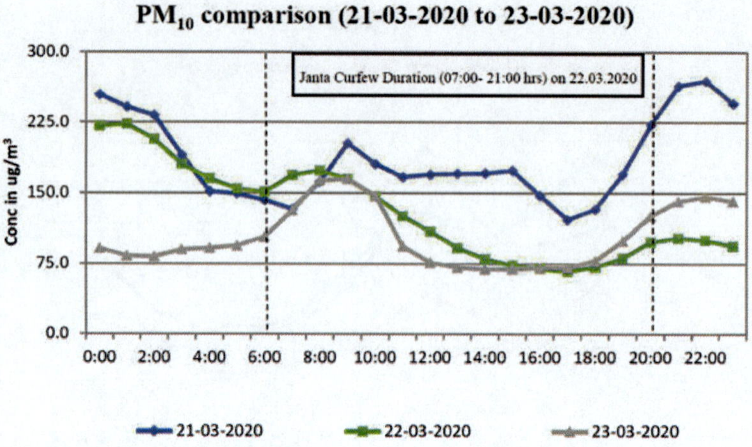

Fig. 9 PM 10 levels during lockdown days [5]

compounds). "A high NOx level can again react with the ozone and mop it up. The ozone that escapes to the cleaner areas has no NOx to further cannibalize it and as a result ozone concentration builds up in these areas."

3.6 Aerosols

Aerosols from human sources like vehicular exhaust or factory exhaust hugely contribute to air pollution and it also affects human health severely. Due to the lockdown, 1.3 billion citizens strictly locked themselves in their houses and this resulted in fewer activities that are mainly responsible for the aerosols generation. Aerosols are tiny liquid and solid particles which remain suspended in the air that leads to a reduction in visibility and can damage the human lungs and heart. After the lockdown, the aerosol level was dropped to a 20-year-low level according to the satellite data published by NASA (Fig. 10).

Given below is the graph showing the decrease in the aerosol level post-lockdown [7] (Fig. 11).

4 Discussions and Summary

The positive effects of the lockdown on the air quality of India have been mentioned in this paper.

As the northern region of India mostly faces poor air quality, data from these cities have been highlighted for a comparative study during this lockdown period. Delhi

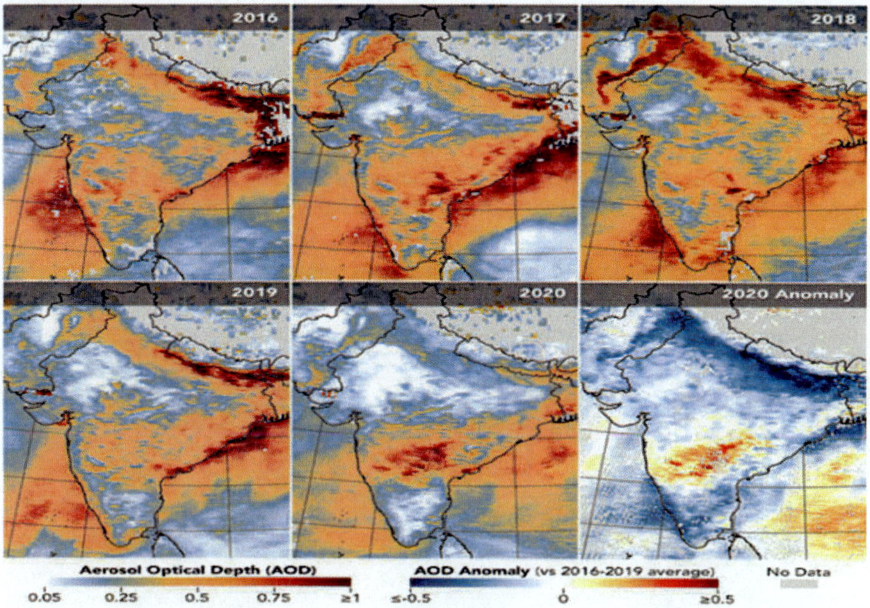

Fig. 10 Aerosol levels in India in past years (NASA image)

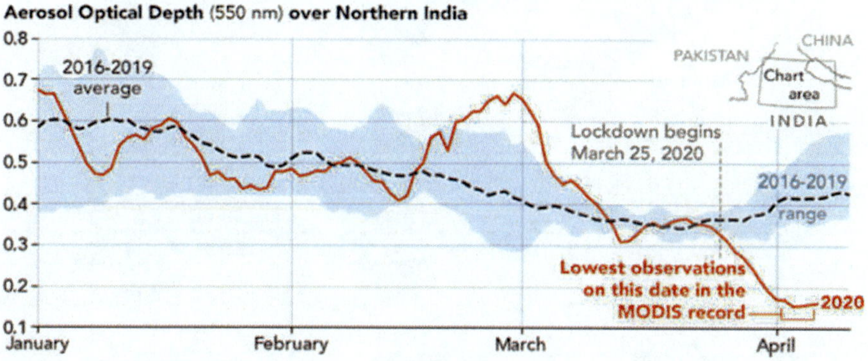

Fig. 11 Aerosol levels in India during lockdown days

a megacity with a population of around 1.9 crores displayed positive results due to the lockdown. As Industrial and vehicular activities came to a halt during this phase, the air quality of many regions of India improved and showed signs of rejuvenation. Amongst all the pollutants that contribute to the poor AQI of India, PM2.5 and PM10 emerged as the major contributors. According to CPCB from the lockdown period of 25th march to 14th April, a significant drop of 46% was recorded for PM2.5 and subsequently 50% for PM10 in Delhi [5]. Also, Mumbai was observed to have levels go as low as 76% for PM2.5 according to the System of Air Quality Weather Forecasting and Researching (SAFAR).

SAFAR also recorded the concentration of other cities like Kolkata and Ahmedabad which recorded a reduction of 30%, Pune by 25%, and Chennai by 11% in PM2.5 levels [8]. From Figs. 12 and 13, we can see the concentration reduction of PM 2.5 and PM10 spiral downwards contributing to a better AQI [9].

Following that a major decrease in NO_2 and CO concentration were recorded by TERI Emission Inventory,2018, of 56% cut for NO_2 and a 37%cut for CO levels in the atmosphere. NO_2 and CO being bi-products of transport and vehicular pollution saw this massive drop in their level due to the restriction of vehicular activity during the lockdown period. According to SAFAR, Pune recorded the highest reduction in No_2 level by 70% followed by Mumbai by 69%, Delhi by 33%, and Ahmedabad by 30% [10].

Similarly reduced concentration levels of NO_2 and CO as reported by [9] due to the halt in vehicular activity.

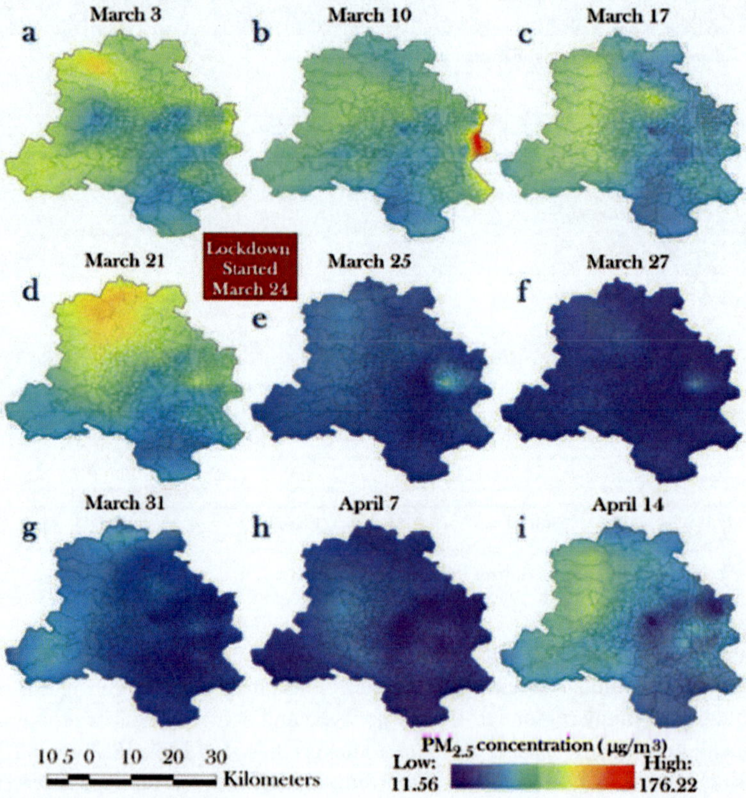

Fig. 12 PM 2.5 Levels in Delhi [9]

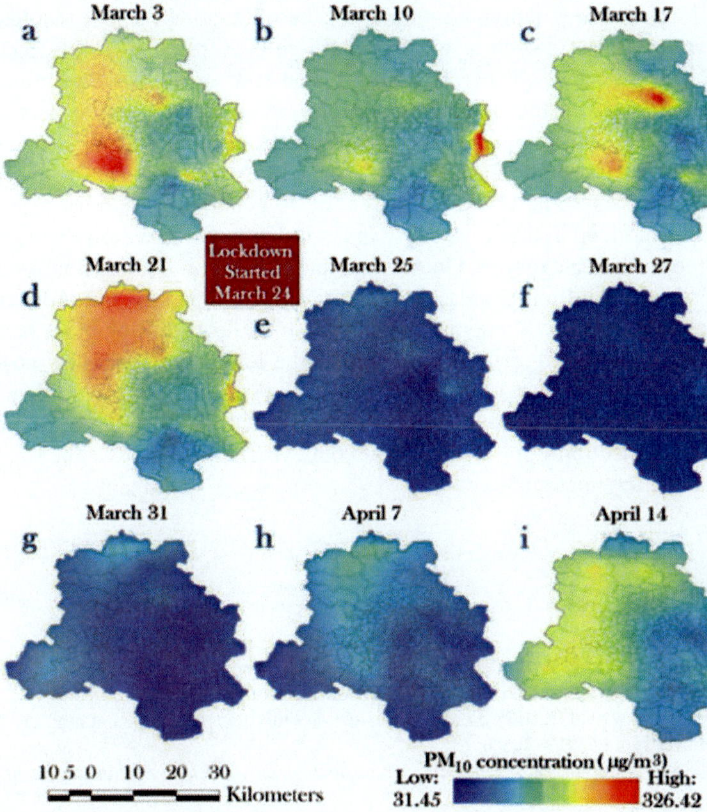

Fig. 13 PM 10 Levels in Delhi [9]

NO discernible difference was noticed in the level of SO_2 before and after the lockdown as the major contributor to SO_2 level is the burning of coal and diesel. With the use of BS-6 fuels, the SO_2 levels were already brought down to a limit as low as possible. Studies [9] show that levels of So_2 start increasing from April onwards. Ozone is a secondary which is mostly consumed by NO to form compounds like NO_2 and NO_3 but as the level of NO_x, compounds have reduced due to low vehicular emission, the accumulations of O_3 have increased significantly in comparison but intolerable limits.

5 Conclusion

Covid-19 is a pandemic that has globally claimed the lives of millions while putting others at risk. This unexpected event has affected the economy of many countries along with forcing many nations to go into a state of standstill. It has resulted in a change in various sectors such as education where everything is transformed into e-learning; next, we have businesses with reduced staff or more job rotations and a significant increase in the demand for Health and Safety in every activity we perform. In all this environment of chaos and pandemonium, nature has benefited from the entire phases of lockdown as it has steadily allowed nature to recover from its damaged state and revive itself for future generations A very significant improvement is noticed in the Air Quality Index of India which was in a severe state and has come to the category of Good-Moderate. India is a country that houses a population of over a billion people; the lockdown has helped the nation to reduce its anthropogenic activity. A major chunk of the total death toll of India is due to diseases contracted due to air pollution, due to the lockdown, India has observed reduced transport and industrial pollution. With the improvement in AQI, we can conclude that India has the potential to improve its AQI for future generations and this will benefit Pollution Control Boards and policymakers to improve the standards for air quality in the upcoming years.

References

1. Singh, R. P., & Chauhan, A. (2020). Impact of lockdown on air quality in India during COVID-19 pandemic. *Air Quality, Atmosphere & Health, 13*(8), 921–928.
2. Venkataraman, C., et al. (2017). Source influence on emission pathways and ambient PM2.5 pollution over India (2015–2050).
3. M.Gupta, A. T., and Kumar, V. (2020). The effect of Covid-19 on Air Pollution in India. *GJESM*.
4. Dekker, I. N., et al. (2019). What caused the extreme CO concentrations during the 2017 high-pollution episode in India? *Atmospheric Chemistry and Physics, 19*(6), 3433–3445.
5. CPCB (2020). Lockdown cuts PM2.5 and PM10 Levels by half in Delhi. *The Hindu*, New Delhi.
6. Anumita Roychowdhury, A. S. (2020, March). Covid-19 flattens peak-hour pollution. *Down to Earth.*
7. Patel, K. (2020). Airborne particle levels plummet in India. Retrieved from https://earthobservatory.nasa.gov/images/146596/airborne-particle-levels-plummet-in-northern-india
8. Borwankar, V. (2020). Air Pollution levels in Mumbai drop by 76%. *Times of India.*
9. Susanta Mahato, S. P. (2020). Effect of lockdown amid COVID-19 pandemic on air quality of the megacity Delhi, India. *Science of The Total Environment.*
10. Chatterjee, B. (2020). Mumbai's Vehicular Pollution dropped by three-fourth during lockdown period. *Hindustan Times.*

A Predictive Consequence Modelling Case Study for Estimating the Dispersion Distance of Methane Gas Vapour Cloud

Bikarama Prasad Yadav, N. A. Siddiqui, and Pranaya Sharma

1 Introduction

The industrial development and population of India are augmenting day by day. With the setting up of new industries and ever-increasing population, there has been an inflating demand for energy in the country. According to Nejat et. al. [1], India stood at the fourth position in terms of oil and gas market globally. In order to meet its domestic and industrial demands, India has been importing a noteworthy amount of natural gas from other countries (22%) [2]. The Indian natural gas market is anticipated to propel by 5.4% per annum for the period of 2007–2030 and standing at 132 bcm by 2030 [3]. The major end users responsible for the increased natural gas demands are fertilizer producers, petrochemicals, city gas distribution and power plants. With the ever-increasing concern for a cleaner and greener fuel for cities in India, city gas distribution is finding its way to align with the need and trend. The City Gas Distribution System is primarily a network of underground natural gas pipelines and the allied paraphernalia. It quenches the demand of end users by supplying gas from a bulk high-pressure transmission pipeline to the service pipes situated at their doors within a single city area [4]. In a research study conducted by Pandian et. al. [5], the leakage of chemicals gives rise to a safety threat for the nearby population and degrades the air quality. The unanticipated release of methane gas can lead to huge number of fatalities and damage to public property.

The CGD system of India is a complex network which employs polyethylene pipe to supply natural gas to end users. Process industries always haunt the organization with the fear of unforeseeable explosion, fire or release of toxic and flammable gases. The City Gas Distribution System is still new in India, and any compromise with the safety aspect during operation and maintenance of the system can prove

B. P. Yadav (✉) · N. A. Siddiqui
University of Petroleum and Energy Studies, Dehradun 248007, India

P. Sharma
Tata Consulting Engineers Limited, Dehradun 411001, India

N. A. Siddiqui et al. (eds.), *Advances in Behavioral Based Safety*,
https://doi.org/10.1007/978-981-16-8270-4_26

to be fatal and disastrous [6]. Accidents like the Nagaram GAIL gas pipeline blast (2014) in Andhra Pradesh had occurred in past in which 6 children were burned alive while 16 people were seriously burned [7]. As per the guidelines laid by the Petroleum and Natural Gas Regulatory Board of India, these polyethylene pipes should be manufactured and installed conforming to the IS 14885 or ISO 4437 standard [8]. The major concern in a city gas distribution network is the leakage of gas from pipeline and valves. The foreseeable causes behind methane gas leakage can be faulty material of construction, corrosion, overpressure, environmental conditions and third party interference [9]. According to the ASME B31.8S, the potential threats to a pipeline have been categorized into 9 categories including external corrosion, internal corrosion and third party/mechanical damage [10].

With the increasing attractiveness of City Gas Distribution System, the Petroleum and Natural Gas Regulatory Board had launched 9th CGD bidding round for constructing network of gas pipelines in 86 geographical areas. The major geographical areas included Dehradun district and GAIL Gas was awarded the license to develop CGD infrastructure in the district. The CGD network will spread over an area of 3088 sq. km connecting 300,001 customers across the district within 8 contract years. The infrastructure will encompass 50 booster stations and 900 km of steel pipeline [8]. It is evident that the pipeline will also be exposed to the aforementioned pipeline threats. Hence, the network needs to be monitored regularly and carefully for the safe operation of the CGD system in the district. It is important that the city gas distribution firms shall work meticulously with other stakeholders in order to prevent any mishappening.

Modelling of leakage scenarios can prove to be very helpful in determining the threat zones, estimation of societal and individual risks, safe evacuation distance and preparation of emergency response plan [11]. In this paper, an attempt has been made to estimate the safe evacuation distance in the event of gas leakage in the city gas distribution network of Dehradun district. However, the study conveys foreseeable risks in advance as city gas distribution network is not yet established in the city. The Areal Location of Hazardous Atmosphere (ALOHA) version 5.4.7.0 has been used to simulate the gas leakage scenario in the district. The topographical details of the district have been entered into the software in order to have an effectual realization of the scene. A hypothetical scenario of the leakage has been assumed at different point of time in the day and seasons to estimate the scale of impact. The outcome of the simulation can be used to quantify safe evacuation distances in the event of any methane gas leakage. The study will also be helpful to have rescue operations in place beforehand and to minimize the degree of impact to the residents living in the proximity.

2 Description of Location

The capital city of Uttarakhand, Dehradun has been selected for the study. The district lies between 290 57′ and 310 2′ north latitudes and 770 35′ and 790 20′

east longitudes. The Dehradun district has been licensed to GAIL Gas Ltd. for the development of city gas distribution network. The area of Dehradun is 3088 sq. km and the population of the district as per the 2011 census was 16,96,694. The location of the release of methane gas is at 30°19'N and 78°002'E co-ordinates.

3 Methodology

3.1 The ALOHA Model of Dispersion

The Areal Location of Hazardous Areas is a freeware software tool developed by the National Oceanic and Atmospheric Administration Office of Response and Restoration (NOAA), Chemical Emergency Preparedness and Prevention Office (CEPPO) and United States Environmental Protection Agency (USEPA). The Gaussian model of dispersion has been used to model the dispersion of methane gas as it is neutrally buoyant and will spread with the downwind direction [12]. The ALOHA air dispersion model can be used to model the hazardous locations associated with the release of methane gas. Also, the model can make calculations about the varying concentration of pollutants with time. The ALOHA's Gaussian dispersion models the releases which do not affect the ambient air flow and are unaffected by gravity. The prediction of concentration of pollutants/ gases with the increasing distance by downwind direction is made by the Gaussian model. The software models continuous, instantaneous and time-dependent releases which give rise to a single cloud of pollutants/ gases.

The model which describes the cloud dispersion is modelled by Palazzi et. al. [13], which models steady-state release of pollutants/ gases for short duration.

$$C(x, y, z, t) = \begin{cases} \frac{\chi}{2}\left\{erf\left(\frac{x}{\sigma_x\sqrt{2}}\right) - erf\left(\frac{x - Ut}{\sigma_x\sqrt{2}}\right)\right\} & (t \leq t_r) \\ \frac{\chi}{2}\left\{erf\left(\frac{x - U(t - t_r)}{\sigma_x\sqrt{2}}\right) - erf\left(\frac{x - Ut}{\sigma_x\sqrt{2}}\right)\right\} & (t_r < t < \infty) \end{cases}$$

where, σ_x, σ_y, σ_z are the dispersion parameters, t_r is the duration of release,

$$\chi(x, y, z, t) = \left(\frac{Q(t)}{U}\right) g_y(x, y) g_z(x, z),$$

χ represents steady-state distribution from a continuous steady-state point source release [17], where,

$$g_y(x, y) = \frac{1}{\sqrt{2\pi}\sigma_y(x)} \exp\left[-\frac{1}{2}\left(\frac{y}{\sigma_y(x)}\right)^2\right]$$

when no inversion is present,

$$g_y(x, z) = \frac{1}{\sqrt{2\pi}\sigma_z(x)} \left\{ \exp\left[-\frac{1}{2} \left(\frac{z - h_s}{\sigma_z(x)} \right)^2 \right] + \exp\left[-\frac{1}{2} \left(\frac{z + h_s}{\sigma_z(x)} \right)^2 \right] \right\}$$

where h_s is the height of release.

3.2 Atmospheric Data Collection

The atmospheric data of Dehradun city have been taken from the report published by the Indian Meteorological Department in 2012 [14]. In order to perform a detailed analysis, the study has been divided into four season, namely winter (December to February), summer (March to May), monsoon (June to September) and post-monsoon (October to November). The values of the atmospheric data have been taken from the Dehradun climate report and averaged for each season. Due to the differences in the atmospheric data (air temperature, relative humidity and wind speed), the simulations have been planned separately for all the four seasons. Further, in order to reduce the error in the study, the maximum air temperature, lowest humidity and highest wind speed are considered. The atmospheric data as per the report of Indian Meteorological Department have been tabulated in Table 1.

3.3 Hypothetical Scenario

An 8-inch diameter mains pipeline has been installed in the district, buried at a depth of 1.0 m below the ground. The mains line carries natural gas and operates at a pressure of 15 kg/cm^2. The majority of the length of the pipeline is constructed from steel. The pipeline is coated with polyethylene coating and cathode protection is applied to prevent from any corrosion. Pipeline markers are placed at every 50 m distance and warning signs are placed throughout the length for identification of the pipeline. Third party damage is being done to the pipeline and the diameter of rupture

Table 1 Atmospheric data (air temperature, relative humidity, wind speed and wind direction) entered in ALOHA

Season	Month	Air temperature (°C)	Relative humidity (%)	Wind speed (kmph)	Wind direction
Winter	Dec–Feb	31.2	20	3.6	NE
Summer	Mar–May	42.8	33	5.4	NE
Monsoon	Jun–Sept	43.7	40	3.6	NE
Post-monsoon	Oct–Nov	33.3	60	3.6	NE

is taken similar to that of the pipeline's diameter. Also, it is assumed that the pipeline is connected to infinite source of methane gas.

3.4 Consequence Analysis

The consequence analysis has been done using ALOHA version 5.4.7.0. In the consequence modelling, distances of toxic area, flammable area and thermal radiation with varying concentrations levels and thermal radiation have been made using the software tool.

3.5 Determination of Level of Concerns (LOCs)

A toxic level of concerns helps in quantifying the threshold concentration of gases which can prove to be harmful for the population inhaling the gas for prolonged periods. In ALOHA, three LOCs can be selected which will be delineated by red, orange and yellow colour. The red zone contains the highest threat. ALOHA utilizes public exposure guidelines for selecting the three tiers of LOCs. The AEGLs, ERPGs and TEELs for a range of chemicals have been encapsulated in the software. However occasionally a situation can arise in which no LOCs have been defined for the chemical or gas under study. ALOHA chooses the LOCs by default when the chemical name and details about the release are entered. For the present study, protective action criteria (PACs) have been selected by ALOHA, for estimating the distance of toxic area of vapour cloud. The software selects the PAC values for a chemical/gas when no AEGL and ERPG values are being defined in the library. For the present study, PAC values are the temporary emergency exposure levels (TEELs) derived by manipulation of the existing data and current LOC's.

3.6 Threat Zone Identification

The computed zones of flammable area, toxic area and thermal radiation estimated in the study were mapped on Google map using Google Earth/ ALOHA.

4 Limitations

The results generated from ALOHA are highly dependent of the quality of the information is fed to it. However, sometimes despite of entering the best available data ALOHA results can be inaccurate and unreliable. The concentration estimates by

ALOHA can be inaccurate at very low wind speeds especially less than 3 miles per hour as the pollutant cloud is unable to mix with the surroundings quickly and ALOHA is unable to predict the concentrations especially near the source. Additionally, ALOHA is unable to estimate the concentrations in very stable atmospheric conditions such as E and F class.

5 Results and discussions

The simulation results have been tabulated in Tables 2 and 3. The threat zones of toxic, flammable and thermal radiation area are mapped in Figs. 1, 2, 3, 4, 5, 6, 7, 8. The transportation of methane is a complex process and any loss of control, can prompt an accident. The simulation results from this study can prove to be very helpful in determining the affected distance over which any injury/damage can reach. Additionally, the results seem to be helpful in construction of an effective emergency response plan to handle such disastrous situation.

Table 2 tabulates the dispersion of toxic and flammable vapour cloud of methane gas post leakage from the CGD pipeline network. It is evident from the simulation results that except summer season the spreading distance of toxic cloud vapour of

Table 2 Threat zone and flammable zone area due to release of methane gas as simulated in ALOHA

Season	Toxic area of vapour cloud		Flammable area of vapour cloud	
	Concentration	Distance (m)	Concentration	Distance (m)
Winter	400,000 ppm (PAC-3)	37	30,000 ppm (60% LEL)	182
	230,000 ppm (PAC-2)	48	5000 ppm (10% LEL)	429
	65,000 ppm (PAC-1)	90		
Summer	400,000 ppm (PAC-3)	30	30,000 ppm (60% LEL)	151
	230,000 ppm (PAC-2)	40	5000 ppm (10% LEL)	357
	65,000 ppm (PAC-1)	74		
Monsoon	400,000 ppm (PAC-3)	37	30,000 ppm (60% LEL)	184
	230,000 ppm (PAC-2)	49	5000 ppm (10% LEL)	433
	65,000 ppm (PAC-1)	91		
Post monsoon	400,000 ppm (PAC-3)	37	30,000 ppm (60% LEL)	183
	230,000 ppm (PAC-2)	49	5000 ppm (10% LEL)	429
	65,000 ppm (PAC-1)	90		

Table 3 Thermal radiation form jet fire as simulated in ALOHA

Season	Thermal radiation from jet fire		
	Radiation energy (kW/sq. m)	Distance (m)	Description
Winter	10	39	Potentially lethal within 60 s
	5	56	2nd degree burns within 60 s
	2	88	Pain within 60 s
Summer	10	37	Potentially lethal within 60 s
	5	53	2nd degree burns within 60 s
	2	83	Pain within 60 s
Monsoon	10	36	Potentially lethal within 60 s
	5	52	2nd degree burns within 60 s
	2	82	Pain within 60 s
Post monsoon	10	37	Potentially lethal within 60 s
	5	53	2nd degree burns within 60 s
	2	84	Pain within 60 s

PAC-3, PAC-2 and PAC-1 level gas concentration is similar for all the other seasons. The dispersion distance of toxic cloud vapour for PAC-3, PAC-2 and PAC-1 level gas concentration was found to be less in summer season, reasonably due to the faster wind speed in the summer season and due to the established fact that faster wind speed increases dilution of the gas. Inferring the simulation results of the winter season, it is apparent that the largest distance to which the toxic area of vapour cloud can spread is 90 m while that of monsoon and post-monsoon season it is anticipated to be 91 m and 90 m, respectively. It is apparent from the Figs. 1, 2, 3, 4, 5, 6, 7, 8,

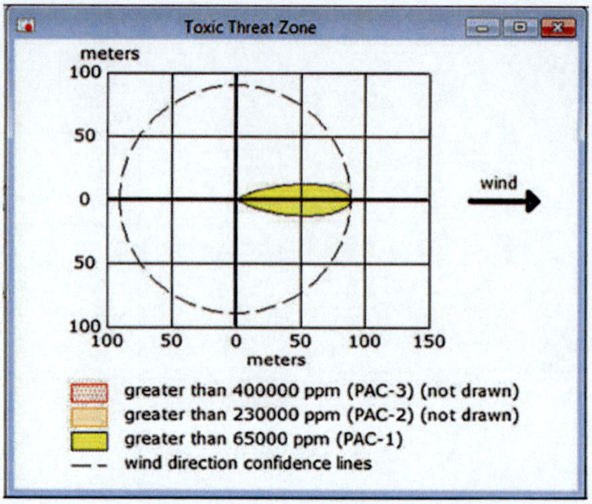

Fig. 1 Toxic threat zone of methane vapour cloud in winter season, Wind speed 2.23mph, Model: ALOHA Gaussian

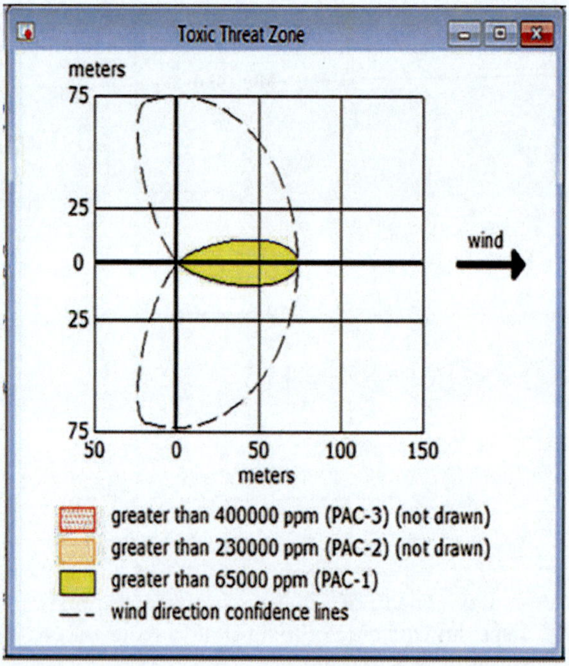

Fig. 2 Toxic threat zone of methane vapour cloud in summer season, wind speed 3.355 mph, Model: ALOHA Gaussian

that the location of release is in a highly populated area and the release of methane gas could possibly affect surrounding residential areas as well as commercial areas.

The flammable area of the vapour cloud of methane is also modelled in order to visualize the distance till the vapour cloud can catch fire. The ALOHA models the

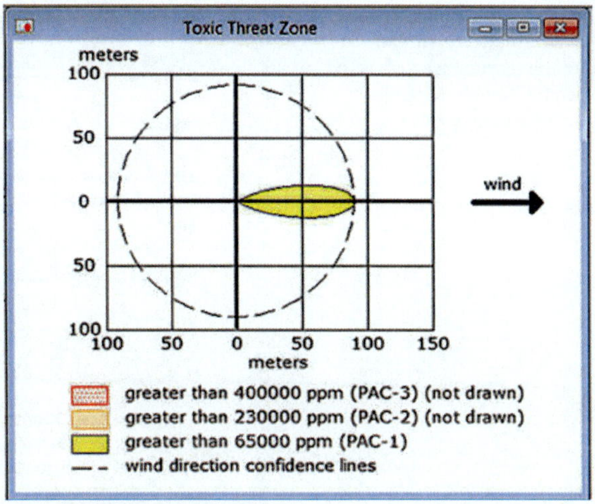

Fig. 3 Toxic threat zone of methane vapour cloud in monsoon season, wind speed 2.23 mph, Model: ALOHA Gaussian

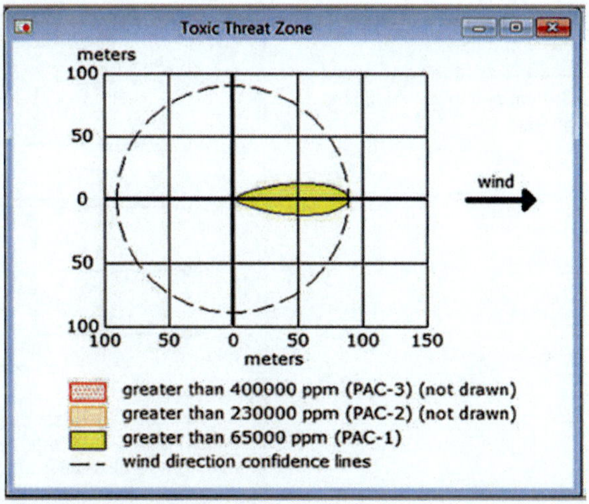

Fig. 4 Toxic threat zone of methane vapour cloud in Post Monsoon, wind speed 2.23 mph, Model: ALOHA Gaussian

flammable areas by considering 60% LEL and 10% LEL, due to the fact that past data have shown that flammable pockets can form even at these concentration. From the simulation results of the winter season, it is evident that flammable area of the vapour cloud having concentration 30,000 ppm (60% LEL) is extended till 182 m, whereas 5000 ppm (10% LEL) is extended till 429 m. The 60% LEL level in summer is found to be covering a distance till 151 m while the 10% LEL level is found till a distance of 357 m from the point of release. The reason of lesser flammable area in summer season is due to the reason that increase in the wind velocity in the summer season dilutes the vapour cloud and restricts it from moving farther. From the modelled results of the monsoon season, the 60% LEL was found to be till a distance of 184 m,

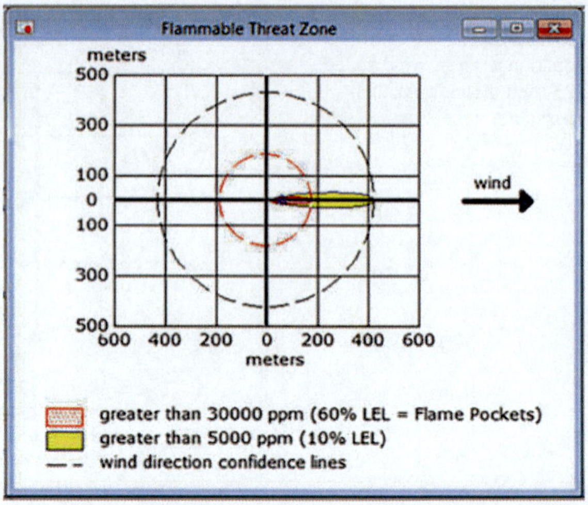

Fig. 5 Flammable threat zone of methane vapour cloud in winter season, Wind speed 2.23mph, Model: ALOHA Gaussian

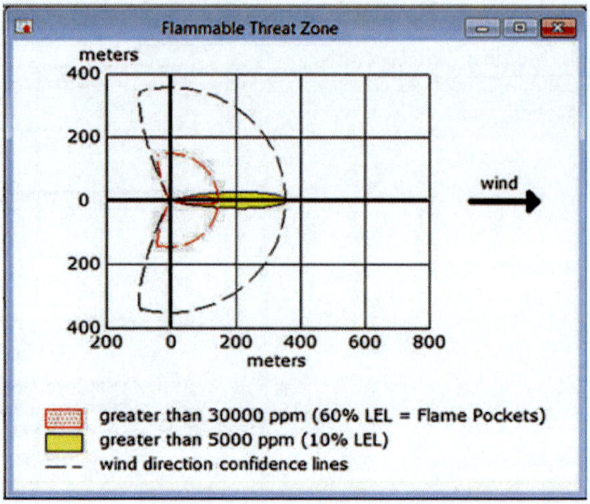

Fig. 6 Flammable threat zone of methane vapour cloud in summer, wind speed 3.355 mph, Model: ALOHA Gaussian

whereas the 10% LEL was found to be at a distance of 433 m. The modelled results for the post-monsoon season depict that the 60% LEL level is at a distance of 183 m, whereas the 10% LEL is at a distance of 429 m.

The simulation of thermal radiation from jet fire in the winter season is found to be largest for all the three levels of radiation energy. This can be attributed to the reason that the radiation energy is a function of atmospheric transmissivity, which is affected by presence of water vapour in the atmosphere. The relative humidity for the winter season is the lowest from rest of the seasons; therefore the distance of radiation energy is largest for all the three levels as the radiation energy experiences less attenuation than the other seasons.

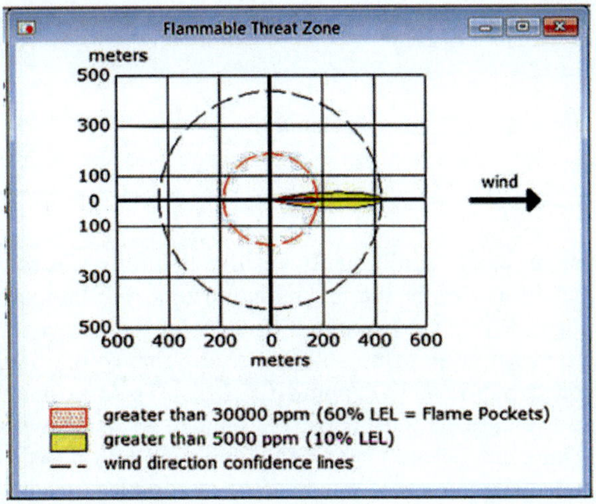

Fig. 7 Flammable threat zone of methane vapour cloud in Monsoon, wind speed 2.23 mph, Model: ALOHA Gaussian

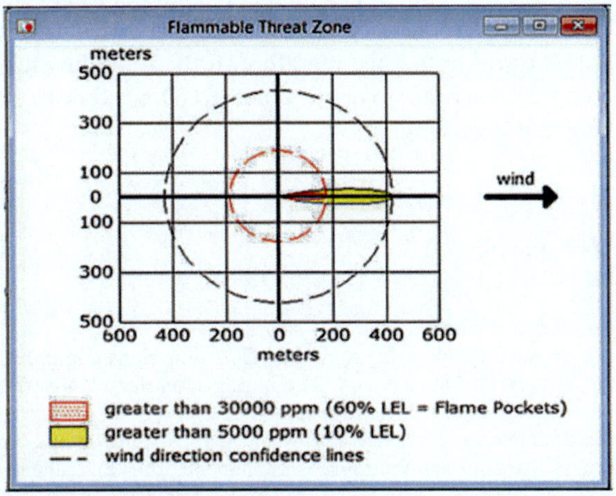

Fig. 8 Flammable threat zone of methane vapour cloud in Post Monsoon, wind speed 2.23 mph, Model: ALOHA Gaussian

The minimum safe evacuation distance to which people are required to be evacuated was found to be 429 m in winter, 357 m in summer, 433 m in monsoon and 429 m in post-monsoon.

6 Conclusion

The analysis of city gas distribution network of Dehradun city using Areal Location of Hazardous Atmospheres suggests that wind velocity was found to be the driving factors behind the variation in the dispersion distance of toxic and flammable area of

Table 4 Minimum safe distance for evacuation

Season	Distance (m)
Winter	429
Summer	357
Monsoon	433
Post-Monsoon	429

vapour cloud. Additionally, relative humidity was found to be the factor governing the distance of the thermal radiation zone. Furthermore, the results of the simulation illustrate that the maximum distance of toxic area of vapour cloud is 91 m having a concentration of 65,000 ppm while that of the flammable area is 433 m having concentration of 5000 ppm (10% LEL). The maximum distance till the thermal radiation will reach causing pain is found to be 88 m having energy of 2 kW/sq.m. These results can be used by CGD companies as a basis of the hazard and risk assessment. The minimum safe evacuation distances are tabulated in Table 4. This paper communicates the risk in advance and it is recommended that the city gas distribution companies and other stakeholders should consider these distances while preparing their emergency response plan and safe evacuation distance. Since, ALOHA has got its own limitations; therefore furthermore investigation is required by the CGD companies in order to qualify their CGD network free from causing any injury to the people it is serving.

References

1. Nejat, P., Jomehzadeh, F., Taheri, M. M., Gohari, M., & Abd Majid, M. Z. (2015). A global review of energy consumption, CO2 emissions and policy in the residential sector (with an overview of the top ten CO_2 emitting countries). *Renewable and Sustainable Energy Reviews, 43*, 843–862.
2. India Energy Statistics. (2013). *Ministry of Statistics and Programme Implementation*. Ministry of Statistics and Programme Implementation, Government of India, New Delhi
3. IEA. (2009). *World Energy Outlook*. Paris: IEA.
4. Chatterjee, R. (2007). Economics of gas distribution—Tariff mechanism for end consumer gas pricing. New Delhi: City Gas India Conference.
5. Pandian, D. J., Mansingh, L., Singhvi, L. K., Tambraparni, S., Sharma, S., & Perrin, A. (2010). City gas India roundtable 2010: Initiatives and challenges. *The Journal for Decision Makers, 35*(4), 61–92.
6. Petroleum and Natural Gas Regulatory Board. (2008). Technical standards and specification including safety standards for city or local natural gas distribution networks) regulations. http://www.pngrb.gov.in/OurRegulation/CGD-Network-GSR612.html
7. Lakshmi, M. R., & Kumar, V. D. (2015). Anthropogenic hazard and disaster relief operations: A case study of GAIL pipeline blaze in east Godavari of AP. *Procedia - Social and Behavioral Sciences, 189*, 198–207.
8. Petroleum and Natural Gas Regulatory Board. (2018). 9th CGD bidding round—Details of the successful bidders Dated: 14.09.2018. http://www.pngrb.gov.in/cgd-network-bid-2018.html
9. ASME. (2018). B31.8S - Managing System Integrity of Gas Pipelines. https://www.asme.org/codes-standards/findcodes-standards/b31-8s-managing-system-integrity-gas-pipelines/2018

10. Han, Z. Y., & Weng, W. G. (2010). An integrated quantitative risk analysis method for NG pipeline network. *Journal of Loss Prevention in the Process Industries, 23*, 428–436.
11. Vidusha, R. S., Sharath, D., Yadav, B. P., & Nandan, A. (2018). Case study on vapour cloud explosions (Buncefield and Jaipur Explosion)—A Review. *Advances in fire and process safety* (pp. 135–147). Singapore: Springer.
12. Jones, R., Lehr, W., Simecek-Beatty, D., & Michael Reynolds, R. (2013). *ALOHA® (Areal locations of hazardous atmospheres) 5.4.4.: Technical documentation.* Seattle, WA: US Department of Commerce, NOAA Technical Memorandum.
13. Palazzi, E., Defaveri, M., Fumarola, G., & Ferraiolo, G. (1982). Diffusion from a steady source of short duration. *Atmospheric Environment, 16*(12), 2785–2790.
14. Sharma, A., Singh, O. P., & Saklani, M. M. (2012). *Climate of Dehradun.* New Delhi: Indian Meteorological Department Ministry of Earth Sciences.

A Study on Improving Safety Through KAIZEN in an FMCG Industry

Sandeep Kumar and P. A. Arun

1 Introduction

Safety KAIZEN is a continuous process of improvement, and its sustainability relies on the maintenance and monitoring of the implemented ideas. Its goal is not only to make a change in the system; its goal is to make an effective change so that the performance can increase.

This study mainly focuses on a specific part of the plant such as packing areas, where high risk exists due to the type of machinery operating in the areas and human interface to it. To achieve the optimum level of safety in the plant, it is necessary to monitor and implement newly introduced ideas by the professionals. This will help in a high standard of decision-making and developing strategies with respect to the business.

So in order to implement and to achieve consistency, two steps need to be followed; these are [1],

- Maintenance of the existing condition on which the company ensures the compliance of current standard of technical and management aspect, which keeps the traditional safety intact, and,
- Improvement of the existing condition—KAIZEN, which pushes further to enable to achieve the zero-tolerance targets.

Safety Kaizen uses 5 Why Techniques to analyze the root cause and to determine the safety measures. Safety Kaizen is a merger of traditional safety as well as modern safety (behavioral-based safety). Kaizen implementation is done by mainly three ways are,

- **Innovation**: which requires research and development to generate a new ideation technique

S. Kumar · P. A. Arun (✉)
School of Engineering, UPES, Dehradun, India
e-mail: arun.pa@ddn.upes.ac.in

N. A. Siddiqui et al. (eds.), *Advances in Behavioral Based Safety*,
https://doi.org/10.1007/978-981-16-8270-4_27

- **Horizontal Deployment:** to implement the same developed technique to other machineries and department of the same plant.
- **Replication**: It is way of taking reference of another person developed technique for their own remedial measures
- **Problem-Solving**: It is nothing but a brainstorming technique where a group of experts sit and analyze the problem, the outcome is then implemented to solving the problem.

For the generation of KAIZEN, ideation can come from anywhere of the worker's hierarchy. Most of the ideation comes from the grass root people like example Machine operators, Maintenance technicians, Fitter-Foreman. These people work day-to-day basis of same job which allows them to understand the kind of difficulty and its countermeasures. Apart from this where there is difficulty to provide any kind of engineering control, KAIZEN also helps in developing administrative control for the respective point of risk as a controlling measure. These administrative controls are intervention procedures that include a safe pathway to follow for those activities having significant risks. This administrative control will allow mitigating the risk as low as reasonably practicable [2–4].

2 Methodology

2.1 Conceptual Flow Chart and Method Statement

To understand the basics of the methodology, it is necessary to develop a roadmap that will depict the initial and final stage of a method in a single alignment. Thereby we developed a Conceptual Flow Chart as depicted below, which gives a firm idea of the further process. It includes the step-by-step process of the Safety KAIZEN methodology which will lead us to our goal. It is based on five stages i.e., 1. Planning and Preparation, 2. Review of Baseline Data, 3. KAIZEN Ideation and Generation, 4. Implementation of KAIZEN, 5. Monitoring and Review (Fig. 1).

2.2 Review Study of Past Recordable Cases and Its Causes

The accidents are resulted by system failures. To understand the cause of these accidents, we took a reference of past two-years accidents statistics. This gave us the idea of all the recordable cases that occurred throughout the Plants. The accident statistics were maintained based on (a) Category Wise, (b) Type Wise, (c) Body Part Wise, and (d) Section wise which allowed us a better understanding of these injuries. The similarity between the statistics of both the working year has recorded highest reportable injury in the packing section. This identifies the sensitivity of this area

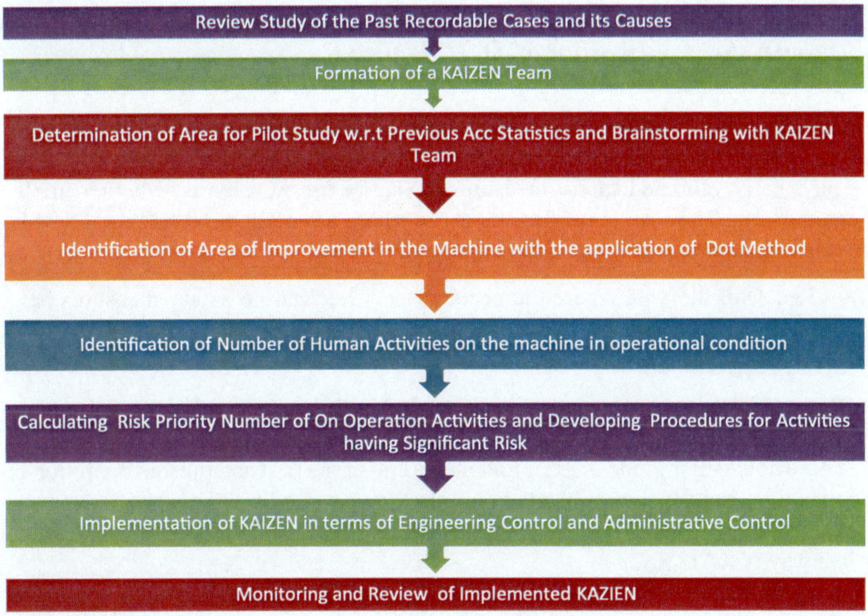

Fig. 1 Methodology adopted

between the machine and the human interface. Through this, we also get to know the cause of these accidents that lead to it.

From this, we can easily determine the sensitive areas of the packing machine which require additional protection. It also helped us to understand the nature of the accident with respect to the parts involved.

2.3 Formation of KAIZEN Team

To make the project more effective, a KAIZEN Team was formed which includes experts of their field. Members of the KAIZEN Team included Production Officer, Maintenance Officer, and their associates, Quality Officer, Fitter-Foreman, and a Packing Machine Operator. Factory Manager was performing as Team Lead. By conducting a series of meeting, we mainly focused on developing KAIZEN for Packing Machines. Here we focused on the parts where the accident occurred and as well as the area where there might be chances of an accident can occur.

To achieve the goal we projected an action plan, this action plan is based on the PDCA cycle. This plan has been developed by keeping in mind to reduce the lost time injuries and increasing additional improvement.

2.4 Identification of an Area of Improvement in the Machine with the Application of DOT Method

In order to improve, a packing machine was selected for Pilot Study. For the Pilot machine, firstly, we observed the parts where the accident already occurred. For this, we physically validated safety measures/KAIZEN for those parts, whether implemented or not. Then we identified the areas which are prone to accidents. For this, we implemented pasted dots on the packing machine,

(i) Red Dot: the area where the accident occurred but no safety measures have been implemented.
(ii) Purple Dot: areas where an accident can happen.

Note: When safety measure is implemented at those places, these dots turn into the green outline as Ex: Red Dot with Green Outline.

These dots will provide a better understanding of the past and present condition of the machine, and we communicated these to the operators and the workers working on the machine. This activity acts as a sensitizing medium to the workmen's which is in a way is KAIZEN.

Most of the Red Dots have already been turned green because of the high risk involved and due to the sensitivity of the area. So we focused on the Purple dots, some of these areas require guard modifications, some of these require additional precautionary measures, and some of them required brainstorming sessions with the KAIZEN team to develop a KAIZEN idea for the respective area. This is how the closure of all the Purple dots took place.

2.5 Identification of Number of Human Activities on the Machine in Operational Condition

With the reference to accident statistics, 80% of it occurred due to the operational interventions hence it was mandatory to analyze all the operational activities and to evaluate its Risk Priority Number. These values will help us to determine the significance of the risk. These activities require to be performed by skilled workmen because it requires a certain level of competency to perform them. Some activities are not completely operational, which means that while performing those activities the workmen can stop the machine and then to restart for ensuring the output.

For the Pilot machine taken into account, we identified a certain number of On Operational Activities out of those some of them have significant risks. Some of these activities are in general which applies to all the packing machines but some of them particularly design specific.

2.6 Intervention Procedure

This is an administrative control measure that has been developed for online activities with high risk, where any sort of engineering control is quite difficult to place. It will help the operator/technician to perform the activity in a way that reduces the significance of the risk. This procedure contains pictorial references as well as the procedural which will allow a better understanding of the workmen. The development of this procedure initially took place during meetings with KAIZEN Team. The inputs of these members were very crucial to develop the intervention procedure. The intervention procedure is made very simple and short so that the operator and technician will understand it easily. It should be specific to the job activity and it should not include anything other than that. The procedure should also consist of all the precautionary measures at the top. In the procedure, responsibility and accountability shall be mentioned clearly. On the pictorial reference, a step-by-step activity shall also be mentioned which will allow another means of graphical communication to the workers.

3 Results and Discussion

After reviewing the previous statistics of recordable cases, we decided to look upon the most accident-prone area that was packing machine areas. It is because of the presence of many machines on the shop floor and human interface which make it most vulnerable. Then we choose a packing machine for Pilot Study, then by using the Dot method, we identified 5 Numbers of Red Dots with green Outlines and 4 numbers of Purple Dots. For all the identified Purple dots, we provided improvement measures, for example, extension of guards, covering of rotating parts, providing adequate locking mechanism. Then we implemented three numbers of KAIZENS which will allow reducing the risk probability. The KAIZENS were in the form of engineering control measures, two of them we implemented in the machine and two of them were modified hand tools. The outcome of the implemented KAIZENS is as followed:

- It will reduce the production loss and unnecessary time loss.
- It will reduce the probability of risk.
- It will increase the proximity distance from the rotating parts.
- It will restrict the operator's hand movements from the hazardous area.

Where engineering control was not feasible, for those we provided administrative control measures, namely, intervention procedure. Firstly we identified the number of human intervention to the packing machine. These intervention activities include routine and non-routine activities. The routine activities were those, the operator used to follow on a daily basis, and non-routines were those like maintenance technician has to work on operational packing machines. Then we calculated the Risk Priority

Number of those on operational activities. According to the RPN, we extracted activities with high significance. The total number of activities with high significance was eight. For all the high significance activities, we developed intervention procedures, these procedures were aligned with the operational activity. At the initial stage of development of the procedure, KAIZEN Team helped it to standardize by conducting series of meetings. In the intervention procedure, it was specific to the step-by-step activity with the responsibility and accountability personnel.

It is necessary to analyze the developed and implemented KAIZEN in order to identify its effectiveness. It should also be noted that this KAIZEN doesn't create any sort of unnecessary hazard to the surroundings. Hence, to assess the quality criteria, certain point needs to be ensured for its adequacy and effectiveness. The number of accepted criteria will allow us to understand its quality. The points mentioned can vary with the KAIZEN implemented within the machine or KAIZEN developed for using as Hand Tools.

- Does it create any obstruction to the surrounding environment?
- Is the operator/technician is getting affected by the implemented KAIZEN?
- Is the Food Safety of the product affected by the implemented KAIZEN?
- Is the operator/technician is clear about before and after scenario?
- Is training and standardization of the KAIZEN is done?
- Is the developed KAIZEN is replicable to other machines
- Is the developed KAIZEN reduces significant risk?

4 Conclusions

This paper is all about increasing the performance of safety by improving the present standard in other sense upgrading the present standard. To make it possible, a Pilot study was done on the packing machine for its improvement which allowed us a path to follow further on the way of increasing performance. The outcome of the Pilot Study gave us an idea to understand the area of improvement. This area of improvement was mitigated through the engineering control methods and where engineering control was not practically possible then administrative control came into light. This Engineering Control is introduced as KAIZENs. To provide administrative control, an intervention procedure was developed for Online Activities with having significant risk. Identification of significant risk was made by developing a Risk Index Value Chart of all the Online Interventions including high risk and low risk. Now the outcome of one Pilot machine has to be replicated to all the packing machines of the shop floor in order to achieve complete effectiveness.

References

1. Karkoszka, T. (2009). Kaizen philosophy a manner of continuous improvement of processes and products. *Journal of Achievements in Materials and Manufacturing Engineering, 35*(2), 197–203.
2. Cox, S., & Flin, R. (1998). Safety culture: Philosopher's stone or man of straw? *Work & stress, 12*(3), 189–201.
3. James, J., Ikuma, L. H., Nahmens, I., & Aghazadeh, F. (2014). The impact of Kaizen on safety in modular home manufacturing. *The International Journal of Advanced Manufacturing Technology, 70*(1–4), 725–734.
4. Ikuma, L. H., Nahmens, I., & James, J. (2011). Use of safety and lean integrated kaizen to improve performance in modular homebuilding. *Journal of construction engineering and management, 137*(7), 551–560.

An Interface for Analysing and Assessing the Inter Personnel Behaviour in Order to Assure Organizational Safety

Bikarama Prasad Yadav, Umang Kumar Yadav, Nikhil Verma, Abhishek Nandan, and P. Mondal

Interface Name

AICOS Analysing Interpersonal Conduct for Organizational Safety

1 Introduction

Indian industry has been observed with the frugal safety culture in past years. Despite the rules, legislations and policies imparted for the industries, it has been seen that the outcome is not at its edge. Also commitment from the top management and efforts may go in vain if there is increase in the accidents in the organization [1]. Researchers have proven that safety has been one of the key factors in the past three decades. Various aspects were introduced during the study to analyse and change the behavioural and psychological aspects of the individual in the organization. Influencing characters play a vital role in analysing the behaviour which includes the safety climate, safety management system, etc. After analysing the various industries, it was concluded that the safety management system has a vital role in industry [2].

In line of the continual improvement, it has been observed that organization learnings need to be an important factor. It is very crucial to understand the relationship between the members to analyse the learning behaviours in order to understand the psychological aspects towards safety [3]. Though the organization has the culture, debate was always there how to describe the safety culture. The previous studies done on the safety models but only key aspects were discussed for the improving the safety culture [4]. Some organization uses visual Prototyping System in which using

B. P. Yadav (✉) · U. K. Yadav · N. Verma · A. Nandan · P. Mondal
University of Petroleum and Energy Studies, Dehradun, India

N. A. Siddiqui et al. (eds.), *Advances in Behavioral Based Safety*,
https://doi.org/10.1007/978-981-16-8270-4_28

cameras, monitoring is done but this cannot help to analyse the behaviour of the individual and can only be effective for the consequences-based analysis [5]. The work environment with positive relationship leads to the generation of positive psychology and positive behaviour and gives the experience for the satisfaction, enrichment and development [6]. A review was conducted on key trends and developments related to safety with respect to psychology, human resource management and behaviour [7]. Though researches have told that the safety climate analysis and improvements have given some good results. It has been recorded that the perceptions and exceptions by looking the actions of supervisor gives the prediction of safety to the workers. The prediction helps to judge the employees motivation to work safely. Studies have also shown that safety measures were done on employee level rather than company level. The most of the previous researchers were associated with the injury rate in the organization and few were related to the hazardous nature of the organization [8].

Identifying and analysing key elements and standard requirements have been considered as major changes in organizational behaviour. Approaches which include BBS or proactive BBS were considered with the help of employee participation (survey and interview sessions) which has brought numerous aspects to reconsider for further implementation [9, 10]. Inclusion of psychological aspect in safety culture improvement has been also introduced as an individual or among different target level of workmen and have been critically analysed for their effectiveness. The best part of it is largely accepted for cultural changes into the organization [11, 12]. A few studies were also conducted on the available evidence on ill effects of occupational health and safety management system (OHSMS) applied for employee health and safety and their related economic outcomes of interventions [13]. In fact, aspects related to social determinants on workplace safety is also one of the consideration largely accepted by researcher in past and facts are proven in the favour [14]. Apart from tools and technique available, practices on guiding documents for improvement of safety culture into the organization through inculcating behaviour-based safety at site level among the employees were also done in the industries [15]. It has been seen that interpersonal risks that can lead to collective learning shows differences in psychological safety from trust and it explains how psychological safety removes interpersonal risks and provides a structured learning process [16]. Some studies focus on the safety as an activity of safety people and labour inspection and have limited their scope traditional changes in organizational safety [17], whereas some study focuses on the present analysis of interpersonal behaviour is based on two interrelated conceptions [18]. For improving process safety culture, identifying observations, generating certain typical characteristics of current situation has given directions that could lead us to solve process safety-related issues in the industries [19]. With the development of chemical industry, the complexity of system has increased along with the handling of highly hazardous chemicals. Hence, implementation of PSM practices and assessing the effectiveness of the same in regards to the minimizing fatality rate and productivity enhancement have become the major concerns [20]. Thereby, designing a framework to ensure that the OH&S managing safety has been carefully constructed and customized to the individual organization as per the requirement [21]. A number of works are included by the researches seeking the information related to incidents in process safety, mainly in pharma industry. They have analysed the consequences along with the factors contributing towards the incidents. They have identified the contributing factors include hazard awareness and their identification; operating procedures; design; safeguards and controls [22]. Work has also highlighted on the challenges coming in process safety including principles of hazard identification and risk assessment are to be reviewed and much focus should be given to include the approaches like engineering and science, life-cycle analysis (LCA), etc. The final HIRA to be conducted by applying all these aspects including the objective of sustainability [23]. These are the few components/indicators are much helpful in managing process safety effectively in major hazardous industries. It also distinguishes 2 dimensions of safety indicators; personal safety versus process safety indicators, which includes leading and lagging indicators as well [24]. Further, two other important aspects include the top management leadership and employee empowerment are considered for TQM (Total Quality Management). Various surveys were conducted in organizations to ensure the job and customer satisfaction in regards to the top management leadership and employee empowerment with improved quality products [25]. Strategies including different levels of contingencies of process industries have stressed on the introduction of organizational-level framework of climate for initiative and psychological safety by using the surveys and data collection methods and further to analyse their results [26]. Researchers have found that psychological safety climate binds the relationship between the task conflict and overall performance. Results showed that task conflict and team performance were interlinked and are associated under situation of high psychological safety [27]. Under certain circumstances, failure-based behaviours in industries were analysed which checks about the role of social capital and psychological safety in developing of failure-based learning behaviours using the help of surveys [28].

2 Methodology

Based on the outcome of the research, the interface will help the organization to analyse the individual behaviour (Top to Bottom) in the organization towards safety (Fig. 1).

A critical analysis of their work place behaviour is done by using available methods. This has given a way out to develop the tool named as Analysing Interpersonal Conduct of Individual for Organizational Safety (AICOS). Towards achieving the outcome, role of primary and secondary data are very important and for that a set of questionnaire includes the level of employees are taken into consideration. In this, their personal behaviours are assessed for organizational safety aspects. Identifying the need of this work and development of tool, it is important to consider the role of line management starting from the CEO or chief of the organization to a common worker who contributes for organizational success. These questionnaire are based on the references available globally in interpersonal behaviours along with standards

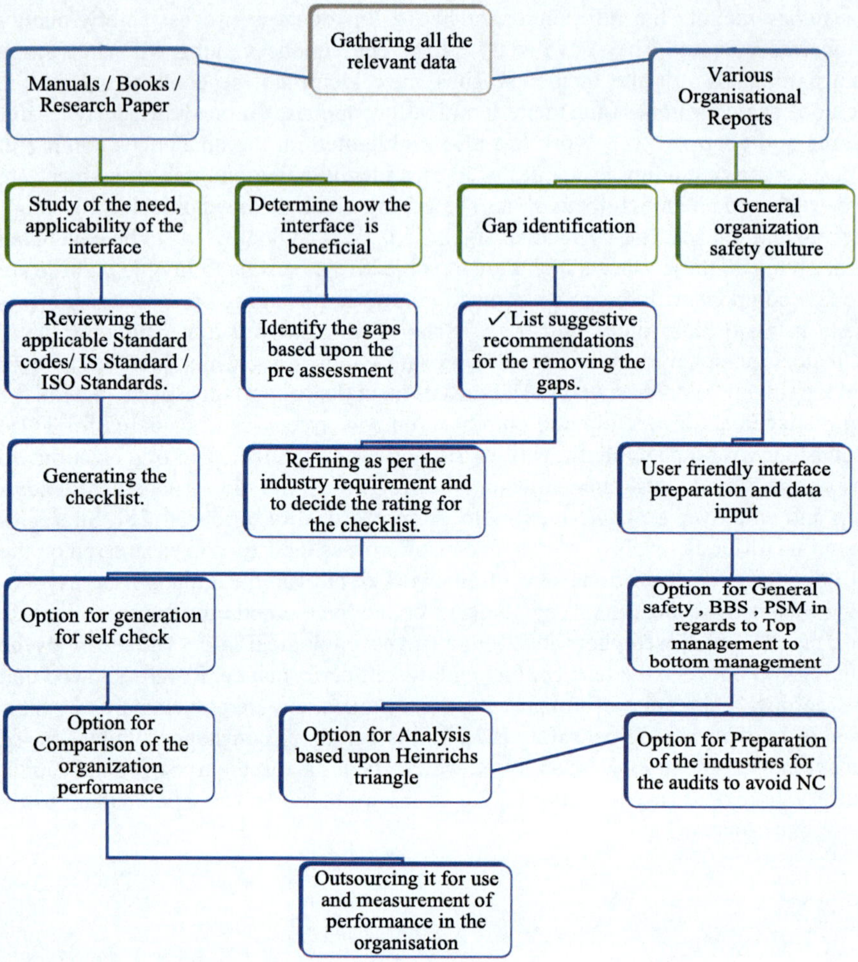

Fig. 1 Methodology

available. Concluding the fact of the findings at workplace, interaction with the help of developed tools is done at various stages avoiding errors and assessed results are verified at workplace based on the current situation available. Assessing the need and its efficacy of interface tool, different organizations are selected for completing the survey and validation.

3 Discussion

3.1 Management Role

Management plays an important role in implementing the BBS at organization level and it is being reflected in organization through reactiveness of the safety management system. As a part of good culture, some companies have positive traits in their management level and do have decisive management talks towards in establishing better work culture and have also given priority for the safety with the production goals [29]. These kind of commitment gives a thought for various dimensions, which are important in developing the safety culture. The trust developed in working environment helps management deems, as safety is important and provides the sense of satisfaction for the employee working.

3.2 Behaviour Based Safety (BBS)

Interestingly, work place behaviour has direct impact not just on the organizational safety culture but it also impacts on the productivity or entire organizational growth. Almost 50% of the unsafe behaviours leading accident in the organization and these facts are known and therefore stressed enough to bring changes to improve the existing condition in last two decades. In fact, behavioural aspects have become a part of standard requirements and witnessed in the guidelines of ISO 45001: 2018. The purpose of these requirements or inclusion into standard requirements are just because of gaining morale and witnessed the changes into the system towards safety system improvements. In fact, these standards are much more helpful in employee participation and sharing their approaches, consultation towards developing a new culture [30]. The good part for the BBS is, it provides a structural and quantitative approaches for prolonged safety culture in the field of psychology and the process relates to reinforcement [10].

3.3 Psychological Safety

Psychological safety helps to reflect the characteristic which resolves the task conflict in contrast to improve the team performance. It can be seen that psychological safety refers to the common belief shared by everyone in the team that they are confident to take the risk. Psychologically safe environment introduces more creativity and decision making without damaging the overall performance. Psychology safety climate helped the way conflict is taken and is managed by the team [10].

3.4 Safety Culture

The culture which the organization persists and needs to be continued towards the safety is the safety culture. Safety attitude is the contribution of management attitude and the communication to make the safety climate. It has been seen that with proper climate, safety can be increased with the reduction of the injury rate which in increase in the participation and compliance. The safety culture is defined as the top down approach used to by the organization approaching the safety management.

3.5 Safety Management System

It has been seen that safety management system is defined as the way organization imparts the safety culture. The change in the top leadership or revised safety management can increase the safety. But it has been seen that impact of the safety management has been ignored in the last years. To assess the quality of the management system, regular auditing needs to be done to identify the gaps.

3.6 Process Safety Management

The PSM is one of the critical aspects in the industry which needs to be covered in pre or during industry changes. The PSM covers small but critical parts of the process which when neglected can cause fire, explosion or toxic release leading to catastrophic event. The PSM checklist will ensure that all necessary safety checks have been done prior starting the process, with a coverage to all the 14 elements of process safety management. Through checklist, the NC's can be easily tracked which can be fulfilled easily and results can be shown to the top management easily rather than going for step by step.

3.7 Heinrich Triangle

Heinrich's expressed that "psychology in accident prevention is a fundamental of great importance"; he believed that psychology lies at root of accident causes and a total of 88% of all industrial accidents are caused by the unsafe acts of persons (Fig. 2).

He originally stated that in a unit group of 330 accidents, 300 result in no injury, 29 result in minor injury and 1 results in a major or lost time case.

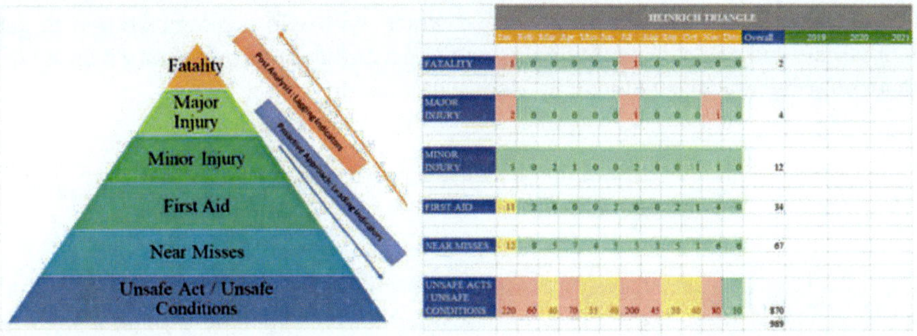

Fig. 2 Heinrich triangle

3.8 Analysing Interpersonal Conduct for Organizational Safety (AICOS)

Assessing the BBS and its implementation are the key areas where one can impart the safety. One of the major challenges faced by the organization is to maintain their assets for the long term so as to implement the successful BBS at the organization. Also, organization faces certain issues which include the commitment, communication, guidelines etc. It also acts as a barrier when there is a continuous breach in implementation. Sometimes, it has been seen that organization commits towards the safety but in ground, things are totally different. Sometimes the procedures are there but they are lack in the implementation phase.

The proposed interface will be effective in analysing the BBS in the organization and to identify the gaps with the recommendations. It will help the organization to prepare themselves for the audits like 45,001: 2018. The Quantification analysis will tell the organization about their current status on BBS. Also, the cross verification methodology will help to check from top management and the bottom management. Based upon the assessment, suitable suggestions will be given which will be acting as a catalyst for implementation. The interface also gives the flexibility by choosing the key areas which are required for the BBS at the site. It also enhances the promotion for the participation and consultation by assessing from top management and the bottom management. It will assess on the basis of various checklist to check the behaviour of people in the organization. Based upon the review done, the score will be generated.

The interface can be used for the yearly review and progress in the organization. Based upon the review, the organization can set their goals accordingly. It is user-friendly and is applicable for various industries. The rating generated by the checklist will make the final scoresheet of the assessment and will also help the organization to prepare for the audit accordingly. The major advantage of the interface that all critical aspects are covered in one place which will help the organization to set their targets and goals. Data for analysing the performance of organization based on incident/accident analysis from the Heinrichs triangle dataset.

3.9 Details About the Links

The links involve three stages of assessment for all areas which includes the initial level, advanced level and final level assessment which include questionnaires in which for each questions, answers will be given based upon proofs. Also at the same levels of checklist, crosschecking with other checklist will also be done to ensure it is under compliance or not. Once the link is filled by the candidate using google forms, it will be analysed and based upon this, result will be concluded.

4 Result and Discussions

The proposed questionnaire-based interface AICOS is intended to judge the analysis of the individual and the organization towards the psychological and behavioural safety. After studying from previous incidents and research works, we tried to do a quantitative analysis by generating relevant questionnaire and assessing them based upon the feedback. Analysis of safety culture and behaviour towards safety was checked with the help of various questionnaire developed; incorporated in the form of an interface. The links to different area questionnaire were circulated so as to gain feedback. The collected responses were analysed basis the inputs as per the rating scale, and the outcome was that the psychological based-questionnaire plays vital role in judging the safety culture and to find the noncompliance at the site at various level of the organization. Based on the input parameters, an idea about where the organization stands can be inferred from the numbers (Table 1, Fig. 3).

- The full compliances are highlighted in **green colour**, the moderate compliances are highlighted in **amber colour** and the major noncompliances are highlighted in **red colour**.
- A total of 25 questions were assessed in stage 1, followed by 25 questions in stage 2 and 20 questions in 3rd stage of assessment.
- All the major noncompliances shall be addressed with utmost priority and resolved to full compliances followed by the moderate level compliances. Organization can set up an acceptable score level separately for different stages and combined too.

Table 1 AICOS checklist details

Analysing interpersonnel conduct for organizational safety (AICOS)				
An interface for analysing the trends of safety behaviour of an individual for an organization; constitutes the questionnaire for personnel employed at worker level, Supervisor level, safety officer, employee level, manager level, as plant head and top management level in the organization. It emphasizes on assessing the key areas and quantify the analysis based upon the survey response; record the findings; calculate score and provide the required recommendations				
		Questionnaire levels	Area	Link
1	Initial stage of assessment	1, 2, 3	Worker	Link 1
		1, 2, 3	Supervisor	Link 2
		1, 2, 3	Safety officer	Link 3
2	Advanced stage of assessment	1, 2, 3	Employee	Link 4
		1, 2, 3	Manager	Link 5
		1, 2, 3	Plant head	Link 6
3	Final stage of assessment	1, 2, 3	Top management	Link 7
		1, 2, 3	ISO 45001	Link 8
		1, 2, 3	PSM	Link 9
Rating criteria: 5—Strongly Agree: The organization is in full compliance with the stated point which shows the impact from top management to the shop floor (Attach proof) 4—Agree: The organization is in compliance with the stated point which shows the impact from top management but needs to be checked as per organization SOP, Shop floor visit or applicable standard. (Attach proof) 3—Neutral: Means Organization is having the structure but not in compliance at the shop floor 2—Strongly Disagree: The organization is not in the compliance with the stated point 1—NA (At the end the points will be highlighted)				

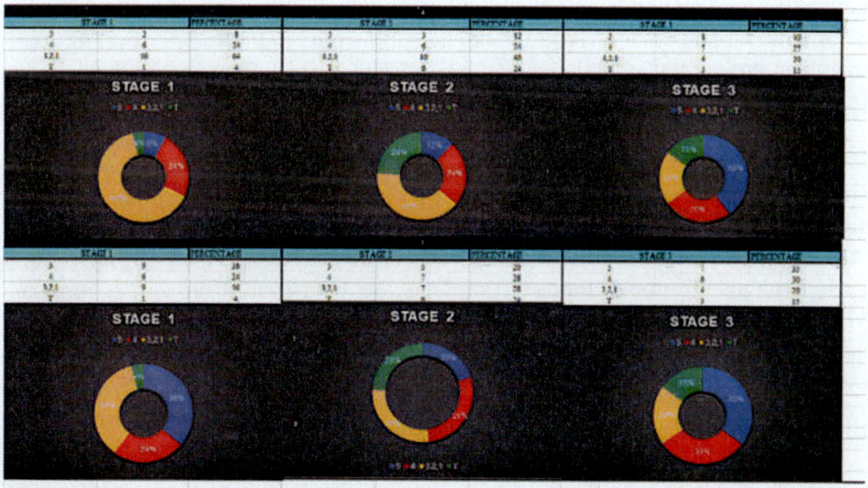

Fig. 3 Comparitive analysis of response received

4.1 Response Comparison 1st Case

Safety Officer 1; Organization 1 (Fig. 4).

- After quantitatively analysing the response received for stage 1 assessment, it was evident that the 4 ID of major noncompliances (24%); 14 ID of moderate compliance (56%); and 5 ID of full compliance were there (20%).
- After quantitatively analysing the response received for stage 2 assessment, it was evident that the 1 ID of major noncompliances (4%); 5 ID of moderate compliance (20%); and 19 ID of full compliance were there (76%).
- After quantitatively analysing the response received for stage 3 assessment, it was evident that the 5 ID of major noncompliances (25%); 8 ID of moderate compliance (40%); and 7 ID of full compliance were there (35%).

Safety Officer 2; Organization 1 (Fig. 5).

- After quantitatively analysing the response received for stage 1 assessment, it was evident that the 11 ID of major noncompliances (44%); 11 ID of moderate compliance (44%); and 3 ID of full compliance were there (12%).
- After quantitatively analysing the response received for stage 2 assessment, it was evident that the 5 ID of major noncompliances (20%); 13 ID of moderate compliance (52%); and 7 ID of full compliance were there (28%).
- After quantitatively analysing the response received for stage 3 assessment, it was evident that the 7 ID of major noncompliances (35%); 9 ID of moderate compliance (45%); and 4 ID of full compliance were there (20%).

Upon comparison of the response mapped, for the very same organization, the number of noncompliances, moderate compliances and full compliances were differencing from far percentages; so as to maintain uniformity of information or floor level implementation of the administered or to be administered safety aspects; the concept of IITS (Information, Instruction, Training, Supervision) must be followed.

The output generated in form of charts gave the organization a clear idea, about the gap identification i.e., at what point they lack and what points are fully complied; hence they can chose upon what best practices can be put through to improve safety culture of the industries. The different stages of questionnaire developed helped in grasping a major loopholes and provide with a stringent outcome and helps to ensure that individuals are not giving the fake details as with each details proof must be attached. The Questionnaire in one way or another emphasizes on the roles and responsibility of the individual from top management to bottom management helps to widen the minds of the individual towards their contribution for safety.

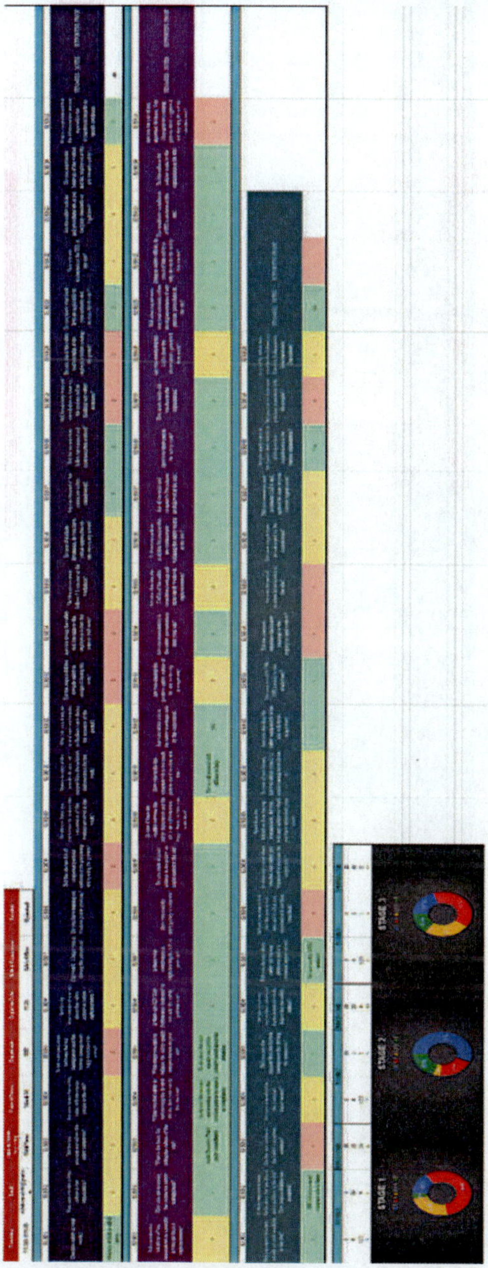

Fig. 4 Response _ SO 1_O1

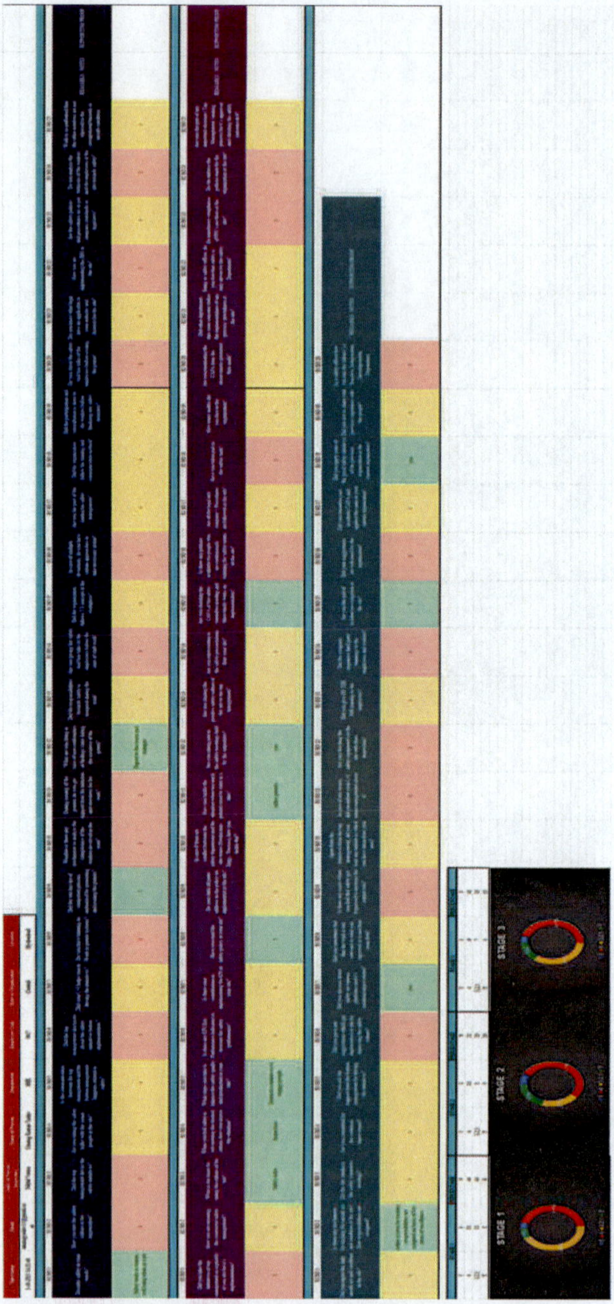

Fig. 5 Response _SO 2_O1

5 Conclusion

This work is helpful in checking the safety behaviour, psychology, culture and attitude of the individual towards safety. Also, where the organization is lacking can be easily understood from the response received. After studying and analysing the feedback revived, it can be seen that the questionnaire is one of the key tools identified to do the compliance at the site and help in irradiating the noncompliance or conflict on different points. The rating system also helps to quantify the score of the industry and gives the area of improvement which can increase the score of the site. The questionnaire also helps to resolve the conflict on certain situation where the top management feels that it is under compliance. Also, it helps the organization and the individual to assess the gaps found and to track them timely. The interface will also give suggestions to remove the non-conformity.

These interface has covered all the aspects related to the safety and at different levels of people perception related to safety at work in the organization. The best part of these interface is, the questionnaire is especially designed to counter check the different aspects for a full proof analysis of individual perception at different levels. This analysis is going to help in change in behavioural aspects of safety for an individual or an organization which could able to change the safety vision of a particular organization. This interface has also given a scope for continual improvement towards and also for robust and easier to usage by an user.

References

1. Larcker, D., & Tayan, B. (2015). *Corporate governance matters: A closer look at organizational choices and their consequences.* Pearson education.
2. Vinodkumar, M., Bhasi, M. (2010). Prevention, safety management practices and safety behaviour: Assessing the mediating role of safety knowledge and motivation. *Accident Analysis & Prevention, 42*(6): 2082–2093.
3. Carmeli, A., et al. (2009). Learning behaviours in the workplace: The role of high-quality interpersonal relationships and psychological safety. *Systems Research and Behavioral Science: The Official Journal of the International Federation for Systems Research, 26*(1), 81–98.
4. Choudhry, R. M., Fang, D., & Mohamed, S. (2007). The nature of safety culture: A survey of the state-of-the-art. *Safety Science, 45*(10): 993–1012.
5. Guo, H. L., Li, H., & Li, V. (2013). VP-based safety management in large-scale construction projects: A conceptual framework. *Automation in Construction, 34*: 16–24.
6. Kark, R., et al. (2009). Alive and creating: The mediating role of vitality and aliveness in the relationship between psychological safety and creative work involvement. *The International Journal of Industrial, Occupational and Organizational Psychology and Behavior, 30*(6), 785–804.
7. Hofmann, D. A., Burke, M. J., & Zohar, D. (2017). 100 years of occupational safety research: From basic protections and work analysis to a multilevel view of workplace safety and risk. *Journal of Applied Psychology, 102*(3): 375.
8. Smith, G. S., et al. (2006). The relationship between safety climate and injury rates across industries: The need to adjust for injury hazards. *Accident Analysis & Prevention, 38*(3), 556–562.

9. Darabont, D. C., Antonov, A. E., & Bejinariu, C. (2017). Key elements on implementing an occupational health and safety management system using ISO 45001 standard. In MATEC web of conferences, EDP Sciences.
10. Li, H., et al. (2015). Proactive behavior-based safety management for construction safety improvement. *Safety Science, 75*, 107–117.
11. Edmondson, A. C., & Lei, Z. (2014). Psychological safety: The history, renaissance, and future of an interpersonal construct. *Annual Review of Organizational Psychology and Organizational Behavior, 1*(1): 23–43.
12. Zhang, M., & Fang, D. (2013). A continuous behavior-based safety strategy for persistent safety improvement in construction industry. *Automation in Construction, 34*: 101–107.
13. Almutairi, N. S., et al. (2019). Effectiveness of intervention occupational safety and health management system on improving safety and health among the laboratory technician/staff. *International Journal of Advanced Scientific Research & Development (IJASRD), 5*(12), 18–32.
14. Nordlöf, H., et al. (2015). Safety culture and reasons for risk-taking at a large steel-manufacturing company: Investigating the worker perspective. *Safety Science, 73*, 126–135.
15. Center for Chemical Process Safety. (2010). *Guidelines for risk based process safety*. Wiley.
16. Edmondson, A. C. (2002). *Managing the risk of learning: Psychological safety in work teams*. Citeseer.
17. Häkkinen, K. (1995). A learning-by-doing strategy to improve top management involvement in safety. *Safety Science, 20*(2–3), 299–304.
18. Feffer, M. (1970). Developmental analysis of interpersonal behavior. *Psychological Review, 77*(3), 197.
19. Knegtering, B., & Pasman, H. J. (2009). Safety of the process industries in the 21st century: A changing need of process safety management for a changing industry. *Journal of Loss Prevention in the Process Industries, 22*(2): 162–168.
20. Kwon, H.-M. (2006) The effectiveness of process safety management (PSM) regulation for chemical industry in Korea. *Journal of Loss Prevention in the Process Industries, 19*(1): 13–16.
21. Makin, A.-M., & Winder, C. (2008). A new conceptual framework to improve the application of occupational health and safety management systems. *Safety Science, 46*(6), 935–948.
22. Maniar, M.S., Kumar, A., & Mentzer, R. A. (2020). Global process safety incidents in the pharmaceutical industry. *Journal of Loss Prevention in the Process Industries, 68*: 104279.
23. Mannan, M. S. et al. (2015). Trends and challenges in process safety.
24. Hopkins, A. (2009). Thinking about process safety indicators. *Safety Science, 47*(4): 460–465.
25. Ugboro, I. O., & Obeng, K. (2005). Top management leadership, employee empowerment, job satisfaction, and customer satisfaction in TQM organizations: an empirical study. *Journal of Quality Management, 5*(2): 247–272.
26. Baer, M. et al. (2003). Innovation is not enough: Climates for initiative and psychological safety, process innovations, and firm performance. *Journal of Organizational Behavior: The International Journal of Industrial, Occupational and Organizational Psychology and Behavior, 24*(1): 45–68.
27. Bradley, B. H., et al. (2012). Reaping the benefits of task conflict in teams: The critical role of team psychological safety climate. *Journal of Applied Psychology, 97*(1), 151.
28. Carmeli, A. (2007). Social capital, psychological safety and learning behaviours from failure in organisations. *Long Range Planning, 40*(1): 30–44.
29. Rundmo, T., & Hale, A. R. (2003). Managers' attitudes towards safety and accident prevention. *Safety Science, 41*(7): 557–574.
30. Kaila, H. L. (2011). Organizational cases on behaviour-based safety (BBS) in India. *The International Journal of Human Resource Management, 22*(10): 2135–2146.